Studies in Systems, Decision and Control

Volume 430

The series "Studies in Systems, Decision and Control" (SSDC) covers both new developments and advances, as well as the state of the art, in the various areas of broadly perceived systems, decision making and control–quickly, up to date and with a high quality. The intent is to cover the theory, applications, and perspectives on the state of the art and future developments relevant to systems, decision making, control, complex processes and related areas, as embedded in the fields of engineering, computer science, physics, economics, social and life sciences, as well as the paradigms and methodologies behind them. The series contains monographs, textbooks, lecture notes and edited volumes in systems, decision making and control spanning the areas of Cyber-Physical Systems, Autonomous Systems, Sensor Networks, Control Systems, Energy Systems, Automotive Systems, Biological Systems, Vehicular Networking and Connected Vehicles, Aerospace Systems, Automation, Manufacturing, Smart Grids, Nonlinear Systems, Power Systems, Robotics, Social Systems, Economic Systems and other. Of particular value to both the contributors and the readership are the short publication timeframe and the world-wide distribution and exposure which enable both a wide and rapid dissemination of research output.

Indexed by SCOPUS, DBLP, WTI Frankfurt eG, zbMATH, SCImago.

All books published in the series are submitted for consideration in Web of Science.

More information about this series at https://link.springer.com/bookseries/13304

Lei Zou · Zidong Wang · Jinling Liang

Communication-Protocol-Based Filtering and Control of Networked Systems

Springer

Lei Zou
College of Information Science
and Technology
Donghua University
Shanghai, China

Zidong Wang
Department of Computer Science
Brunel University London
London, Middlesex, UK

Jinling Liang
School of Mathematics
Southeast University
Nanjing, Jiangsu, China

ISSN 2198-4182 ISSN 2198-4190 (electronic)
Studies in Systems, Decision and Control
ISBN 978-3-030-97511-1 ISBN 978-3-030-97512-8 (eBook)
https://doi.org/10.1007/978-3-030-97512-8

Mathematics Subject Classification: 34H05, 37N35, 49J15, 58E25, 93Bxx, 93Exx

This Springer imprint is published by the registered company Springer Nature Switzerland AG
The registered company address is: Gewerbestrasse 11, 6330 Cham, Switzerland

This book is dedicated to the Dream Dynasty, consisting of a group of diligent people who have enjoyed intensive research into analysis and synthesis of networked systems ...

Preface

In recent years, theoretical and practical research on networked systems (NSs) has attracted growing attention due primarily to their successful applications in an extensive range of fields. In the NS, signal transmissions among system components (e.g. controllers, sensors, filters and actuators) are implemented through a shared communication network, thereby improving the system reliability and reducing the maintenance cost. In such a network-based communication, signal transmissions might be failed due to data collisions if multiple network nodes try to access the channel simultaneously. Accordingly, various communication agreements are introduced to restrain the network accesses for all the network nodes. These agreements are referred to communication protocols. The utilization of communication protocols has a great impact on signal transmissions, which further complicates the dynamical behaviors of NSs. In this case, the traditional filtering and control schemes can no longer guarantee satisfactory performance for NSs subject to communication protocols. Consequently, it is of practical significance to establish new techniques for the filtering and control of NSs subject to communication protocols.

This book is concerned with the communication-protocol-based filtering and control problems for several classes of discrete-time NSs. Particularly, the communication protocols under consideration contain the Round-Robin (RR) protocol, Try-Once-Discard (TOD) protocol and Random Access (RA) protocol. The content of this book can be divided into two parts, where the first part (Chaps. 2–7) studies the filter design methodologies and the second part (Chaps. 8–10) investigates the controller design methodologies. These results provide a framework dealing with the controller/filter design, stability analysis and performance analysis for different NSs (e.g. nonlinear systems, time-varying systems, time-delay systems, complex networks and multi-agent systems) subject to different communication protocols. Some techniques including the backward Riccati difference equations, minimum mean square error estimation theory, mathematical induction method, linear matrix inequalities and optimization approaches are employed to handle the filtering and control issues with specific performance requirements.

The compendious framework and description of this book are given as follows. Chapter 1 introduces the recent advances on communication-protocol-based filtering and control problems of NSs and the outline of this book. Chapter 2 studies the ultimately bounded filtering problem for time-delay complex networks under the effects of RR protocol. Chapter 3 is concerned with the finite-horizon $\mathscr{H}_\infty$ filtering issue of nonlinear time-varying systems with high-rate communication network and the RA protocol scheduling. Chapter 4 extends the results in Chap. 3 to the finite-horizon $\mathscr{H}_\infty$ fault estimation issue subject to the RA protocol. Chapter 5 handles the communication-protocol-based set-membership filtering problem for time-varying systems with mixed-time-delays. Chapter 6 deals with the recursive filtering problem for time-varying systems subject to the RA protocol scheduling. Chapter 7 addresses the ultimately bounded filtering issue of communication-based train control systems subject to the p-persistent CSMA protocol. In Chap. 8, we consider the finite-horizon $\mathscr{H}_\infty$ control problem subject to the RA protocol. Chapter 9 discusses the ultimately bounded control problem of nonlinear systems subject to the TOD protocol scheduling and uniform quantization effects. In Chap. 10, the finite-horizon $\mathscr{H}_\infty$ consensus control problem is investigated for time-varying multi-agent systems subject to the RA protocol. Chapter 11 provides the conclusion of this book and suggests some possible research topics related to the results proposed in this book.

This book is a research monograph whose intended audience is graduate and postgraduate students as well as researchers. The background required of the reader is the knowledge of basic Lyapunov stability theory, basic stochastic process, basic optimal estimation theory and matrix theory.

Shanghai, China — Lei Zou
London, UK — Zidong Wang
Nanjing, China — Jinling Liang

Acknowledgements

We would like to express our sincere thanks to those who have been devoted to the research in this book. Particular thanks are given to Professor Huijun Gao from Harbin Institute of Technology of China, Professor Donghua Zhou from Shandong University of Science and Technology of China, Professor Xiaohui Liu from Brunel University London of the U.K. and Professor Qing-Long Han from Swinburne University of Technology of Australia. We also extend our thanks to many colleagues who have offered support and encouragement throughout this research effort. In particular, we would like to acknowledge the contributions and friendly support from Guoliang Wei, Bo Shen, Hongli Dong, Xiao He, Lifeng Ma, Derui Ding, Jun Hu, Yurong Liu, Liang Hu, Yang Liu, Nianyin Zeng, Sunjie Zhang, Qinyuan Liu, Fan Wang, Wenying Xu and Hang Geng. Last but not the least, we are especially grateful to our families for their never-ending understanding, unfailing encouragement and never-ending support when it was most required.

The writing of this book was supported in part by the National Natural Science Foundation of China under Grants 61703245, 61873148, 61933007, 61873058 and 61673141, the China Postdoctoral Science Foundation under Grant 2018T110702, the European Union's Horizon 2020 Research and Innovation Programme under Grant 820776 (INTEGRADDE), the Royal Society of the UK and the Alexander von Humboldt Foundation of Germany. The support of these organizations is much acknowledged.

Contents

List of Notations

$\mathbb{R}^n$	The n-dimensional Euclidean space
$\mathbb{R}^{n \times m}$	The set of all $n \times m$ real matrices
$\mathbb{N}^+$	The set of non-negative integers
$\mathbb{N}^-$	The set of negative integers
$\mathbb{N}$	The set of integers
$\|A\|$	The spectral norm of the matrix A
$\|A\|_F$	The Frobenius norm of the matrix A
A^T	The transpose of the matrix A
A^{-1}	The inverse of the matrix A
$A^\dagger$	The Moore–Penrose pseudo inverse of the matrix A
I	An identity matrix of compatible dimension
0	A zero matrix of compatible dimension
$\mathbf{1}_n$	An n dimensional column vector with all ones
$\text{Prob}(\cdot)$	The occurrence probability of the event "·"
$\mathbb{E}\{x\}$	The expectation of the stochastic variable x
$\mathbb{E}\{x\|y\}$	The expectation of the stochastic variable x conditional on y
$\lambda_{\max}\{A\}$	The largest eigenvalue of a square matrix A
$\lambda_{\min}\{A\}$	The smallest eigenvalue of a square matrix A
$\text{diag}\{\dots\}$	The block-diagonal matrix
$l_2[0, \infty)$	The space of square summable sequences
$l_2([0, N], \mathbb{R}^n)$	The space of the square-summable n-dimensional vector functions over the interval $[0, N]$
$\text{tr}\{A\}$	The trace of a matrix A
$X > Y$	The $X - Y$ is positive definite, where X and Y are real symmetric matrices
$X \geq Y$	The $X - Y$ is positive semi-definite, where X and Y are real symmetric matrices
$\|x\|$	The Euclidean norm of a vector x
$\delta(a)$	The Kronecker delta function that equals 1 if $a = 0$ and equals 0 otherwise

mod (a, b)	The unique non-negative remainder on division of the integer a by the positive integer b
$\lfloor a \rfloor$	The largest integer not greater than a
$\otimes$	The Kronecker product of matrices
$\circ$	The Hadamard product of matrices
"$*$"	An ellipsis for terms induced by symmetry in symmetric block matrices

Chapter 1
Introduction

In recent years, the communication-protocol-based synthesis and analysis issues have gained substantial research interest owing mainly to their significance in networked systems. The core characteristic of networked systems is the utilization of network-based communication among different system components. In such network-based communication, signal transmissions via the communication channel are implemented according to certain agreements, namely, the communication protocols. These communication protocols would give rise to certain complex yet important impact on the system performance (e.g. stability). The necessity of designing the filtering/control strategies arises naturally in situations where the effects induced by communication protocols are inevitable due to the network-based communication. As such, the communication-protocol-based filtering and control problems for different networked systems serve as very interesting, imperative yet challenging topics.

1.1 Research Background

The research on networked systems has attracted a vast amount of interest in the past several decades [1–10]. The most typical characteristic of networked systems is the utilization of network-based communication technology, under which the data exchange among different system components (e.g. controllers, sensors, filters and actuators) is implemented via the shared communication network. Different from the traditional point-to-point communication technology, such network-based communication technology possesses numerous advantages (e.g. simple installation, reduced hardwire, low cost and high reliability) [11]. Accordingly, networked systems have achieved successful applications in an extensive range of fields such as smart vehicles,

L. Zou et al., *Communication-Protocol-Based Filtering and Control of Networked Systems*, Studies in Systems, Decision and Control 430,
https://doi.org/10.1007/978-3-030-97512-8_1

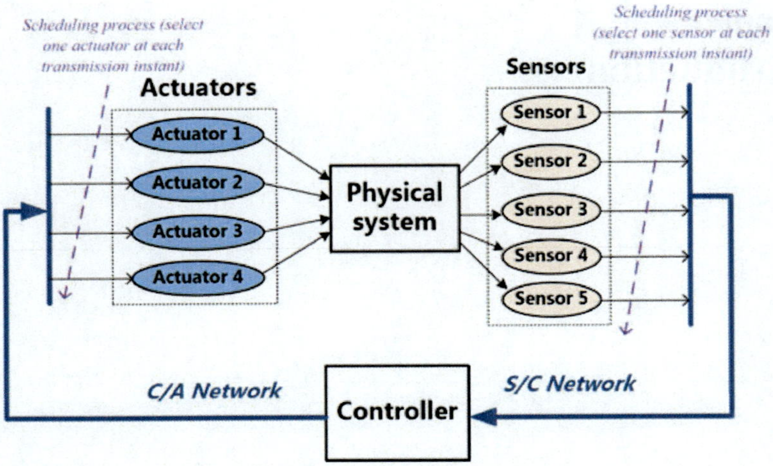

Fig. 1.1 A typical networked system with two communication networks

smart grids, environmental monitoring, industrial automation and intelligent transportation. By now, networked systems have become the focus of intensive research in signal processing and control communities. A typical networked system is shown in Fig. 1.1, where the signal transmission between the controller and plant is implemented through two communication channels: the sensor-to-controller (S/C) channel and the controller-to-actuator (C/A) channel. Nevertheless, the utilization of network-based communication would also give rise to certain special phenomena that might degrade the system performance. Such network-induced phenomena include, but are not limited to, transmission delays [12–14], packet dropouts [15–17], data packet disorder [18–20], signal quantization [21–25], communication protocol scheduling [26–28] and channel fading effects [29, 30]. As such, the synthesis and analysis issues of networked systems subject to various network-induced phenomena have attracted an ever-increasing research interest; see, for example, [31–38] and the references therein. For instance, in [39], the finite-time tracking control problem has been investigated for a type of networked system with quantized inputs. The distributed and centralized estimation issues have been studied in [40] for networked systems with missing measurements.

Among various network-induced phenomena, communication protocol scheduling is one of the most typical phenomena and would lead to a particularly significant impact on the performance of networked systems. Such a phenomenon is mainly caused by the limited communication capability of the underlying channel in the networked system. More specifically, in a typical network-based communication process, data transmissions would unavoidably suffer from data collisions in case of simultaneous multiple accesses to the shared communication network (as shown in Fig. 1.2). In order to prevent such data collisions from occurring, an effective way is to restrain the network accesses according to the so-called communication protocols by guaranteeing that only one network node is permitted to transmit its data at each

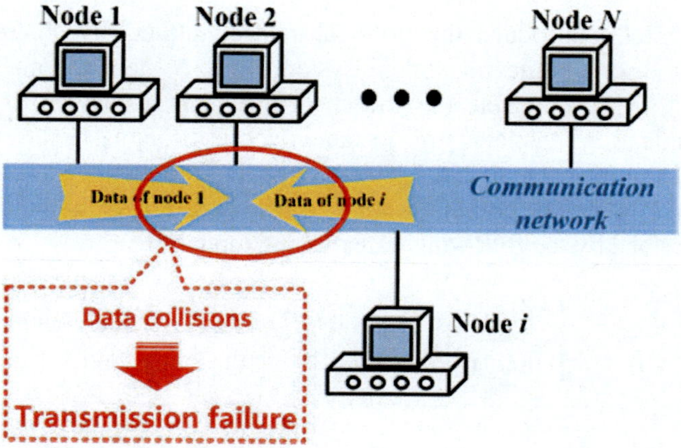

Fig. 1.2 Data collisions in case of simultaneous multiple accesses

transmission instant through the shared communication channel [41]. Communication protocols are capable of orchestrating the transmission order of all the network nodes and thereby generating certain scheduling behaviors that inevitably complicate the analysis and synthesis issues of networked systems. By now, there are three kinds of extensively investigated communication protocols in the literature; for example, the Try-Once-Discard (TOD) protocol [42], the Round-Robin (RR) protocol [27] and the Random Access (RA) protocol [30, 43].

1.2 Theoretical Frameworks

In networked systems, communication protocols are in fact a type of agreement with the aim to regulate the signal transmissions over shared channels. Accordingly, communication protocols would have a dramatic effect on the dynamical behaviors of networked systems. So far, the analysis and synthesis issues subject to different communication protocols have attracted an ever-increasing research interest. It is worth mentioning that the dynamical behaviors of networked systems subject to communication protocols are mainly affected by two aspects: the scheduling behaviors of communication protocols and the signal compensation methods. The former one determines which network node is selected to transmit its data at each transmission instant, while the latter one prescribes how to compensate the data corresponding to the network nodes that have not been selected to transmit data. Nevertheless, the so-called protocol-induced effects can be regarded as the impacts of the corresponding scheduling behavior and the signal compensation method on the networked system. Generally, there are two signal compensation methods widely adopted in practical applications: the zero-order holder (ZOH) method [44] and the zero-input (ZI) method [45].

Let us briefly introduce the protocol-induced effects by a simple example. Consider a discrete-time networked system with N network nodes labeled as $\{1, 2, \ldots, N\}$. Let ξ_k denote the chosen node which is allowed to get access to the communication channel at time instant k, $y_{i,k}$ be the signal of the ith node before transmitted at time instant k and $\bar{y}_{i,k}$ be the signal of the ith node after transmitted at time instant k. Then, the communication subject to certain communication protocol can be described by the following transmission model:

$$\bar{y}_k = \begin{cases} \urcorner(\xi_k)y_k + \big(I - \urcorner(\xi_k)\big)\bar{y}_{k-1}, & \text{the ZOH method} \\ \urcorner(\xi_k)y_k, & \text{the ZI method}, \end{cases} \tag{1.1}$$

where

$$\urcorner(\xi_k) \triangleq \operatorname{diag}\{\delta(\xi_k - 1)I, \delta(\xi_k - 2)I, \ldots, \delta(\xi_k - N)I\},$$
$$\bar{y}_k \triangleq \begin{bmatrix} \bar{y}_{1,k}^T \ \bar{y}_{2,k}^T \cdots \bar{y}_{N,k}^T \end{bmatrix}^T, \quad y_k \triangleq \begin{bmatrix} y_{1,k}^T \ y_{2,k}^T \cdots y_{N,k}^T \end{bmatrix}^T,$$

and $\delta(\xi_k - i) \in \{0, 1\}$ is a Kronecker delta function (i.e. $\delta(\xi_k - i) = 1$ holds if $\xi_k = i$ and $\delta(\xi_k - i) = 0$ otherwise). In this model, the scheduling behavior of the underlying communication protocol is characterized by the time-varying variable ξ_k. The main research topic of networked systems subject to communication protocols is to analyze the effects of the transmission model (1.1) on system performance and to design the corresponding controllers, filters or fault estimators according to the system dynamics and the transmission model (1.1), i.e. the analysis and synthesis problems of networked systems under the transmission model (1.1).

To date, there are three main different theoretical frameworks available in the literature dealing with the analysis and synthesis problems of networked systems subject to various communication protocols, namely, switched-system-based (SWB) framework [46], impulsive-hybrid-system-based (IHSB) framework [47] and switched-time-delay-based (STDB) framework [27]. Generally speaking, the SWB framework is always adopted to settle the discrete-time systems subject to communication protocols. Let's take the control problem of discrete-time networked system with a certain communication protocol for example. Consider a discrete-time networked system with N sensor nodes. ξ_k is the selected sensor node obtaining access to the communication network at time instant k. The networked system is characterized as follows:

$$\begin{cases} x_{k+1} = \digamma(x_k, u_k, \omega_k) \\ y_k = \beth(x_k, \nu_k), \end{cases} \tag{1.2}$$

where x_k, y_k, u_k, ω_k and ν_k are the system state, measurement output, control input, process noise and measurement noise at time instant k, respectively. $\digamma(\cdot, \cdot, \cdot)$ and $\beth(\cdot, \cdot)$ are two real-valued functions. Under the effects of communication protocol, the transmission model can be described by (1.1). Then, letting the control input be $u_k = \daleth(\bar{y}_k)$, the dynamics of the closed-loop system can be described by the following difference equation:

$$x_{k+1} = \begin{cases} \bar{F}_{\xi_k}(x_k, \bar{y}_{k-1}, \omega_k, \nu_k), & \text{the ZOH method}, \\ F_{\xi_k}(x_k, \omega_k, \nu_k), & \text{the ZI method}. \end{cases} \tag{1.3}$$

Obviously, the closed-loop system (1.3) can be regarded as a switched system and the switching law is determined by ξ_k which represents the scheduling behavior of the underlying communication protocol. As such, the performance analysis and controller design issues of such a networked system can be implemented based on the SWB framework. More details regarding the SWB framework can be found in [46, 48].

The IHSB framework is always employed to handle the continuous-time systems with sampled measurements and communication protocol scheduling effects. For such kinds of systems, the corresponding transmission model can be described by an impulsive system. Then, the resulted system dynamics could be represented by an impulsive switched system in which the system state would switch its value at every sampling instant. One of the representative works of the IHSB framework can be found in [49]. Both the SWB framework and IHSB framework are capable of dealing with networked systems subject to three widely studied communication protocols (i.e. the RR protocol, TOD protocol and RA protocol). Compared with such two frameworks, the STDB framework is developed to cope with the analysis and synthesis issues of networked systems with protocol scheduling and communication delays (or the delay effects induced by the sampling mechanism). In the STDB framework, the transmission model under the protocol scheduling is described by a delayed measurement model with switching parameters. Based on such a transmission model, the resulted system dynamics could be modeled by a switched time-delay system. More details regarding the STDB framework can be found in [27]. So far, a rich body of literature has appeared on the analysis and synthesis issues subject to various communication protocols based on these aforementioned frameworks; see, for example, [26, 49–53]. The relevant research on networked systems subject to communication protocols have the main focused attention on (1) the stability analysis of networked systems subject to various communication protocols, (2) the communication-protocol-based control problem of networked systems, (3) the communication-protocol-based filtering (or state estimation) problem of networked systems and (4) the communication-protocol-based fault diagnosis problem of networked systems.

1.3 Stability Analysis Subject to Protocol Scheduling

The performance analysis issue has gained a lot of research attention for networked systems subject to protocol scheduling effects since the pioneering works [54–56].

Stability is one of the most investigated performance indices of dynamical systems [57]. The scheduling effects of communication protocols would lead to an enormous impact on the stability of networked systems [41]. In [56], the stability analysis issue has been studied for continuous-time linear time-invariant (LTI) systems subject to the effects of RR protocol and TOD protocol, respectively. Furthermore, a maximum allowable transfer interval (MATI) has been proposed to guarantee the global exponential stability (GES) of a networked system subject to the TOD protocol. In [58], an improved MATI has been obtained to guarantee the GES compared with [56] based on the hybrid system method. The input–output $\mathscr{L}_p$ stability issue has been investigated in [59] for continuous-time nonlinear systems subject to RR protocol and TOD protocol, respectively. The results have been extended to the continuous-time nonlinear systems subject to hybrid communication protocols in [60]. Based on the IHSB framework, the $\mathscr{L}_p$ stability problem has been addressed in [49] for continuous-time networked systems with communication protocol scheduling, time-varying transmission intervals and communication delays, where the trade-offs between the MATI, maximally allowable delay (MAD) and performance gains have been provided.

The study on networked systems with RA protocol has been first reported in [47], where sufficient conditions have been acquired to guarantee the $\mathscr{L}_p$ stability for continuous-time networked systems. For discrete-time networked systems, the mean-square stability has been analyzed based on the SWB framework in [51] with RR, TOD and RA protocols, respectively. Generally speaking, there are two different stochastic processes describing the scheduling behavior of the RA protocol. The first one is the independent and identically distributed (i.i.d.) sequence of random variables, which has been first introduced in [47]. The other one is the discrete-time Markov chain, which has been first adopted in [51]. The choice between such two stochastic processes is dependent on the actual communication channel. In [61], the authors have studied the exponential mean-square stability of networked systems subject to two different RA scheduling models (i.e. the i.i.d. sequence of random variables and the discrete-time Markov chain) respectively based on the IHSB framework. In [62], the $\mathscr{L}_p$ stability issue has been addressed for nonlinear networked systems with TOD protocol scheduling by using the small gain theorem.

The exponential mean-square stability analysis issue has been considered for networked systems with two kinds of RR protocols in [63], where two Markov chains have been employed to model the packet dropouts. In [64], the stability problem has been investigated based on the STDB framework for discrete-time networked systems with RR scheduling, constant communication delays and nonuniform sampling scheme. The authors have extended the results in [27], where the exponential stability has been considered for networked systems with time-varying communication delays and RR scheduling. Then, the authors have discussed the stability issues subject to TOD protocol scheduling effects for continuous-time networked systems and discrete-time networked systems in [65] and [52], respectively, where the corresponding stability criteria have been derived by using the hybrid-delayed-system-based approach. In [66], the mean-square stability has been discussed based on the STDB framework for a class of stochastic networked systems with protocol scheduling effects, where it has been shown that the TOD protocol would lead to a larger MATI compared with the RR protocol. It should be noted that the TOD protocol is designed based on the "competitive" principle. As such, it is sometimes difficult for certain network nodes to obtain sufficient opportunities in getting access to the communication channel. In this case, some "improved" TOD protocols have been developed to ensure that every node is eventually assigned with the network access opportunity within a finite window of time. In [60], a so-called constant-penalty TOD (CP-TOD) communication protocol has been introduced based on the mechanism of "silent-time" and the $\mathscr{L}_p$ stability of the networked system subject to the CP-TOD scheduling has been studied. The input-to-state stability in probability has been studied in [67] for nonlinear stochastic systems under quantization effects and communication protocols in virtue of the switched Lyapunov function method.

1.4 Communication-Protocol-Based Filtering and Control

The filtering and control problems are two fundamental research topics in industrial automation community. In order to evaluate the control and filtering performance, various control and filtering methods have been developed. These control and filtering schemes can be categorized into several groups according to the considered systems and noises as shown in Table 1.1.

As discussed in Sect. 1.2, the closed-loop system dynamics of a networked system is largely dependent on the protocol-induced effects. As such, the controller/filter design of a networked system should take the protocol-induced effects into consideration in order to achieve the desired performance. In this section, we would like to review the communication-protocol-based control and filtering problems for different systems.

1.4.1 Communication-Protocol-Based Filtering and Control of Linear Time-Invariant Systems

The control problem of linear time-invariant (LTI) system is a hot research topic that has attracted quite a lot of attention [34, 68–70]. The protocol-induced effects lead to an enormous impact on the closed-loop system. Consider a typical linear time-invariant system of the following form:

$$\begin{cases} x_{k+1} = Ax_k + Bu_k + E\omega_k \\ y_k = Cx_k + Dv_k. \end{cases}$$

Let the control input be $u_k = K\bar{y}_k$. Then, based on the transmission model (1.1), the dynamics of the closed-loop system subject to the ZOH method can be described as follows:

Table 1.1 The control and filtering methods for different systems with different noises

Noises	Systems				
	Control problems				
	Linear time-invariant systems	Nonlinear systems	Uncertain systems	Time-varying systems	Multi-agent systems
Energy-bounded noises	$\mathscr{H}_\infty$ control, energy-to-peak control			Finite-horizon $\mathscr{H}_\infty$ control	$\mathscr{H}_\infty$ consensus control
Norm-bounded noises	Ultimate-bounded control, l_2 control			Ultimate-bounded control	Bounded consensus control
Stochastic noises	LQG control, variance-constrained control			Error-constrained control	Mean-square consensus control
Noises	Systems				
	Filtering problems				
	Linear time-invariant systems	Nonlinear systems	Uncertain systems	Time-varying systems	Networked systems over sensor networks
Energy-bounded noises	$\mathscr{H}_\infty$ filtering, energy-to-peak filtering			Finite-horizon $\mathscr{H}_\infty$ filtering	Distributed $\mathscr{H}_\infty$ filtering
Norm-bounded noises	Ultimate-bounded filtering, l_2 filtering			Set-membership filtering	Distributed ultimate-bounded filtering
Stochastic noises	Variance-constrained filtering			Recursive filtering	Distributed recursive filtering

$$\begin{bmatrix} x_{k+1} \\ \bar{y}_k \end{bmatrix} = \begin{bmatrix} A + BK\daleth(\xi_k)C & BK\big(I - \daleth(\xi_k)\big) \\ \daleth(\xi_k)C & I - \daleth(\xi_k) \end{bmatrix} \begin{bmatrix} x_k \\ \bar{y}_{k-1} \end{bmatrix} + \begin{bmatrix} E & BK\daleth(\xi_k)D \\ 0 & \daleth(\xi_k)D \end{bmatrix} \begin{bmatrix} \omega_k \\ \nu_k \end{bmatrix}.$$

The dynamics of the closed-loop system subject to the ZI method can be described as follows:

$$x_{k+1} = \big(A + BK\daleth(\xi_k)C\big)x_k + E\omega_k + BK\daleth(\xi_k)D\nu_k.$$

Obviously, the above two difference equations are in fact linear time-invariant systems with certain switching behaviors. The main purpose of the communication-protocol-based control problem of time-invariant systems is to design the controller parameters subject to such switching systems. In [71], a co-design strategy of TOD protocol scheduling and controller has been developed for a class of LTI systems, where the desired controller parameter has been acquired by solving a set of matrix inequalities. The results have been extended to the co-design problem of TOD protocol and controller for linear time-delay networked systems in [72], where the desired controller has been developed by using linear matrix inequality (LMI) techniques. Considering the input saturation effects, the control problem has been addressed in [73] for discrete-time systems with network-induced delays and TOD protocol scheduling by using the delay-dependent Lyapunov–Krasovskii method. The results have been extended to the control problem subject to RR protocol scheduling in [74], where the corresponding stability has been analyzed based on the STDB framework. The quantized control problem has been investigated in [75] under the RR protocol scheduling, large communication delays and time-varying sampling intervals.

Considering the random packet dropouts in the communication channel, the optimal control problem has been studied in [76] for linear time-invariant systems with RA protocol scheduling. The co-design of controller and protocol scheduling strategy has been investigated for linear networked systems with random packet losses in [77], where the resulting closed-loop networked systems have achieved a minimal decay rate. $\mathscr{H}_\infty$ control issue is one of the research focused in the field of networked control systems. In [78], the $\mathscr{H}_\infty$ control problem has been considered for linear systems with RR protocol scheduling and measurement missing effects. The control problem of medium-constrained vehicular networks has been studied in [30], where a shared communication channel with protocol scheduling and channel fading effects has been considered. Considering that the statistical properties of the RA protocol scheduling are partly unknown, the output feedback controller has been designed in [79] for linear networked systems by solving a set of LMIs. In [80], a so-called most regular binary sequences protocol has been considered and the corresponding controller has been designed by using piecewise Lyapunov functional and the average dwell time technique.

Filtering (or state estimation) is one of the most studied fundamental issues in signal processing and control communities [81–84]. For the filtering issue of a typical networked system, the filtering process is implemented based on the received measurement data, which are largely affected by the adopted communication. Similar to the communication-protocol-based control problems, the filtering error systems under the effects of communication protocols can be modeled by linear time-invariant systems with certain switching behaviors. The main purpose of the communication-protocol-based filtering problem of time-invariant systems is to design the filter parameters subject to such switching systems. In [85], the moving-horizon estimation problem has been studied for a class of linear time-delay systems under the RR protocol, where the developed estimation strategy has achieved a satisfactory performance by using a lifting-based design framework. The $\mathscr{H}_\infty$ filtering problem subject to certain communication protocol has been investigated in [86], where a continuous-time linear networked system with a TOD protocol has been considered. Multi-rate sampling scheme is a widely utilized method in various practical systems to achieve the relatively low resource consumption. In [87], the $\mathscr{H}_\infty$ filter has been constructed for multi-rate multi-sensor systems subject to the p-persistent CSMA protocol (a special RA protocol), where the multi-rate system is converted to a single-rate system by adopting the lifting method.

1.4.2 Communication-Protocol-Based Filtering and Control of Nonlinear Systems

The control and filtering issues of nonlinear systems have always been two of the most challenging issues in the past decades [88]. Considering the communication-protocol-based control problem for a typical nonlinear system described by (1.2), the dynamics of the closed-loop systems subject to different signal compensation methods can be modeled, respectively, by

$$\begin{bmatrix} x_{k+1} \\ \bar{y}_k \end{bmatrix} = \hat{F}_{\xi_k}\left(\begin{bmatrix} x_k \\ \bar{y}_{k-1} \end{bmatrix}, \omega_k, \nu_k\right), \quad \text{the ZOH method,}$$

and

$$x_{k+1} = \check{F}_{\xi_k}\big(x_k, \omega_k, \nu_k\big), \quad \text{the ZI method.}$$

Similarly, for the communication-protocol-based filtering problems, the corresponding filtering error systems can also be modeled by such nonlinear switching systems. Compared with the linear systems, the dynamical behaviors of nonlinear systems are much more complex, thereby leading to difficulties in performance analysis and controller/filter design issues of nonlinear systems. So far, there are some representative methods dealing with the control and filtering problem of nonlinear systems; see, for example, [47, 49, 89–92]. With the rapid development of networked systems, the research on nonlinear networked systems has gained considerable research interests owing to the potential applications in modern industry. The control problem of nonlinear systems subject to protocol scheduling is an important topic in such a research. In [89], the tracking control problem has been addressed for nonlinear networked systems subject to communication protocols, where the controller has been designed based on the IHSB framework. The reliable control problem has been investigated in [90] for nonlinear networked systems with TOD protocol scheduling in the presence of actuator faults, where the remote observer-based reliable controller has been designed by using Lyapunov–Krasovskii functional method and some matrix manipulations.

FlexRay is a deterministic communication protocol that is widely employed in automotive control. Under the effects of FlexRay, the transmission order of network nodes is orchestrated according to the pre-set communication cycles composed of a static segment and a dynamic segment that are periodically repeated. In [93], an emulation controller has been developed for nonlinear networked systems subject to FlexRay scheduling based on the IHSB framework. Considering the case that the transmission delays are larger than the transmission intervals, the predictive control problem has been studied in [94] for nonlinear systems subject to communication protocols. Neural network is an effective tool in dealing with the nonlinear systems. In [91, 92], the neural-network-based output-feedback control problem has been investigated for nonlinear systems subject to communication protocols, where the adaptive dynamic programming method has been employed to handle the controller design issue. In [95], a neural-network-based adaptive controller has been designed for bilateral teleoperation systems with multiple slaves under Round-Robin scheduling protocol. The Takagi–Sugeno (T–S) fuzzy modeling method is one of the most widely employed approaches to handle the nonlinear systems. In [96], the fuzzy model predictive control problem has been considered for discrete-time T–S fuzzy systems under a so-called event-triggering-based TOD protocol.

The filter (or state estimator) design issue for nonlinear systems is a hot yet important topic in signal processing community. The existence of communication protocol would further complicate the filter design task. In [97], the state estimator has been designed for a type of nonlinear networked systems with communication protocols based on the IHSB framework and the small gain theorem. Measurement outliers might occur in system operation due to various reasons such as sensor malfunction, large non-Gaussian noises and cyber-attacks. Noting that the measurement outliers would pose serious threats to the filtering process, in [98], the $\mathscr{H}_\infty$ filtering issue has been addressed for a type of nonlinear stochastic systems with RR protocol scheduling. Noting that the dynamics of certain complex systems can be modeled by nonlinear systems, the corresponding $\mathscr{H}_\infty$ filters have been proposed in [99, 100] for nonlinear complex networks and genetic regulatory networks. The well-known artificial neural network is a special nonlinear system whose nonlinearity is determined by the underlying neuron activation functions. In [101], the finite-time state estimation issue has been investigated for delayed artificial neural networks, where a shared communication network with the RA protocol has been adopted for the signal transmissions between the sensors and estimator.

Quantization and missing measurement phenomena are two widely studied network-induced complexities in networked systems. The simultaneous existence of protocol scheduling effects and other network-induced complexities would further complicate the analysis and design issues of networked systems. In [102], a networked recursive filter has been constructed for a type of nonlinear stochastic systems subject to uniform quantization, missing measurements and RR protocol scheduling effects. Sliding mode observer is an effective estimator dealing with the uncertain nonlinear systems. In [103], the sliding model observer design issue has been considered for a type of discrete nonlinear time-delay systems subject to the RA protocol scheduling, where sufficient conditions for the existence of the desired sliding model observer have been proposed in terms of the feasibility of a minimization problem. By introducing a special extended dissipative property index, the generalized state estimation problem has been considered in [104] for Markovian-coupled networks under the RR protocol scheduling, where the Markovian-coupled networks have been modeled by a nonlinear complex network with a Markov stochastic process. Moving-horizon estimation is an effective scheme to tackle nonlinear systems. The communication-protocol-based moving-horizon estimation problem has been studied in [105] for nonlinear networked systems by using a special robust-based scheme.

1.4.3 Communication-Protocol-Based Filtering and Control of Uncertain Systems

Uncertainties serve as a class of important complexities for system modeling, which describe the parameter changes in system dynamics [106, 107]. There are two kinds of uncertainties widely studied in the literature, namely, the norm-bounded uncertainties [108, 109] and the polytopic uncertainties [110, 111]. The existence of uncertainties would pose serious threats to the system performance (e.g. the stability of a closed-loop system). As such, the control and filtering problems subject to uncertainties have gained an ongoing research interest for various systems; see, for example, [112–114]. It is worth mentioning that, for a networked system, the utilization of communication protocol would further complicate the control and filter design of uncertain systems. Considering the communication-protocol-based control problem for a networked system with norm-bounded uncertainties, the dynamics of the closed-loop systems subject to different signal compensation methods can be described by the following models:

$$\begin{cases} \begin{bmatrix} x_{k+1} \\ \bar{y}_k \end{bmatrix} = \Re_{\xi_k}\left(\begin{bmatrix} x_k \\ \bar{y}_{k-1} \end{bmatrix}, \omega_k, \nu_k, \Delta_k \right), & \text{the ZOH method,} \\ x_{k+1} = \bar{\Re}_{\xi_k}(x_k, \omega_k, \nu_k, \Delta_k), & \text{the ZI method.} \end{cases}$$

Obviously, the utilization of communication protocols would affect the stability and robustness of closed-loop systems. To data, the communication-protocol-based control and filtering problems of uncertain systems have begun to stir some initial research interest. For instance, in [115], a resilient robust model predictive controller has been developed for polytopic uncertain systems subject to the TOD protocol scheduling by using an average dwell-time-based approach. The results have been extended to the model predictive control problem of polytopic uncertain systems with event-triggered mechanism and Round-Robin protocol scheduling in [116]. In [117], a robust $\mathscr{H}_2/\mathscr{H}_\infty$ model predictive control problem has been studied for polytopic uncertain systems subject to the TOD protocol scheduling based on the SWB framework.

In [118], the dynamic output-feedback robust model predictive control problem has been studied for polytopic uncertain systems with RR protocol, where a model-dependent observer has been developed to estimate the system states. The authors have then extended their results to the robust model predictive control problem for polytopic uncertain systems with state saturation nonlinearities under Round-Robin protocol scheduling in [119], in which the state saturation nonlinearities have been reformulated into the sum of finite number of linearities. The sliding mode control is an effective tool to handle the uncertainties by forcing the state trajectories subject to parameter perturbations to reach certain sliding manifolds. In [28], the authors have developed an $\mathscr{H}_\infty$ sliding mode controller for discrete-time systems with norm-bounded uncertainties subject to the RA protocol scheduling, where the reachability of the sliding mode dynamics has been guaranteed by using a token-dependent stochastic Lyapunov function. The results have been extended to the sliding mode control problem for cyber-physical switched systems with RR protocol scheduling in [120], where the input-to-state stability in probability has been analyzed based on the linear matrix inequality method.

In practical applications, sometimes parameter uncertainties may occur in a random fashion. In [121], the robust $\mathscr{H}_\infty$ state estimation problem has been considered for two-dimensional systems with randomly occurring uncertainties, signal quantization effects and RR protocol scheduling. The results have then been extended to the $\mathscr{H}_\infty$ control problem of a type of two-dimensional system with randomly occurring uncertainties, nonlinearities and unknown time-delays subject to RA protocol scheduling effects in [122]. It is worth noting that uncertainties can describe not only the parameters changes in plant dynamics but also the parameters perturbations of the transmission behavior (e.g. the statistical properties of RA protocol scheduling). In [123], a remote $\mathscr{H}_\infty$ state estimator has been formulated for time-delay neural networks under the scheduling of RA protocol, where the scheduling behavior of the RA protocol is governed by a Markov chain whose transmission probability is uncertain. For certain complex systems, the inner coupling strengths of nodes might suffer from some variations, which result in the so-called uncertain inner coupling. In [124], a resilient set-membership state estimator has been constructed for a type of time-varying complex network with sensor saturation, uncertain inner coupling and distributed delays subject to the RR protocol. The communication-protocol-based filtering problem has been addressed in [125] for nonlinear systems with stochastic uncertainties, where two resource-saving unscented Kalman filters have been developed.

1.4.4 Communication-Protocol-Based Filtering and Control of Time-Varying Systems

In practical applications, a large number of systems are subject to certain time-varying parameters variations, and thereby leading to a rich body of research works concerning time-varying systems [126]. Considering the communication-protocol-based control problem of a time-varying system, the closed-loop system can be described as follows:

$$\begin{cases} \begin{bmatrix} x_{k+1} \\ \bar{y}_k \end{bmatrix} = \hat{\Re}_{\xi_k}\left(\begin{bmatrix} x_k \\ \bar{y}_{k-1} \end{bmatrix}, \omega_k, \nu_k, k \right), & \text{the ZOH method,} \\ x_{k+1} = \mathfrak{R}_{\xi_k}\left(x_k, \omega_k, \nu_k, k\right), & \text{the ZI method.} \end{cases}$$

The aim of communication-protocol-based control problems is to design the desired controllers for such time-varying systems subject to different switching behaviors. Some representative results concerning the communication-protocol-based filtering problem for time-varying systems are listed as follows. In [127], the communication-protocol-based finite-horizon $\mathscr{H}_\infty$ state estimation issue has been first investigated for time-varying artificial neural networks with component-based distributed delays. The results have been extended to the communication-protocol-based finite-horizon $\mathscr{H}_\infty$ state estimation in [128] for delayed memristive neural networks, where the impacts induced by the memristors have been modeled by some time-varying uncertain terms. In [129], the communication-protocol-based set-membership filtering problem has been studied for time-varying nonlinear systems with censored measurements. The results have been extended to the non-fragile set-membership filtering problem for sensor-saturated memristive neural networks with TOD protocol in [130] and for delayed memristive neural networks with quantization and TOD protocol in [131]. The quantized finite-horizon $\mathscr{H}_\infty$ filtering problem has been studied in [132] for multi-rate systems with RA protocol. The results have then been extended to the fusion estimation problem in [133] for multi-rate linear repetitive processes subject to the TOD protocol and, to the communication-protocol-based finite-horizon $\mathscr{H}_\infty$ state estimation problem in [134] for stochastic coupled networks with random inner couplings. Piecewise linear system is a special time-varying system that has received considerable research attention during the past decades. In [135], the set-membership filtering problem has been considered for piecewise linear systems with censored measurements under the effects of RR protocol scheduling.

In practical systems, the controllers and filters might sometimes suffer from certain perturbations. In order to guarantee the desired system performance, the so-called resilient scheme has been adopted in the controllers and filters design processes. In [136], a near-optimal resilient controller has been developed for networked time-varying systems with gain perturbations, state saturations and additive nonlinearities, where the RA protocol has been utilized to schedule the signal transmissions between the sensors and the controller. Event-triggered transmission is an effective approach to reduce the transmission frequency. The noncooperative event-triggered control problem has been proposed in [137] for time-varying networked systems under the RR protocol scheduling. Furthermore, the developed control strategy has been applied to the load frequency control problem in circuit systems to show the effectiveness of the proposed method.

Note that the communication-protocol-based control and filter issues are very important for complex networks since it is quite crucial to guarantee the efficient communication among nodes. The communication-protocol-based $\mathscr{H}_\infty$ filtering issues of time-varying complex networks have been addressed in [138] and [139] with random coupling strengths and state saturations, respectively, where the corresponding filters have been designed by using the backward recursive Riccati difference equations method and the recursive linear matrix inequality technique. Similar communication-protocol-based $\mathscr{H}_\infty$ estimation method has been employed to deal with the genetic regulatory networks in [140], where the genetic regulatory networks have been modeled by time-varying systems with nonlinearities. The filtering error covariance is important to evaluate the estimation accuracy under stochastic noises. By minimizing the traces of error covariance (or the upper bounds of error covariance), a communication-protocol-based Kalman filters and communication-protocol-based extended Kalman filter have been designed in [141, 142] for linear time-varying systems and nonlinear time-varying systems, respectively. In [143], a recursive filter has been developed for time-varying nonlinear complex networks under RA protocol scheduling effects. A recursive full information estimator has been designed in [144] for time-varying systems subject to the RR scheduling by solving a minimization problem, where the upper bound of the norm of the state estimation error has been achieved. The recursive fusion estimation issue has been considered in [145] for a class of time-varying state-saturated complex networks under the effects of RA protocol.

1.4.5 Communication-Protocol-Based Filtering and Control of Distributed Networked Systems

For the communication-protocol-based control problem of a distributed system, the dynamics of node i can be described by

$$\begin{cases} \begin{bmatrix} x_{i,k+1} \\ \bar{y}_{i,k} \end{bmatrix} = \hat{\Im}\big(x_{i,k}, \bar{y}_{i,k-1}, \mathbf{x}_{i,k}, \omega_{i,k}, v_{i,k}, \boldsymbol{v}_{i,k}, \xi_{i,k}\big), & \text{the ZOH method,} \\ x_{i,k+1} = \Im\big(x_{i,k}, \mathbf{x}_{i,k}, \omega_{i,k}, v_{i,k}, \boldsymbol{v}_{i,k}, \xi_{i,k}\big), & \text{the ZI method,} \end{cases}$$

where $\mathbf{x}_{i,k}$, $\boldsymbol{v}_{i,k}$ denote the state information and the measurement noise information of the neighbor nodes for node i; $\xi_{i,k}$ is the scheduling behavior of the transmissions between the node i and its neighbor nodes. Obviously, such communication-protocol-based control problem for distributed networked systems are more challenging compared with the centralized networked systems. The main difficulty to deal with such control problems is induced by the complex scheduling behavior described by $\xi_{i,k}$. Similarly, for the communication-protocol-based distributed filtering problems, the corresponding filtering error systems can also be modeled by such distributed systems with complex scheduling behavior. The communication-protocol-based distributed filtering problem has been first studied in [146], where the RR protocol has been adopted to schedule the signal transmissions among sensor nodes. The results have been further extended to the finite-time distributed state estimation issue in [147] for nonlinear systems over sensor networks with RR protocol scheduling effects and fading channels. Considering the case that the network-based communication suffers from cyber-attacks, the communication-protocol-based distributed secure filtering problem has been investigated in [148] for linear networked systems in the STDB framework. Considering the filtering problem of time-varying state-saturated systems, a communication-protocol-based distributed recursive filtering scheme has been developed in [149], where a special matrix simplification approach has been utilized to tackle the sensor network topology's sparseness issue. The distributed set-membership filtering problem has been studied in [150] for nonlinear systems subject to RR and RA protocols over sensor networks.

The resilient method is capable of reducing the impact of the system performance on the gain perturbations of the controller (or filter). The distributed resilient control and filter problems have been addressed, respectively, in [151] and [152] subject to communication protocols. Considering the time-varying multi-agent systems, the finite-horizon $\mathscr{H}_\infty$ consensus control problem has been studied in [153] subject to RR protocol.

1.5 Communication-Protocol-Based Fault Diagnosis

In real-world applications, practical systems usually suffer from the threats of faults that might take place in various system components. As such, the research on fault diagnosis has attracted significant attention; see, for example, [154–157] and the references therein. Generally speaking, the research works of fault diagnosis include the fault detection, fault estimation and fault isolation problems. The fault detection and isolation problems of networked systems have been first reported in [53], where the RA protocol has been employed to schedule the transmission between the sensors and the remote fault detection and isolation filter. Furthermore, a finite frequency stochastic H_- performance requirement has been achieved by using the linear matrix inequality method. The results have been extended to the $\mathscr{H}_\infty$ fault detection issue with RA protocol in [158] and the frequency-dependent fault detection issue with TOD protocol in [159]. The robust $\mathscr{H}_\infty$ fault detection issue has been considered in [160] for nonlinear 2-D systems with so-called randomly occurring linear fractional uncertainties under the effects of RR protocol. The authors have then extended the results in [161], where an $\mathscr{H}_\infty$ fuzzy fault detection filter has been developed for networked fuzzy systems with multiplicative noises subject to the RR protocol scheduling.

Considering the phenomenon of missing measurements with uncertain occurrence probabilities, the communication-protocol-based fault detection has been studied in [162] for time-delay systems with missing measurements under uncertain missing probabilities. The authors have then extended their results in [163], where a robust fault detection filter has been developed for nonlinear systems with data drift and randomly occurring faults subject to TOD protocol. In [50], the fault detection has been considered over a finite-frequency domain for time-delay networked systems with RR protocol scheduling, in which both the required H_- and $\mathscr{H}_\infty$ performance indices have been achieved by using an improved Kalman–Yakubovich–Popov lemma. Considering the gain variations in the filter, the non-fragile $\mathscr{H}_\infty$ fault detection problem has been proposed in [164] for fuzzy systems subject to the RA protocol by adopting the strong centralized stochastic analysis technique and the matrix calculation method. In [165], the distributed fault estimation problem has been investigated for a type of delayed complex networks subject to the RR protocol, where a set of unknown input observers has been designed to decouple the external disturbance from the estimation process as much as possible.

Recently, the fault detection and estimation problems of time-varying systems have received an increasing amount of attention. The fault detection problem has been investigated in [166] and [167] for time-varying systems subject to TOD protocol and RR protocol, respectively. In [168], the finite-horizon $\mathscr{H}_\infty$ problem has been studied for a type of nonlinear time-varying systems with randomly occurring faults subject to the RR protocol. The main idea of fault estimation is to reconstruct the desired fault information (e.g. the size and shape of the fault) based on the available measurements. The information about the fault could help to improve the reliability of control systems. In [169], the communication-protocol-based fault estimation problem has been studied for time-varying systems with randomly occurring sensor nonlinearities where the desired fault estimator has been designed by using the recursive linear matrix inequality technique.

1.6 Outline

In this book, we aim to investigate the communication-protocol-based filtering and control problems for several chasses of networked systems. The organization structure of this book is shown in Fig. 1.3 and the outline of this book is given as follows:

- In this chapter, we first introduce the research background, motivations and research problems about the filtering and control problems of networked systems under the effects of communication protocols. The research progresses of communication-protocol-based filtering and control problems are reviewed. Then, the outline of this book is listed.
- Chapter 2 investigates the ultimately bounded filtering problem for a class of time-delay complex networks where the RR protocol is adopted to schedule the signal transmissions between sensors and the remote filter. By using a switch-based approach, the resultant filtering error dynamics is modeled by a periodic parameter-switching system with time delays. The mean-square ultimate boundedness of the filtering error is analyzed by constructing a novel Lyapunov-like functional and employing the stochastic analysis approach. The desired filter gains are derived by solving a convex problem. Within the established framework, the developed filtering scheme is specialized to two kinds of complex networks.
- Chapter 3 considers the finite-horizon $\mathscr{H}_\infty$ filtering problem for a class of nonlinear time-varying systems with a high-rate communication network and the RA protocol scheduling effects. The high-rate communication will give rise to the over-sampling of transmission signals (i.e. *multiple transmissions* will occur between each two adjacent sampling instants of the sensors). Under the scheduling effects of the RA protocol, the signal transmissions between sensors and filter will be regulated according to certain scheduling behavior (which can be described by a Markov chain). A special mapping technology is employed to formulate the transmission model subject to the effects induced by the high-rate communication and RA protocol scheduling. The desired filter parameters are obtained by solving a

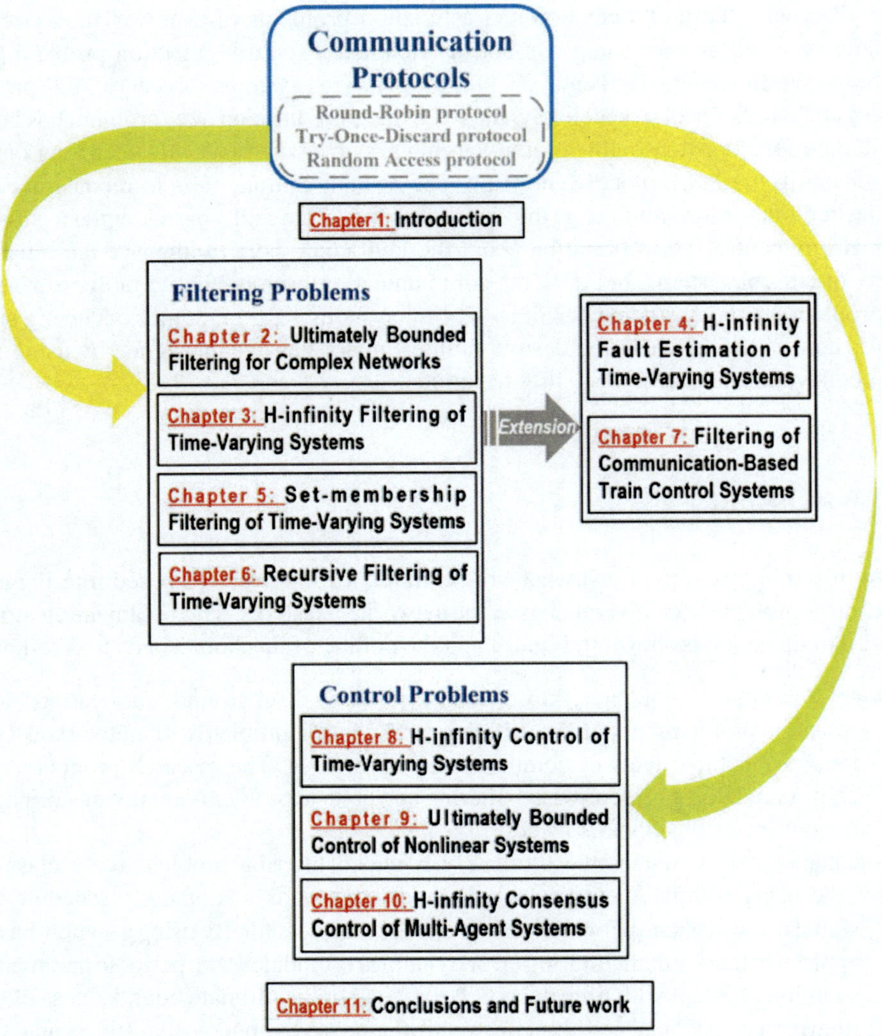

Fig. 1.3 Organization of this book

set of recursive matrix inequalities to ensure the required $\mathscr{H}_\infty$ performance of the resultant filtering error dynamics.

- Chapter 4 is concerned with the finite-horizon $\mathscr{H}_\infty$ fault estimation issue of time-varying systems subject to the RA protocol scheduling. Actually, the fault estimation studied in this chapter can be regarded as the extension of the results addressed in Chap. 3. In this chapter, the communication between the sensors and the remote fault estimator is regulated by the RA protocol. In order to characterize the effects of RA protocol, a sequence of i.i.d. random variables is employed to model the scheduling behavior of the signal transmission process. The faults under consideration are assumed to be either incipient faults or abrupt faults, which are the most common faults in industrial processes. By applying the recursive-matrix-inequality-based method that is introduced in Chap. 3, a finite-horizon $\mathscr{H}_\infty$ fault estimator is developed to generate the desired estimates of states and faults.

- In Chap. 5, the set-membership filtering issue is studied for a class of time-varying systems with mixed-time-delays subject to RR protocol and TOD protocol. By resorting to the recursive-matrix-inequality-based approach, sufficient conditions are obtained for the time-varying filter to guarantee that the derived state estimate of the given system is confined to a certain ellipsoidal region at each time instant. Within the established theoretical framework, two optimization problems are addressed to calculate the desired filter parameters for different underlying protocols.
- In Chap. 6, we deal with the recursive filtering problem for time-varying systems subject to the RA protocol scheduling. The desired state estimate is calculated recursively by minimizing the trace of the filtering error covariance. Under the scheduling behavior of the RA protocol, the resultant filtering error is characterized by a time-varying system with a stochastic parameter matrix. Furthermore, a special effort is proposed to address the boundedness of the filtering error covariance. According to the statistics properties of the RA protocol and the system parameters, the uniform lower and upper bounds of the filtering error covariance are derived.
- In Chap. 7, we further investigate the filter design issue for the so-called communication-based train control systems subject to the p-persistent CSMA protocol. The train under consideration is described by n cars linked by couplers, whose dynamic behavior is characterized by a sequence of Newton's motion equations. The wireless communication between the train and wayside access points is regulated according to the so-called p-persistent CSMA protocol, under which the signal transmission behavior is described by a Bernoulli distributed variable. Two optimization problems are solved to derive the desired filter gains according to different requirements.
- In Chap. 8, we focus our attention on the finite-horizon $\mathscr{H}_\infty$ control problem for time-varying systems under the scheduling of RA protocol. An observer-based controller is developed to handle the control task. The signal transmissions between the plant and remote controller are implemented via two communication networks where two RA protocols are employed to determine the sensor and actuator obtaining accesses to the networks. The mapping technique is then utilized to describe the joint scheduling effects of two RA protocols. The desired controller parameters are calculated by solving two coupled backward recursive difference equations. Furthermore, the impact of the RA protocol on the characteristic of the controller parameters is analyzed.
- Chapter 9 considers the ultimately bounded control problem of a class of nonlinear systems subject to the TOD protocol scheduling and uniform quantization effects. Under the scheduling of TOD protocol, the sensor node getting access to the communication channel is determined by a quadratic selection principle. A novel nonlinear-observer-based controller is constructed to ensure that the dynamics of the closed-loop system is ultimately bounded in mean square subject to the

norm-bounded disturbance and quantization error. The desired controller gain matrices are calculated by using the linear-matrix-inequality-based approach.

- Chapter 10 discusses the finite-horizon $\mathscr{H}_\infty$ consensus control for time-varying multi-agent systems subject to the RA protocol scheduling. A scheduling model is introduced for multi-agent systems by extending the original RA protocol scheduling model for centralized networked systems. By using the mapping technology and the Hadamard product, the closed-loop system would be described by a time-varying system with a stochastic parameter matrix. The desired time-varying controller parameters are computed by solving two coupled backward recursive difference equations (RDEs).
- Chapter 11 draws the conclusions and points out some potential research topics related to the results presented in this book.

References

1. Liu, L., Ma, L., Zhang, J., Bo, Y.: Distributed non-fragile set-membership filtering for nonlinear systems under fading channels and bias injection attacks. Int. J. Syst. Sci. **52**(6), 1192–1205 (2021)
2. Hu, J., Zhang, H., Liu, H., Yu, X.: A survey on sliding mode control for networked control systems. Int. J. Syst. Sci. **52**(6), 1129–1147 (2021)
3. Zhang, L., Nguang, S.K., Quyang, D., Yan, S.: Synchronization of delayed neural networks via integral-based event-triggered scheme. IEEE Tans. Neural Netw. Learn. Syst. **31**(12), 5092–5102 (2020)
4. Xu, Y., Wu, Z.-G., Pan, Y.-J.: Event-based dissipative filtering of Markovian jump neural networks subject to incomplete measurements and stochastic cyber-attacks. IEEE Trans. Cybern. **51**(3), 1370–1379 (2021)
5. Leong, A.S., Quevedo, D.E.: Kalman filtering with relays over wireless fading channels. IEEE Trans. Autom. Control **61**(6), 1643–1648 (2016)
6. Caballero-Águila, R., Hermoso-Carazo, A., Linares-Pérez, J.: Distributed fusion filters from uncertain measured outputs in sensor networks with random packet losses. Inf. Fusion **34**, 70–79 (2017)
7. Heemels, W.P.M.H., Donkers, M.C.F., Teel, A.R.: Periodic event-triggered control for linear systems. IEEE Trans. Autom. Control **58**(4), 847–861 (2013)
8. Cheng, P., Qi, Y., Xin, K., Chen, J., Xie, L.: Energy-efficient data forwarding for state estimation in multi-hop wireless sensor networks. IEEE Trans. Autom. Control **61**(5), 1322–1327 (2016)
9. Hu, J., Liang, J., Chen, D., Ji, D., Du, J.: A recursive approach to non-fragile filtering for networked systems with stochastic uncertainties and incomplete measurements. J. Frankl. Inst. **352**(5), 1946–1962 (2015)
10. Li, Y., Liu, S., Zhao, D., Shi, X., Cui, Y.: Event-triggered fault estimation for discrete time-varying systems subject to sector-bounded nonlinearity: a Krein space based approach. Int. J. Robust Nonlinear Control **31**(11), 5360–5380 (2021)
11. Peng, C., Yang, T.C.: Event-triggered communication and H_∞ control co-design for networked control systems. Automatica **49**(5), 1326–1332 (2013)
12. Peng, C., Han, Q.-L.: On designing a novel self-triggered sampling scheme for networked control systems with data losses and communication delays. IEEE Trans. Ind. Electron. **63**(2), 1239–1248 (2016)

13. Wang, Y.-W., Zhang, W.-A., Dong, H., Zhu, J.-W.: Generalized extended state observer based control for networked interconnected systems with delays. Asian J. Control **20**(3), 1253–1262 (2018)
14. Li, T., Zhang, W.-A., Yu, L.: Improved switched system approach to networked control systems with time-varying delays. IEEE Trans. Control Syst. Technol. **27**(6), 2711–2717 (2019)
15. Han, H., Zhang, X., Zhang, W.: Robust distributed model predictive control under actuator saturations and packet dropouts with time-varying probabilities. IET Control Theory Appl. **10**(5), 534–544 (2016)
16. Chen, B., Zhang, W.-A., Yu, L.: Distributed fusion estimation with missing measurements, random transmission delays and packet dropouts. IEEE Trans. Autom. Control **59**(7), 1961–1967 (2014)
17. Lu, R., Peng, H., Liu, S., Xu, Y., Li, X.-M.: Reliable l_2-l_∞ filtering for fuzzy Markov stochastic systems with sensor failures and packet dropouts. IET Control Theory Appl. **11**(14), 2195–2203 (2017)
18. Lian, B., Zhang, Q., Li, J.: Integrated sliding mode control and neural networks based packet disordering prediction for nonlinear networked control systems. IEEE Trans. Neural Netw. Learn. Syst. **30**(8), 2324–2335 (2019)
19. Zhang, F., Zhang, Q., Li, J.: Networked control for T-S fuzzy descriptor systems with network-induced delay and packet disordering. Neurocomputing **275**, 2264–2278 (2018)
20. Liu, A., Zhang, W.-A., Yu, L., Liu, S., Chen, M.Z.Q.: New results on stabilization of networked control systems with packet disordering. Automatica **52**, 255–259 (2015)
21. Zhao, Z., Wang, Z., Zou, L., Guo, J.: Set-membership filtering for time-varying complex networks with uniform quantisations over randomly delayed redundant channels. Int. J. Syst. Sci. **51**(16), 3364–3377 (2020)
22. Ren, W., Xiong, J.: Tracking control of nonlinear networked and quantized control systems with communication delays. IEEE Trans. Autom. Control **65**(8), 3685–3692 (2020)
23. Wang, F., Zhang, L., Zhou, S., Huang, Y.: Neural network-based finite-time control of quantized stochastic nonlinear systems. Neurocomputing **362**, 195–202 (2019)
24. Shen, H., Men, Y., Wu, Z.-G., Cao, J., Lu, G.: Network-based quantized control for fuzzy singularly perturbed semi-Markov jump systems and its application. IEEE Trans. Circuits Syst. I: Regul. Pap. **66**(3), 1130–1140 (2019)
25. Liu, A., Yu, L., Zhang, W., Chen, M.: Moving horizon estimation for networked systems with quantized measurements and packet dropouts. IEEE Trans. Circuits Syst. I: Regul. Pap. **60**(7), 1823–1834 (2013)
26. Zou, L., Wang, Z., Hu, J., Gao, H.: On $\mathscr{H}_\infty$ finite-horizon filtering under stochastic protocol: dealing with high-rate communication networks. IEEE Trans. Autom. Control **62**(9), 4884–4890 (2017)
27. Liu, K., Fridman, E., Hetel, L.: Stability and L_2-gain analysis of Networked Control Systems under Round-Robin scheduling: a time-delay approach. Syst. Control Lett. **61**(5), 666–675 (2012)
28. Song, J., Wang, Z., Niu, Y.: On H_∞ sliding mode control under stochastic communication protocol. IEEE Trans. Autom. Control **64**(5), 2174–2181 (2019)
29. Hu, B., Wang, Y., Orlik, P.V., Koike-Akino, T., Guo, J.: Co-design of safe and efficient networked control systems in factory automation with state-dependent wireless fading channels. Automatica **105**, 334–346 (2019)
30. Guo, G., Eang, L.: Control over medium-constrained vehicular networks with fading channels and random access protocol: a networked systems approach. IEEE Trans. Veh. Technol. **64**(8), 3347–3358 (2015)
31. Chen, Y., Chen, Z., Chen, Z., Xue, A.: Observer-based passive control of non-homogeneous Markov jump systems with random communication delays. Int. J. Syst. Sci. **51**(6), 1133–1147 (2020)
32. Zhang, L., Gao, H., Kaynak, O.: Network-induced constraints in networked control systems-a survey. IEEE Trans. Ind. Inf. **9**(1), 403–416 (2013)

33. Zhang, L., Shi, Y., Chen, T., Huang, B.: A new method for stabilization of networked control systems with random delays. IEEE Trans. Autom. Control **50**(8), 1177–1181 (2005)
34. Zhang, X.-M., Han, Q.-L., Yu, X.: Survey on recent advances in networked control systems. IEEE Trans. Ind. Inf. **12**(5), 1740–1752 (2016)
35. Ge, X., Yang, F., Han, Q.-L.: Distributed networked control systems: a brief overview. Inf. Sci. **380**, 117–131 (2017)
36. Zhang, X.-M., Han, Q.-L., Zhang, B.-L.: An overview and deep investigation on sampled-data-based event-triggered control and filtering for networked systems. IEEE Trans. Ind. Inf. **13**(1), 4–16 (2017)
37. Tabuada, P.: Event-triggered real-time scheduling of stabilizing control tasks. IEEE Trans. Autom. Control **52**(9), 1680–1685 (2007)
38. Dimarogonas, D.V., Frazzoli, E., Johansson, K.H.: Distributed event-triggered control for multi-agent systems. IEEE Trans. Autom. Control **57**(5), 1291–1297 (2012)
39. Wang, F., Chen, B., Lin, C., Zhang, J., Meng, X.: Adaptive neural network finite-time output feedback control of quantized nonlinear systems. IEEE Trans. Cybern. **48**(6), 1839–1848 (2018)
40. Caballero-Águila, R., García-Garrido, I., Linares-Pérez, J.: Information fusion algorithms for state estimation in multi-sensor systems with correlated missing measurements. Appl. Math. Comput. **226**, 548–563 (2014)
41. Dačić, D.B., Nešić, D.: Quadratic stabilization of linear networked control systems via simultaneous protocol and controller design. Automatica **43**(7), 1145–1155 (2007)
42. Zou, L., Wang, Z., Han, Q.-L., Zhou, D.: Ultimate boundedness control for networked systems with Try-Once-Discard protocol and uniform quantization effects. IEEE Trans. Autom. Control **62**(12), 6582–6588 (2017)
43. Zhu, K., Hu, J., Liu, Y., Alotaibi, N.D., Alsaadi, F.E.: On ℓ_2-ℓ_∞ output-feedback control scheduled by stochastic communication protocol for two-dimensional switched systems. Int. J. Syst. Sci. (in press). https://doi.org/10.1080/00207721.2021.1914768
44. Sun, X.-M., Liu, G.-P., Wang, W., Rees, D.: L_2-gain of systems with input delays and controller temporary failure: zero-order hold model. IEEE Trans. Control Syst. Technol. **19**(3), 699–706 (2011)
45. Schenato, L.: To zero or to hold control inputs with lossy links? IEEE Trans. Autom. Control **54**(5), 1093–1099 (2009)
46. Donkers, M.C.F., Heemels, W.P.M.H., van de Wouw, N., Hetel, L.: Stability analysis of networked control systems using a switched linear systems approach. IEEE Trans. Autom. Control **56**(9), 2101–2115 (2011)
47. Tabbara, M., Nešić, D.: Input-output stability of networked control systems with stochastic protocols and channels. IEEE Trans. Autom. Control **53**(5), 1160–1175 (2008)
48. Bauer, N.W., Donkers, M.C.F., van de Wouw, N., Heemels, W.P.M.H.: Decentralized observer-based control via networked communication. Automatica **49**(7), 2074–2086 (2013)
49. Heemels, W.P.M.H., Teel, A.R., van de Wouw, N., Nešić, D.: Networked control systems with communication constraints: tradeoffs between transmission intervals, delays and performance. IEEE Trans. Autom. Control **55**(8), 1781–1796 (2010)
50. Ju, Y., Wei, G., Ding, D., Zhang, S.: Fault detection for discrete time-delay networked systems with round-robin protocol in finite-frequency domain. Int. J. Syst. Sci. **50**(13), 2409–2497 (2019)
51. Donkers, M.C.F., Heemels, W.P.M.H., Bernardini, D., Bemporad, A., Shneer, V.: Stability analysis of stochastic networked control systems. Automatica **48**(5), 917–925 (2012)
52. Liu, K., Seuret, A., Fridman, E., Xia, Y.: Improved stability conditions for discrete-time systems under dynamic network protocols. Int. J. Robust Nonlinear Control **28**(15), 4479–4499 (2018)
53. Long, Y., Yang, G.-H.: Fault detection and isolation for networked control systems with finite frequency specifications. Int. J. Robust Nonlinear Control **24**(3), 495–514 (2014)
54. Walsh, G.C., Ye, H.: Scheduling of networked control systems. IEEE Control Syst. Mag. **21**(1), 57–65 (2001)

55. Zhang, W., Branicky, M.S., Phillips, S.M.: Stability of networked control systems. IEEE Control Syst. Mag. **21**(1), 84–99 (2001)
56. Walsh, G.C., Ye, H., Bushbell, L.G.: Stability analysis of networked control systems. IEEE Trans. Control Syst. Technol. **10**(3), 438–446 (2002)
57. Yu, H., Hao, F., Chen, T.: A uniform analysis on input-to-state stability of decentralized event-triggered control systems. IEEE Tans. Autom. Control **64**(8), 3423–3430 (2019)
58. Carnevale, D., Teel, A.R., Nešić, D.: A Lyapunov proof of an improved maximum allowable transfer interval of networked control systems. IEEE Trans. Autom. Control **52**(5), 892–897 (2007)
59. Nešić, D., Teel, A.R.: Input-output stability properties of networked control systems. IEEE Trans. Autom. Control **49**(10), 1650–1667 (2004)
60. Tabbara, M., Nešić, D., Teel, A.R.: Stability of wireless and wireline networked control systems. IEEE Trans. Autom. Control **52**(9), 1615–1630 (2007)
61. Liu, K., Fridman, E., Johansson, K.H.: Networked control with stochastic scheduling. IEEE Trans. Autom. Control **61**(11), 3071–3076 (2015)
62. Zhou, L., Wei, Y.: Stability analysis of networked control systems based on $\mathscr{L}_p$ properties. Int. J. Control Autom. Syst. **13**(2), 390–397 (2015)
63. Xu, Y., Su, H., Pan, Y., Wu, Z., Xu, W.: Stability analysis of network control systems with round-robin scheduling and packet dropouts. J. Frankl. Inst. **350**(8), 2013–2027 (2013)
64. Liu, K., Fridman, E., Hetel, L., Richard, J.: Sampled-data stabilization via Round-Robin scheduling: a direct Lyapunov–Krasovskii approach. In: Proceedings of the 18th IFAC World Congress, Milano, Italy, pp. 1459–1464 (2011)
65. Liu, K., Fridman, E., Hetel, L.: Networked control systems in the presence of scheduling protocols and communication delays. SIAM J. Control Optim. **53**(4), 1768–1788 (2015)
66. Antunes, D., Hespanha, J.P., Silvestre, C.: Stochastic networked control systems with dynamic protocols. Asian J. Control **17**(1), 99–110 (2015)
67. Li, B., Wang, Z., Han, Q.-L., Liu, H.: Input-to-state stabilization in probability for nonlinear stochastic systems under quantization effects and communication protocol. IEEE Trans. Cybern. **49**(9), 3242–3254 (2019)
68. Ding, D., Han, Q.-L., Wang, Z., Ge, X.: A survey on model-based distributed control and filtering for industrial cyber-physical systems. IEEE Trans. Ind. Inf. **15**(5), 2483–2499 (2019)
69. Zhang, X.-M., Han, Q.-L., Ge, X., Ding, D., Ding, L., Yue, D., Peng, C.: Networked control systems: a survey of trends and techniques. IEEE/CAA J. Automatica Sinica **7**(1) (2020)
70. Ding, D., Han, Q.-L., Xiang, Y., Ge, X., Zhang, X.-M.: A survey on security control and attack detection for industrial cyber-physical systems. Neurocomputing **275**, 1674–1683 (2018)
71. Zhou, C., Du, M., Chen, Q.: Co-design of dynamic scheduling and H_∞ control for networked control systems. Appl. Math. Comput. **218**(21), 10767–10775 (2012)
72. Zhou, C., Lu, H., Ren, J., Chen, Q.: Co-design of dynamic scheduling and quantized control for networked control systems. J. Frankl. Inst. **352**(10), 3988–4003 (2015)
73. Liu, K., Fridman, E.: Discrete-time network-based control under Try-Once-discard protocol and actuator constraints. In: Proceedings of the 2014 European Control Conference (ECC), Strasbourg, France, pp. 442–447 (2014)
74. Liu, K., Fridman, E.: Discrete-time network-based control under scheduling and actuator constraints. Int. J. Robust Nonlinear Control **25**(12), 1816–1830 (2015)
75. Liu, K., Fridman, E., Johansson, K.H., Xia, Y.: quantized control under Round-Robin communication protocol. IEEE Trans. Ind. Electron. **63**(7), 4461–4471 (2016)
76. Zhu, C., Guo, G., Yang, B., Wang, Z.: Networked optimal control with random medium access protocol and packet dropouts. Math. Probl. Eng. **2015**, art. no. 105416 (2015)
77. Zhang, W., Yu, L., Feng, G.: Stabilization of linear discrete-time networked control systems via protocol and controller co-design. Int. J. Robust Nonlinear Control **25**(16), 3072–3085 (2015)
78. Ishii, H.: H_∞ control with limited communication and message losses. Syst. Control Lett. **57**(4), 322–331 (2008)

79. Zhang, J., Peng, C., Fei, M.-R., Tian, Y.-C.: Output feedback control of networked systems with a stochastic communication protocol. J. Frankl. Inst. **354**(9), 3838–3853 (2017)
80. Guo, G., Wen, S.: Protocol sequence and control co-design for a collection of networked control systems. Int. J. Robust Nonlinear Control **26**(3), 489–508 (2016)
81. Chen, T.: Robust state estimation for power systems via moving horizon strategy. Sustain. Energy Grids Netw. **10**, 46–54 (2017)
82. Zou, L., Wen, T., Wang, Z., Chen, L., Roberts, C.: State estimation for communication-based train control systems with CSMA protocol. IEEE Trans. Intell. Transp. Syst. **20**(3), 843–854 (2019)
83. Ge, X., Han, Q.-L., Zhang, X.-M., Ding, L., Yang, F.: Distributed event-triggered estimation over sensor networks: a survey. IEEE Trans. Cybern. (in press). https://doi.org/10.1109/TCYB.2019.2917179
84. Chen, B., Yu, L., Zhang, W.-A., Wang, H.: Distributed $\mathscr{H}_\infty$ fusion filtering with communication bandwidth constraints. Signal Process. **96**, Part B, 284–289 (2014)
85. Zou, L., Wang, Z., Han, Q.-L., Zhou, D.: Moving horizon estimation for networked time-delay systems under Round-Robin protocol. IEEE Trans. Autom. Control **64**(12), 5191–5198 (2019)
86. Zhang, J., Peng, C.: Networked H_∞ filtering under a weighted TOD protocol. Automatica **107**, 333–341 (2019)
87. Shen, Y., Wang, Z., Shen, B., Alsaadi, F.E.: H_∞ filtering for multi-rate multi-sensor systems with randomly occurring sensor saturations under the p-persistent CSMA protocol. IET Control Theory Appl. **14**(10), 1255–1265 (2020)
88. Mao, J., Sun, Y., Yi, X., Liu, H., Ding, D.: Recursive filtering of networked nonlinear systems: a survey. Int. J. Syst. Sci. **52**(6), 1110–1128 (2021)
89. Postoyan, R., van de Wouw, N., Nešić, D., Heemels, W.P.M.H.: Tracking control for nonlinear networked control systems. IEEE Trans. Autom. Control **59**(6), 1539–1554 (2014)
90. Ahmadi, A., Salmasi, F.R.: Observer-based reliable control for Lipschitz nonlinear networked control systems with quadratic protocol. Int. J. Control Autom. Syst. **13**(3), 753–763 (2015)
91. Ding, D., Wang, Z., Han, Q.-L.: Neural-network-based output-feedback control with stochastic communication protocols. Automatica **106**, 221–229 (2019)
92. Ding, D., Wang, Z., Han, Q.-L., Wei, G.: Neural-network-based output-feedback control under Round-Robin scheduling protocols. IEEE Trans. Cybern. **49**(6), 2372–2384 (2019)
93. Wang, W., Nešić, D., Postoyan, R.: Emulation-based stabilization of networked control systems implemented on FlexRay. Automatica **59**, 73–83 (2015)
94. Sun, X.-M., Liu, K.-Z., Wen, C., Wang, W.: Predictive control of nonlinear continuous networked control systems with large time-varying transmission delays and transmission protocols. Automatica **64**, 76–85 (2016)
95. Li, Y., Zhang, K., Liu, K., Johansson, R., Yin, Y.: Neural-network-based adaptive control for bilateral teleoperation with multiple slaves under Round-Robin scheduling protocol. Int. J. Control **94**(6), 1461–1474 (2021)
96. Dong, Y., Song, Y., Wang, J., Zhang, B.: Dynamic output-feedback fuzzy MPC for Takagi-Sugeno fuzzy systems under event-triggering-based try-once-discard protocol. Int. J. Robust Nonlinear Control **30**(4), 1394–1416 (2020)
97. Liu, K.-Z., Wang, R., Li, Y.: Observer design for nonlinear networked control systems with variable transmission delays and protocols based on a hybrid system technique. Neurocomputing **173**, 2115–2120 (2016)
98. Fu, H., Dong, H., Han, F., Shen, Y., Hou, N.: Outlier-resistant H_∞ filtering for a class of networked systems under Round-Robin protocol. Neurocomputing **403**, 133–142 (2020)
99. Wan, X., Wang, Z., Wu, M., Liu, X.: H_∞ state estimation for discrete-time nonlinear singularly perturbed complex networks under the Round-Robin protocol. IEEE Trans. Neural Netw. Learn. Syst. **30**(2), 415–426 (2019)
100. Shen, H., Men, Y., Cao, J., Park, J.H.: H_∞ filtering for fuzzy jumping genetic regulatory networks with Round-Robin protocol: a hidden-Markov-model-based approach. IEEE Trans. Fuzzy Syst. **28**(1), 112–121 (2020)

101. Alsaadi, F.E., Luo, Y., Liu, Y., Wang, Z.: State estimation for delayed neural networks with stochastic communication protocol: the finite-time case. Neurocomputing **281**, 86–95 (2018)
102. Mao, J., Ding, D., Wei, G., Liu, H.: Networked recursive filtering for time-delayed nonlinear stochastic systems with uniform quantisation under Round-Robin protocol. Int. J. Syst. Sci. **50**(4), 871–884 (2019)
103. Chen, S., Guo, J., Ma, L.: Sliding mode observer design for discrete nonlinear time-delay systems with stochastic communication protocol. Int. J. Control Autom. Syst. **17**(7), 1666–1676 (2019)
104. Shen, H., Huo, S., Cao, J., Huang, T.: Generalized state estimation for Markovian coupled networks under Round-Robin protocol and redundant channels. IEEE Trans. Cybern. **49**(4), 1292–1391 (2019)
105. Zou, L., Wang, Z., Han, Q.-L., Zhou, D.: Moving horizon estimation of networked nonlinear systems with random access protocol. IEEE Trans. Syst. Man Cybern. Syst. **51**(5), 2937–2948 (2021)
106. Li, Q., Liang, J.: Dissipativity of the stochastic Markovian switching CVNNs with randomly occurring uncertainties and general uncertain transition rates. Int. J. Syst. Sci. **51**(6), 1102–1118 (2020)
107. Tan, H., Shen, B., Peng, K., Liu, H.: Robust recursive filtering for uncertain stochastic systems with amplify-and-forward relays. Int. J. Syst. Sci. **51**(7), 1188–1199 (2020)
108. Shen, B., Wang, Z., Tan, H.: Guaranteed cost control for uncertain nonlinear systems with mixed time-delays: the discrete-time case. Eur. J. Control **40**, 62–67 (2018)
109. Hu, J., Wang, Z., Liu, G.-P., Zhang, H.: Variance-constrained recursive state estimation for time-varying complex networks with quantized measurements and uncertain inner coupling. IEEE Trans. Neural Netw. Learn. Syst. **31**(6), 1955–1967 (2020)
110. Shi, Y., Peng, X.: Fault detection filters design of polytopic uncertain discrete-time singular Markovian jump systems with time-varying delays. J. Frankl. Inst. **357**(11), 7343–7367 (2020)
111. Rao, H., Xu, Y., Zhang, B., Yao, D.: Robust estimator design for periodic neural networks with polytopic uncertain weight matrices and randomly occurred sensor nonlinearities. IET Control Theory Appl. **12**(9), 1299–1305 (2018)
112. Zhang, X.-M., Han, Q.-L., Ge, X.: A novel finite-sum inequality-based method for robust H_∞ control of uncertain discrete-time Takagi-Sugeno fuzzy systems with interval-like time-varying delays. IEEE Trans. Cybern. **48**(9), 2569–2582 (2018)
113. Caballero-Águila, R., Hermoso-Carazo, A., Linares-Pérez, J.: Centralized, distributed and sequential fusion estimation from uncertain outputs with correlation between sensor noises and signal. Int. J. Gen. Syst. **48**(7), 713–737 (2019)
114. Basin, M.V., Maldonado, J.J.: Optimal controller for uncertain stochastic linear systems with Poisson noises. IEEE Trans. Ind. Inf. **10**(1), 267–275 (2014)
115. Zhu, K., Song, Y., Ding, D.: Resilient RMPC for polytopic uncertain systems under TOD protocol: a switched system approach. Int. J. Robust Nonlinear Control **28**(16), 5103–5117 (2018)
116. Zhu, K., Song, Y., Ding, D., Wei, G., Liu, H.: Robust MPC under event-triggered mechanism and Round-Robin protocol: an average dwell-time approach. Inf. Sci. **457–458**, 126–140 (2018)
117. Song, Y., Wang, Z., Ding, D., Wei, G.: Robust H_2/H_∞ model predictive control for linear systems with polytopic uncertainties under weighted MEF-TOD protocol. IEEE Trans. Syst. Man Cybern. Syst. **49**(7), 1470–1481 (2019)
118. Wang, J., Song, Y., Wei, G.: Dynamic output-feedback RMPC for systems with polytopic uncertainties under Round-Robin protocol. J. Frankl. Inst. **356**(4), 2421–2439 (2019)
119. Wang, J., Song, Y., Wei, G., Dong, Y.: Robust model predictive control for polytopic uncertain systems with state saturation nonlinearities under Round-Robin protocol. Int. J. Robust Nonlinear Control **29**(6), 2188–2202 (2019)
120. Zhao, H., Niu, Y., Jia, T.: Security control of cyber-physical switched systems under Round-Robin protocol: input-to-state stability in probability. Inf. Sci. **508**, 121–134 (2020)

121. Li, D., Liang, J., Wang, F.: H_∞ state estimation for two-dimensional systems with randomly occurring uncertainties and Round-Robin protocol. Neurocomputing **349**, 248–260 (2019)
122. Li, D., Liang, J., Wang, F., Ren, X.: Observer-based H_∞ control of two-dimensional delayed networks under the random access protocol. Neurocomputing **401**, 353–363 (2020)
123. Li, J., Wang, Z., Dong, H., Fei, W.: Delay-distribution-dependent state estimation for neural networks under stochastic communication protocol with uncertain transition probabilities. Neural Netw. **130**, 143–151 (2020)
124. Chen, D., Yang, N., Hu, J., Du, J.: Resilient set-membership state estimation for uncertain complex networks with sensor saturation under Round-Robin protocol. Int. J. Control Autom. Syst. **17**(12), 3035–3046 (2019)
125. Liu, S., Wang, Z., Chen, Y., Wei, G.: Protocol-based unscented Kalman filtering in the presence of stochastic uncertainties. IEEE Trans. Autom. Control **65**(3), 1303–1309 (2020)
126. Dong, H., Bu, X., Hou, N., Liu, Y., Alsaadi, F.E., Hayat, T.: Event-triggered distributed state estimation for a class of time-varying systems over sensor networks with redundant channels. Inf. Fusion **36**, 243–250 (2017)
127. Zhao, Z., Wang, Z., Zou, L., Wang, Z.: Finite-horizon H_∞ state estimation for artificial neural networks with component-based distributed delays and stochastic protocol. Neurocomputing **321**, 169–177 (2018)
128. Liu, H., Wang, Z., Fei, W., Li, J., Alsaadi, F.E.: On finite-horizon H_∞ state estimation for discrete-time delayed memristive neural networks under stochastic communication protocol. Inf. Sci. **555** (2021)
129. Li, J., Wei, G., Ding, D., Li, Y.: Set-membership filtering for discrete time-varying nonlinear systems with censored measurements under Round-Robin protocol. Neurocomputing **281**, 20–26 (2018)
130. Hu, J., Yang, Y., Liu, H., Chen, D., Du, J.: Non-fragile set-membership estimation for sensor-saturated memristive neural networks via weighted try-once-discard protocol. IET Control Theory Appl. **14**(13), 1671–1680 (2020)
131. Yang, Y., Hu, J., Chen, D., Wei, Y., Du, J.: Non-fragile suboptimal set-membership estimation for delayed memristive neural networks with quantization via maximum-error-first protocol. Int. J. Control Autom. Syst. **18**(7), 1904–1914 (2020)
132. Liu, S., Wang, Z., Wang, L., Wei, G.: On quantized $\mathfrak{H}_\infty$ filtering for multi-rate systems under stochastic communication protocols: the finite-horizon case. Inf. Sci. **456**, 211–223 (2018)
133. Shen, Y., Wang, Z., Shen, B., Alsaadi, F.E., Alsaadi, F.E.: Fusion estimation for multi-rate linear repetitive processes under weighted try-once-discard protocol. Inf. Fusion **55**, 281–291 (2020)
134. Chen, Y., Wang, Z., Wang, L., Sheng, W.: Finite-horizon H_∞ state estimation for stochastic coupled networks with random inner couplings using Round-Robin protocol. IEEE Trans. Cybern. **51**(3), 1204–1215 (2021)
135. Li, X., Han, F., Hou, N., Dong, H., Liu, H.: Set-membership filtering for piecewise linear systems with censored measurements under Round-Robin protocol. Int. J. Syst. Sci. **51**(9), 1578–1588 (2020)
136. Yuan, Y., Wang, Z., Zhang, P., Liu, H.: Near-optimal resilient control strategy design for state-saturated networked systems under stochastic communication protocol. IEEE Trans. Cybern. **49**(8), 3155–3167 (2019)
137. Yuan, Y., Zhang, P., Wang, Z., Chen, Y.: Noncooperative event-triggered control strategy design with Round-Robin protocol: applications to load frequency control of circuit systems. IEEE Trans. Ind. Electron. **67**(3), 2155–2166 (2020)
138. Shen, L., Niu, Y., Zou, L., Liu, Y., Alsaadi, F.E.: Finite-horizon state estimation for time-varying complex networks with random coupling strengths under Round-Robin protocol. J. Frankl. Inst. **355**(15), 7417–7442 (2018)
139. Wang, D., Wang, Z., Shen, B., Li, Q.: H_∞ finite-horizon filtering for complex networks with state saturations: the weighted try-once-discard protocol. Int. J. Robust Nonlinear Control **29**(7), 2096–2111 (2019)

140. Wan, X., Wang, Z., Han, Q.-L., Wu, M.: A recursive approach to quantized H_∞ state estimation for genetic regulatory networks under stochastic communication protocols. IEEE Trans. Neural Netw. Learn. Syst. **30**(9), 2840–2852 (2019)
141. Zou, L., Wang, Z., Han, Q.-L., Zhou, D.: Recursive filtering for time-varying systems with random access protocol. IEEE Trans. Autom. Control **64**(12), 720–727 (2019)
142. Liu, S., Wang, Z., Hu, J., Wei, G.: Protocol-based extended Kalman filtering with quantization effects: the Round-Robin case. Int. J. Robust Nonlinear Control (in press). https://doi.org/10.1002/rnc.5205
143. Zhang, H., Hu, J., Liu, H., Yu, X., Liu, F.: Recursive state estimation for time-varying complex networks subject to missing measurements and stochastic inner coupling under random access protocol. Neurocomputing **346**, 49–57 (2019)
144. Zou, L., Wang, Z., Han, Q.-L., Zhou, D.: Full information estimation for time-varying systems subject to Round-Robin scheduling: a recursive filter approach. IEEE Trans. Syst. Man Cybern. Syst. **51**(3), 1904–1916 (2021)
145. Alsaadi, F.E., Wang, Z., Wang, D., Alsaadi, F.E., Alsaade, F.W.: Recursive fusion estimation for stochastic discrete time-varying complex networks under stochastic communication protocol: the state-saturated case. Inf. Fusion **60**, 11–19 (2020)
146. Ugrinovskii, V., Fridman, E.: A Round-Robin type protocol for distributed estimation with H_∞ consensus. Syst. Control Lett. **69**, 103–110 (2014)
147. Xu, Y., Lu, R., Shi, P., Li, H., Xie, S.: Finite-time distributed state estimation over sensor networks with Round-Robin protocol and fading channels. IEEE Trans. Cybern. **48**(1), 336–345 (2018)
148. Liu, K., Guo, H., Zhang, Q., Xia, Y.: Distributed secure filtering for discrete-time systems under Round-Robin protocol and deception attacks. IEEE Trans. Cybern. **50**(8), 3571–3580 (2020)
149. Shen, B., Wang, Z., Wang, D., Liu, H.: Distributed state-saturated recursive filtering over sensor networks under Round-Robin protocol. IEEE Trans. Cybern. **50**(8), 3605–3615 (2020)
150. Chen, S., Ma, L., Ma, Y.: Distributed setmembership filtering for nonlinear systems subject to round-robin protocol and stochastic communication protocol over sensor networks. Neurocomputing **385**, 13–21 (2020)
151. Yuan, Y., Shi, M., Guo, L., Yang, H.: A resilient consensus strategy of near-optimal control for state-saturated multiagent systems with round-robin protocol. Int. J. Robust Nonlinear Control **29**(10), 3200–3216 (2019)
152. Shen, L., Niu, Y., Gao, M.: Distributed resilient filtering for time-varying systems over sensor networks subject to Round-Robin/stochastic protocol. ISA Trans. **87**, 55–67 (2019)
153. Song, J., Han, F., Fu, H., Liu, H.: Finite-horizon distributed H_∞-consensus control of time-varying multi-agent systems with Round-Robin protocol. Neurocomputing **364**, 219–226 (2019)
154. Wang, Y., Ding, S.X., Xu, D., Shen, B.: An $\mathscr{H}_\infty$ fault estimation scheme of wireless networked control systems for industrial real-time applications. IEEE Trans. Control Syst. Technol. **22**(6), 2073–2086 (2014)
155. Ding, S.X., Shen, B., Wang, Z., Zhong, M.: A fault detection scheme for linear discrete-time systems with an integrated online performance evaluation. Int. J. Control **87**(12), 2511–2521 (2014)
156. Li, J., Wei, G., Ding, D., Zhang, S.: Event-triggered fault detection for switched systems with time-varying sojourn probabilities. Int. J. Robust Nonlinear Control **29**(18), 6463–6482 (2019)
157. Yang, H., Yin, S.: Actuator and sensor fault estimation for time-delay Markov jump systems with application to wheeled mobile manipulators. IEEE Trans. Ind. Inf. **16**(5), 3222–3232 (2019)
158. Long, Y., Yang, G.-H.: Fault detection filter design for stochastic networked control systems. Int. J. Robust Nonlinear Control **25**(3), 443–460 (2015)
159. Long, Y., Park, J.H., Ye, D.: Frequency-dependent fault detection for networked systems under uniform quantization and try-once-discard protocol. Int. J. Robust Nonlinear Control **30**(2), 787–803 (2020)

160. Luo, Y., Wang, Z., Wei, G., Alsaadi, F.E.: H_∞ fuzzy fault detection for uncertain 2-D systems under Round-Robin scheduling protocol. IEEE Trans. Syst. Man Cybern. Syst. **47**(8), 2172–2184 (2017)
161. Luo, Y., Wang, Z., Wei, G.: Fault detection for fuzzy systems with multiplicative noises under periodic communication protocols. IEEE Trans. Fuzzy Syst. **26**(4), 2384–2395 (2017)
162. Chen, W., Hu, J., Yu, X., Chen, D.: Protocol-based fault detection for discrete delayed systems with missing measurements: the uncertain missing probability case. IEEE Access **6**, 76616–76626 (2018)
163. Chen, W., Hu, J., Yu, X., Chen, D., Du, J.: Robust fault detection for nonlinear discrete systems with data drift and randomly occurring faults under weighted try-once-discard protocol. Circuits Syst. Signal Process. **39**(1), 111–137 (2020)
164. Ren, W., Sun, S., Huo, F., Lu, Y.: Nonfragile H_∞ fault detection for fuzzy discrete systems under stochastic communication protocol. Optimal Control Appl. Methods (in press). https://doi.org/10.1002/oca.2674
165. Gao, M., Zhang, W., Sheng, L., Zhou, D.: Distributed fault estimation for delayed complex networks with Round-Robin protocol based on unknown input observer. J. Frankl. Inst. **357**(13), 8678–8702 (2020)
166. Ju, Y., Wei, G., Ding, D., Liu, S.: A novel fault detection method under weighted try-once-discard scheduling over sensor networks. IEEE Trans. Control Netw. Syst. **7**(3), 1489–1499 (2020)
167. Gao, M., Yang, S., Sheng, L., Zhou, D.: Fault diagnosis for time-varying systems with multiplicative noises over sensor networks subject to Round-Robin protocol. Neurocomputing **346**, 65–72 (2019)
168. Fu, H., Dong, H., Song, J., Hou, N., Li, G.: Fault estimation for time-varying systems with Round-Robin protocol. Kybernetika **56**(1), 107–126 (2020)
169. Dong, H., Hou, N., Wang, Z., Liu, H.: Finite-horizon fault estimation under imperfect measurements and stochastic communication protocol: dealing with finite-time boundedness. Int. J. Robust Nonlinear Control **29**(1), 117–134 (2019)

Chapter 2
Ultimately Bounded Filtering for Complex Networks under Round-Robin Protocol

The classical view of complex networks stems from the percolation and random graph theory [1]. In the past decades, complex networks have attracted an ever-growing research interest due mainly to their strong capacity to model a wide range of systems with web-like structures. Examples of complex networks include the well-known Internet, brain networks, biological networks, neural networks and social networks. A typical complex network is composed of a number of interconnected dynamical units subject to certain topology. The connections in a complex network stand for the information exchanges among different nodes. Early research works on complex networks have focused on the statistical mechanics of the network topology by applying the graph theory. Subsequently, since the discovery of small-world property [2] and scale-free property [3], particular research attention has been devoted to the dynamical behaviors for complex networks. Different from the individual unit, the dynamical behaviors of complex networks are largely influenced by the network topology, and this has led to the particular difficulty in analyzing the dynamical properties.

Filtering (or state estimation) is a fundamental research topic for dynamics analysis of complex networks since the acquisition of accurate state information is often imperative for many practical tasks such as synchronization and pining control. Nevertheless, the exact state of a complex network is usually inaccessible but only its measurement output is made available to the users, and this necessitates the development of filtering techniques according to certain performance requirements. These filtering techniques include, but are not limited to, the well-known minimum mean square error (MMSE) filtering, ultimately bounded filtering, $\mathscr{H}_\infty$ filtering, set-membership filtering, energy-to-peak filtering and finite-time filtering.

On another research front, owing to the rapid development of network-based communication, the research on the filtering problem of complex network with network-induced effects has attracted quite a lot attention. Such network-induced effects would

L. Zou et al., *Communication-Protocol-Based Filtering and Control of Networked Systems*, Studies in Systems, Decision and Control 430,
https://doi.org/10.1007/978-3-030-97512-8_2

bring some new challenging issues in the filtering problem of complex networks. In this chapter, we intend to study the ultimately bounded filtering problem for a class of time-delay complex networks where the RR protocol is adopted to schedule the signal transmissions between sensors and the filter. By using a switch-based approach, the dynamics of the filtering error is modeled by a periodic parameter-switching system with time-delays. The purpose of the problem addressed is to design the filter such that the filtering error is exponentially ultimately bounded with a certain asymptotic upper bound in mean square. Furthermore, such a bound is subsequently minimized by the designed filter parameters. A novel Lyapunov-like functional is employed to deal with the dynamics analysis of the filtering error. Sufficient conditions are established to guarantee the ultimate boundedness of the filtering error in mean square by applying the stochastic analysis approach. Then, the desired filter gains are characterized by solving a convex problem. Finally, a numerical example is given to illustrate the effectiveness of the filter design scheme.

2.1 Problem Formulation

Consider the following time-delay complex network consisting of N nodes:

$$\begin{cases} x_{i,k+1} = f(x_{i,k}) + g(x_{i,k-\tau}) + \sum_{j=1}^{N} w_{i,j}\Gamma x_{j,k} + B_i\omega_k + E_i\nu_k \\ y_{i,k} = C_i x_{i,k} + D_i\nu_k \\ z_{i,k} = M_i x_{i,k}, \quad i = 1, 2, \ldots, N \\ x_{i,j} = \psi_{i,j}, \quad j = -\tau, -\tau+1, \ldots, -1, 0 \end{cases} \tag{2.1}$$

where $x_{i,k} \in \mathbb{R}^n$, $y_{i,k} \in \mathbb{R}^q$ and $z_{i,k} \in \mathbb{R}^m$ are, respectively, the state vector, the measurement output and the output vector to be estimated for the ith node. $\omega_k \in \mathbb{R}^{r_1}$ is the process noise with zero-mean and known variance $R = \mathscr{R}\mathscr{R}^T \geq 0$. $\nu_k \in \mathbb{R}^{r_2}$ denotes the exogenous disturbance input. $f(\cdot)$ and $g(\cdot)$ are nonlinear vector-valued functions satisfying certain conditions given later. $\psi_{i,j}$ $(j = -\tau, -\tau+1, \ldots, -1, 0)$ are the initial conditions. The constant matrices B_i, C_i, D_i, E_i and M_i are known with appropriate dimensions. $\Gamma = \text{diag}\{\gamma_1, \gamma_2, \ldots, \gamma_n\} > 0$ is an inner-coupling matrix linking the jth state variable if $\gamma_j \neq 0$, and $W = [w_{i,j}]_{N\times N}$ is the coupled configuration matrix of the network with $w_{i,j} \geq 0$ $(i \neq j)$ but not all zero. In this chapter, we consider the case that the topology of the complex network is weight balanced. Thus, the coupling configuration matrix $W = [w_{i,j}]_{N\times N}$ satisfies

$$\sum_{j=1}^{N} w_{i,j} = \sum_{j=1}^{N} w_{j,i} = 0 \quad (i = 1, 2, \ldots, N). \tag{2.2}$$

Remark 2.1 The model (2.1) includes the term of the time-delay $g(x_{i,k-\tau})$. As is well known, time-delays commonly occur in many dynamical systems including microwave oscillators, electronics, biological systems and hydraulic systems [4–6]. The existence of time-delays could lead to the deterioration of the system performance or even cause the instability of the system [7, 8]. The inclusion of the time-delay term will bring additional difficulties in the analysis.

Assumption 2.1 $f(\cdot)$ and $g(\cdot)$ are two sector-bounded nonlinear function satisfying the following constraints:

$$\begin{aligned}&(f(x) - f(z) - U_1(x-z))^T (f(x) - f(z) - U_2(x-z)) \leq 0, \ f(0) = 0,\\ &(g(x) - g(z) - L_1(x-z))^T (g(x) - g(z) - L_2(x-z)) \leq 0, \ g(0) = 0,\end{aligned} \tag{2.3}$$

where U_i and L_i $(i = 1, 2)$ are real matrices of appropriate dimensions.

Assumption 2.2 The exogenous disturbance input v_k is a norm-bounded vector satisfying the constraint $\|v_k\| \leq \bar{v}$, where $\bar{v}$ is a known constant.

For notation simplicity, we let

$$\begin{aligned}
x_k &= \left[x_{1,k}^T \ x_{2,k}^T \ \cdots \ x_{N,k}^T\right]^T, \quad y_k = \left[y_{1,k}^T \ y_{2,k}^T \ \cdots \ y_{N,k}^T\right]^T,\\
B &= \left[B_1^T \ B_2^T \ \cdots \ B_N^T\right]^T, \ F(x_k) = \left[f^T(x_{1,k}) \ f^T(x_{2,k}) \ \cdots \ f^T(x_{N,k})\right]^T,\\
D &= \left[D_1^T \ D_2^T \ \cdots \ D_N^T\right]^T, \ G(x_k) = \left[g^T(x_{1,k}) \ g^T(x_{2,k}) \ \cdots \ g^T(x_{N,k})\right]^T,\\
z_k &= \left[z_{1,k}^T \ z_{2,k}^T \ \cdots \ z_{N,k}^T\right]^T, \ C = \mathrm{diag}\{C_1, \ C_2, \ \ldots, \ C_N\},\\
E &= \left[E_1^T \ E_2^T \ \cdots \ E_N^T\right]^T, \ M = \mathrm{diag}\{M_1, \ M_2, \ \ldots, \ M_N\}.
\end{aligned}$$

By using the matrix Kronecker product, the complex dynamical network (2.1) can be rewritten in a compact form as follows:

$$\begin{cases} x_{k+1} = F(x_k) + G(x_{k-\tau}) + (W \otimes \Gamma)x_k + B\omega_k + Ev_k,\\ y_k = Cx_k + Dv_k,\\ z_k = Mx_k. \end{cases} \tag{2.4}$$

In this chapter, we aim to design a filter for the complex network (2.4) where the communication between the remote filter and the complex network (2.4) is implemented via a shared network. The filtering problem under consideration is described in Fig. 2.1. Next, let us introduce the proposed network and filter.

2.1.1 Signal Transmissions over the Communication Network

In this chapter, we consider the case that only one node is physically allowed to transmit measurement data at each transmission instant due to the communication

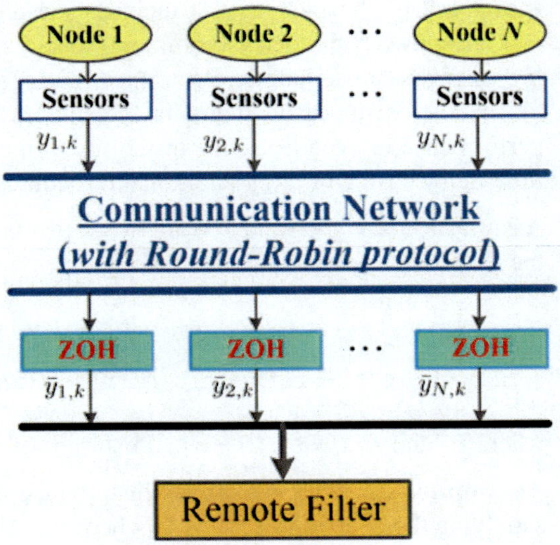

Fig. 2.1 The filtering problem of a complex network subject to the RR protocol

constraints. The RR protocol is adopted to schedule the signal transmissions over such a communication network, based on which all the network nodes are allocated with the access opportunities to the communication network in a fixed order.

Let the first transmission instant of the ith node be i. When all the nodes are given the access one by one in a circle, we say that a round has been finished. Then, the transmission sequence of all the nodes in the $(\gamma+1)$th $(\gamma = 0, 1, 2, \ldots)$ round can be described as

$$y_{1,\gamma N+1},\ y_{2,\gamma N+2},\ y_{3,\gamma N+3},\ \ldots,\ y_{N,\gamma N+N},$$

which implies that the measurement output of the ith node, $y_i(k)$, is transmitted at the kth instant if and only if $\mathrm{mod}(k-i, N) = 0$.

Next, we consider the signals received by the filter. In this chapter, the zero-order holder (ZOH) strategy is adopted to compensate the signals received by the filter. Let $\bar{y}_k = \left[\bar{y}_{1,k}^T\ \bar{y}_{2,k}^T\ \cdots\ \bar{y}_{N,k}^T\right]^T \in \mathbb{R}^{qN}$ denote the measurement output after transmission. The update of $\bar{y}_{i,k}$ $(i = 1, 2, \ldots, N)$ subject to the RR protocol according to the scheme depicted in Fig. 2.1 is given as follows:

$$\bar{y}_{i,k} = \begin{cases} y_{i,k}, & \text{if } \mathrm{mod}\ (k-i, N) = 0 \text{ and } k > 0, \\ \bar{y}_{i,k-1}, & \text{otherwise.} \end{cases} \tag{2.5}$$

For the purpose of simplicity, we assume that $\bar{y}_l = 0$ for $l = -\tau_M, -\tau_M + 1, \ldots, -1, 0$. Then we define the following update matrix

$$\Phi_i = \mathrm{diag}\{\delta(i-1)I, \delta(i-2)I, \ldots, \delta(i-N)I\}, \tag{2.6}$$

where $\delta(\cdot) \in \{0, 1\}$ is the Kronecker delta function. In such a case, we have the following proposition.

Proposition 2.1 *Consider the communication scheduling described by (2.5). It can be derived that the update of $\bar{y}_k$ can be expressed as*

$$\bar{y}_k = \Phi_{\hbar(k)} y_k + (I - \Phi_{\hbar(k)})\bar{y}_{k-1}, \tag{2.7}$$

where $\hbar(k) = \text{mod}(k-1, N) + 1$ denotes the selected node obtaining access to the communication network.

Proof $\forall k \in \mathbb{N}^+$, it can be seen that there exists a pair of non-negative integers $q_1 \in \mathbb{N}$, $q_2 \in \mathbb{N}^+$ satisfying $0 \le q_2 \le N-1$ and $k = q_1 N + q_2$. Hence, it can be concluded from (2.5) that

$$\bar{y}_k = \begin{cases} \Phi_N y_k + (I - \Phi_N)\bar{y}_{k-1}, & \text{if } q_2 = 0, \\ \Phi_{q_2} y_k + (I - \Phi_{q_2})\bar{y}_{k-1}, & \text{otherwise.} \end{cases}$$

Noting that

$$\hbar(k) = \begin{cases} N, & \text{if } q_2 = 0, \\ q_2, & \text{otherwise,} \end{cases}$$

Equation (2.7) follows immediately and the proof is complete.

Remark 2.2 There are generally two signal compensation methods widely adopted in practical applications to generate the value of $\bar{y}_{i,k}$ when node i is not selected to transmit its measurement data: the ZOH strategy and ZI strategy. The choice between these strategies is dependent on the actual system performance requirement. In this chapter, the zero-order holder strategy is employed to compensate the values in $\bar{y}_k$, and we point out that the main results can be easily extended to the zero-input strategy where the update of $\bar{y}_k$ is written as $\bar{y}_k = \Phi_{\hbar(k)} y_k$.

Remark 2.3 The update law (2.7) of $\bar{y}_k$ is derived based on the communication scheduling (2.5). It is easy to see that the update law (2.7) can be easily extended to the traditional filtering issue for complex networks (where all the network nodes are allowed to transmit their measurement data at each transmission instant simultaneously). In this case, the update matrix $\Phi_{\hbar(k)}$ in (2.7) would be replaced by the identity matrix I, and the update law (2.7) would be rewritten as $\bar{y}_k = y_k$.

Denoting $\bar{x}_k = \left[x_k^T \ \bar{y}_{k-1}^T\right]^T$, the complex network (2.4) with a RR protocol can be represented as follows:

$$\begin{cases} \bar{x}_{k+1} = \bar{A}_{\hbar(k)}\bar{x}_k + \bar{F}(\bar{x}_k) + \bar{G}(\bar{x}_{k-\tau}) + \bar{B}\omega_k + \bar{E}_{\hbar(k)}\nu_k, \\ \bar{y}_k = \bar{C}_{\hbar(k)}\bar{x}_k + \bar{D}_{\hbar(k)}\nu_k, \\ z_k = \bar{M}\bar{x}_k, \end{cases} \tag{2.8}$$

where

$$\bar{A}_{\hbar(k)} = \begin{bmatrix} (W \otimes \Gamma) & 0 \\ \Phi_{\hbar(k)} C & I - \Phi_{\hbar(k)} \end{bmatrix}, \quad \bar{B} = \begin{bmatrix} B \\ 0 \end{bmatrix}, \quad \bar{C}_{\hbar(k)} = \begin{bmatrix} \Phi_{\hbar(k)} C & I - \Phi_{\hbar(k)} \end{bmatrix},$$

$$\bar{D}_{\hbar(k)} = \Phi_{\hbar(k)} D, \quad \bar{G}(\bar{x}_{k-\tau(k)}) = \begin{bmatrix} G(x_{k-\tau(k)}) \\ 0 \end{bmatrix}, \quad \bar{M} = \begin{bmatrix} M & 0 \end{bmatrix},$$

$$\bar{E}_{\hbar(k)} = \begin{bmatrix} E \\ \Phi_{\hbar(k)} D \end{bmatrix}, \quad \bar{F}(\bar{x}_k) = \begin{bmatrix} F(x_k) \\ 0 \end{bmatrix}.$$

2.1.2 Filter Structure

In order to estimate the states of the complex network (2.4) with a RR protocol, we construct the filter on the node i as follows:

$$\begin{cases} \hat{x}_{1,i,k+1} = f(\hat{x}_{1,i,k}) + K_{1,i}\Big(\bar{y}_{i,k} - \delta(\hbar(k)-i)C_i\hat{x}_{1,i,k} \\ \qquad - \big(1-\delta(\hbar(k)-i)\big)\hat{x}_{2,i,k}\Big) + g(\hat{x}_{1,i,k-\tau}) + \sum_{j=1}^{N} w_{i,j}\Gamma\hat{x}_{1,j,k}, \\ \hat{x}_{2,i,k+1} = \delta(\hbar(k)-i)C_i\hat{x}_{1,i,k} + \big(1-\delta(\hbar(k)-i)\big)\hat{x}_{2,i,k} + K_{2,i}\Big(\bar{y}_{i,k} \\ \qquad - \delta(\hbar(k)-i)C_i\hat{x}_{1,i,k} - \big(1-\delta(\hbar(k)-i)\big)\hat{x}_{2,i,k}\Big), \\ \hat{z}_{i,k} = M_i\hat{x}_{i,k}, \qquad (i = 1, 2, \ldots, N), \end{cases} \tag{2.9}$$

where $\hat{x}_{1,i,k} \in \mathbb{R}^n$ is the estimate of the network state $x_{i,k}$, $\hat{x}_{2,i,k} \in \mathbb{R}^q$ is the estimate of $\bar{y}_{i,k-1}$, $\hat{z}_{i,k} \in \mathbb{R}^m$ is the estimate of the output $z_{i,k}$, $K_{1,i} \in \mathbb{R}^{n\times q}$ and $K_{2,i} \in \mathbb{R}^{q\times q}$ are the filter parameters to be designed.

By setting

$$\hat{x}_k = \begin{bmatrix} \hat{x}_{1,1,k}^T & \cdots & \hat{x}_{1,N,k}^T & \hat{x}_{2,1,k}^T(k) & \cdots & \hat{x}_{2,N,k}^T \end{bmatrix}^T,$$
$$\hat{z}_k = \begin{bmatrix} \hat{z}_{1,k}^T & \hat{z}_{2,k}^T & \cdots & \hat{z}_{N,k}^T \end{bmatrix}^T,$$

we rewrite the filter for the complex network (2.9) in the following compact form:

$$\begin{cases} \hat{x}_{k+1} = \bar{A}_{\hbar(k)}\hat{x}_k + \bar{F}(\hat{x}_k) + \bar{G}(\hat{x}_{k-\tau}) + K(\bar{y}_k - \bar{C}_{\hbar(k)}\hat{x}_k) \\ \hat{z}_k = \bar{M}\hat{x}_k \\ \hat{x}_j = j = -\tau_M, -\tau_M + 1, \ldots, -1, 0 \end{cases} \tag{2.10}$$

in which

$$K = \begin{bmatrix} K_x^T & K_y^T \end{bmatrix}^T, \quad K_x = \mathrm{diag}\{K_{1,1}, K_{1,2}, \ldots, K_{1,N}\}, \quad K_y = \mathrm{diag}\{K_{2,1}, K_{2,2}, \ldots, K_{2,N}\}.$$

Letting $e_k = \bar{x}_k - \hat{x}_k$ and $\tilde{z}_k = z_k - \hat{z}_k$ be the filtering error and output filtering error, the filtering error dynamics for the complex network can be obtained from (2.4) and (2.10) as follows:

$$\begin{cases} e_{k+1} = (\bar{A}_{\hbar(k)} - K\bar{C}_{\hbar(k)})e_k + \tilde{F}(e_k) + \bar{B}\omega_k + \tilde{G}(e_{k-\tau}) \\ \qquad\quad + (\bar{E}_{\hbar(k)} - K\bar{D}_{\hbar(k)})\nu_k, \\ \tilde{z}_k = \bar{M}e_k, \end{cases} \tag{2.11}$$

where

$$\tilde{F}(e_k) = \bar{F}(\bar{x}_k) - \bar{F}(\hat{x}_k), \tilde{G}(e_{k-\tau(k)}) = \bar{G}(\bar{x}_{k-\tau(k)}) - \bar{G}(\hat{x}_{k-\tau(k)}).$$

Now, let us introduce the following definition which is necessary to give our main results.

Definition 2.1 [9] The dynamics of the filtering error e_k (i.e. the solution of system (2.11)) is said to be exponentially ultimately bounded in mean square if there exist constants $\alpha > 0$, $\beta > 0$, $\delta > 0$ such that

$$\mathbb{E}\{\|e_k\|^2\} \le \alpha^k \beta + \delta, \tag{2.12}$$

where $\alpha \in [0, 1)$, $\beta > 0$ and $\delta > 0$. We denote α and δ as the decay rate and the asymptotic upper bound of $\mathbb{E}\{\|e_k\|^2\}$, respectively.

The purpose of the problem addressed in this chapter is to design a filter for the complex network (2.1) such that the following requirements are met simultaneously:

(1) the dynamics of the filtering error e_k is exponentially ultimately bounded in mean square subject to the process noise ω_k and the exogenous disturbance ν_k;
(2) an asymptotic upper bound of the output filtering error $\tilde{z}_k$ exists and is minimized by the designed filter gain matrices $K_{1,i}$ and $K_{2,i}$ $(i = 1, 2, \ldots, N)$.

Remark 2.4 It is noted from (2.1) that: (i) the additive stochastic noise ω_k is considered which leads to the non-zero error variance; and (ii) the exogenous deterministic disturbance ν_k which further leads to the derivation of the possible equilibrium. As such, we aim to study the ultimate boundedness (rather than the stability) of the filtering error in mean square. To avoid unnecessarily large bound, we will not only obtain the decay rate for the error dynamics to converge to a bounded set but also minimize the bound obtained.

2.2 Main Results

In this section, a sufficient condition is presented to guarantee the ultimate boundedness of the filtering error e_k in mean square. Furthermore, the desired filter gain matrices are derived by solving a minimization problem subject to a set of matrix inequality constraints.

Theorem 2.1 *Let the filter parameter K be given. Assume that there exist $N+2$ positive definite matrices $P_{1,i}$ ($i = 1, 2, \ldots, N$), P_2, $P_3 \in \mathbb{R}^{N(n+q)\times N(n+q)}$, four positive scalars λ_i ($i = 1, 2, 3$) and ρ satisfying*

$$\hat{\Xi}_i = \begin{bmatrix} \chi_{11} & \chi_{12} \\ * & \chi_{22} \end{bmatrix} < 0, \tag{2.13}$$

$$P_2 + \tau P_3 \leq \rho P_3, \tag{2.14}$$

where

$$\chi_{11} = \begin{bmatrix} \chi_{11}^{11} & \chi_{11}^{12} & \chi_{11}^{13} \\ * & \chi_{11}^{22} & 0 \\ * & * & \chi_{11}^{33} \end{bmatrix}, \quad \chi_{12} = \begin{bmatrix} \chi_{12}^{11} & \chi_{12}^{12} \\ \chi_{12}^{21} & \chi_{12}^{22} \\ 0 & 0 \end{bmatrix}, \quad \chi_{22} = \begin{bmatrix} \chi_{22}^{11} & \chi_{22}^{12} \\ * & \chi_{22}^{22} \end{bmatrix},$$

$$\mathfrak{U}_1 = \begin{bmatrix} \mathrm{diag}_N\{U_1\} & 0 \\ 0 & 0 \end{bmatrix}, \quad \mathfrak{U}_2 = \begin{bmatrix} \mathrm{diag}_N\{U_2\} & 0 \\ 0 & 0 \end{bmatrix}, \quad \mathfrak{L}_1 = \begin{bmatrix} \mathrm{diag}_N\{L_1\} & 0 \\ 0 & 0 \end{bmatrix},$$

$$\mathfrak{L}_2 = \begin{bmatrix} \mathrm{diag}_N\{L_2\} & 0 \\ 0 & 0 \end{bmatrix}, \quad \bar{U} = \mathfrak{U}_1 + \mathfrak{U}_2, \quad \bar{L} = \mathfrak{L}_1 + \mathfrak{L}_2, \quad \hat{U} = \mathfrak{U}_1^T \mathfrak{U}_2 + \mathfrak{U}_2^T \mathfrak{U}_1,$$

$$\hat{L} = \mathfrak{L}_1^T \mathfrak{L}_2 + \mathfrak{L}_2^T \mathfrak{L}_1, \quad \chi_{12}^{21} = P_{1,i+1}, \quad \chi_{12}^{22} = P_{1,i+1}\left(\bar{E}_{\hbar(k)} - K\bar{D}_{\hbar(k)}\right),$$

$$\chi_{11}^{11} = \left(\bar{A}_i - K\bar{C}_i\right)^T P_{1,i+1}\left(\bar{A}_i - K\bar{C}_i\right) - \left(1 - \frac{1}{\rho}\right) P_{1,i} - \lambda_1 \hat{U} - \lambda_2 \hat{L},$$

$$\chi_{11}^{12} = \lambda_1 \bar{U}^T + \left(\bar{A}_i - K\bar{C}_i\right)^T P_{1,i+1}, \quad \chi_{11}^{13} = \lambda_2 \bar{L}^T, \quad \chi_{11}^{22} = -2\lambda_1 I + P_{1,i+1},$$

$$\chi_{12}^{11} = \left(\bar{A}_i - K\bar{C}_i\right)^T P_{1,i+1}, \quad \chi_{11}^{33} = -2\lambda_2 I + P_2 + \tau P_3, \quad \chi_{22}^{11} = -P_2 + P_{1,i+1},$$

$$\chi_{12}^{12} = \left(\bar{A}_i - K\bar{C}_i\right)^T P_{1,i+1}\left(\bar{E}_{\hbar(k)} - K\bar{D}_{\hbar(k)}\right), \quad \chi_{22}^{12} = P_{1,i+1}\left(\bar{E}_{\hbar(k)} - K\bar{D}_{\hbar(k)}\right),$$

$$\chi_{22}^{22} = \left(\bar{E}_{\hbar(k)} - K\bar{D}_{\hbar(k)}\right)^T P_{1,i+1}\left(\bar{E}_{\hbar(k)} - K\bar{D}_{\hbar(k)}\right) - \lambda_3 I,$$

with $P_{1,N+1} = P_{1,1}$ for all $i \in \{1, 2, \ldots, N\}$. Then, the dynamics of the filtering error is ultimately bounded in mean square subject to the process noise $\omega(k)$ and exogenous disturbance $\nu(k)$.

Proof To analyze the ultimate boundedness of the filtering error $e(k)$, we choose the following Lyapunov-like functional:

$$V_k = V_{1,k} + V_{2,k} + V_{3,k}, \tag{2.15}$$

where

$$V_{1,k} = e_k^T P_{1,\hbar(k)} e_k, \quad V_{2,k} = \sum_{l=k-\tau}^{k-1} \tilde{G}^T(e_l) P_2 \tilde{G}(e_l),$$

$$V_{3,k} = \sum_{j=-\tau}^{-1} \sum_{l=k+j}^{k-1} \tilde{G}^T(e_l) P_3 \tilde{G}(e_l).$$

The corresponding difference of V_k along the trajectory of system (2.11) can be calculated as follows:

$$\mathbb{E}\{\Delta V_k\} = \sum_{i=1}^{3} \mathbb{E}\{\Delta V_{i,k}\}, \tag{2.16}$$

where

$$\begin{aligned}
&\mathbb{E}\{\Delta V_{1,k}\} = \mathbb{E}\{e_{k+1}^T P_{1,\hbar(k+1)} e_{k+1} - e_k^T P_{1,\hbar(k)} e_k\} \\
=&\mathbb{E}\Big\{e_k^T \big((\bar{A}_{\hbar(k)} - K\bar{C}_{\hbar(k)})^T P_{1,\hbar(k+1)} (\bar{A}_{\hbar(k)} - K\bar{C}_{\hbar(k)}) - P_{1,\hbar(k)}\big) e_k \\
&+ \tilde{F}^T(e_k) P_{1,\hbar(k+1)} \tilde{F}(e_k) + 2e_k^T (\bar{A}_{\hbar(k)} - K\bar{C}_{\hbar(k)})^T P_{1,\hbar(k+1)} \tilde{F}(e_k) \\
&+ 2e_k^T (\bar{A}_{\hbar(k)} - K\bar{C}_{\hbar(k)})^T P_{1,\hbar(k+1)} \tilde{G}(e_{k-\tau}) + 2e_k^T (\bar{A}_{\hbar(k)} \\
&- K\bar{C}_{\hbar(k)})^T P_{1,\hbar(k+1)} (\bar{E}_{\hbar(k)} - K\bar{D}_{\hbar(k)}) \nu_k + 2\tilde{F}^T(e_k) P_{1,\hbar(k+1)} \tilde{G}(e_{k-\tau}) \\
&+ 2\tilde{F}^T(e_k) P_{1,\hbar(k+1)} (\bar{E}_{\hbar(k)} - K\bar{D}_{\hbar(k)}) \nu_k + 2\tilde{G}^T(e_{k-\tau}) P_{1,\hbar(k+1)} (\bar{E}_{\hbar(k)} \\
&- K\bar{D}_{\hbar(k)}) \nu_k + \tilde{G}^T(e_{k-\tau}) P_{1,\hbar(k+1)} \tilde{G}(e_{k-\tau}) + \nu_k^T (\bar{E}_{\hbar(k)} \\
&- K\bar{D}_{\hbar(k)})^T P_{1,\hbar(k+1)} (\bar{E}_{\hbar(k)} - K\bar{D}_{\hbar(k)}) \nu_k + \mathrm{tr}\{\omega_k^T B^T P_{1,\hbar(k+1)} B \omega_k\}\Big\} \\
=&\mathbb{E}\left\{\xi_k^T \Xi_{\hbar(k+1)} \xi_k - e_k^T P_{1,\hbar(k)} e_k\right\} + \mathrm{tr}\{\mathscr{R}^T B^T P_{1,\hbar(k+1)} B \mathscr{R}\},
\end{aligned}$$

$$\mathbb{E}\{\Delta V_{2,k}\} = \mathbb{E}\{\tilde{G}^T(e_k) P_2 \tilde{G}(e_k) - \tilde{G}^T(e_{k-\tau}) P_2 \tilde{G}(e_{k-\tau})\},$$

$$\mathbb{E}\{\Delta V_{3,k}\} = \mathbb{E}\left\{\tau \tilde{G}^T(e_k) P_3 \tilde{G}(e_k) - \sum_{l=k-\tau}^{k-1} \tilde{G}^T(e_l) P_3 \tilde{G}(e_l)\right\},$$

with

$$\begin{aligned}
\Xi_{\hbar(k+1)} &= \bar{\chi}^T P_{1,\hbar(k+1)} \bar{\chi}, \ \bar{\chi} = \left[(\bar{A}_{\hbar(k)} - K\bar{C}_{\hbar(k)}) \ I \ 0 \ I \ (\bar{E}_{\hbar(k)} - K\bar{D}_{\hbar(k)})\right], \\
\xi_k &= \left[e_k^T \ \tilde{F}^T(e_k) \ \tilde{G}(e_k) \ \tilde{G}^T(e_{k-\tau}) \ \nu_k^T\right]^T.
\end{aligned}$$

According to the above derivation, we have

$$\mathbb{E}\{\Delta V_k\} \leq \mathbb{E}\left\{\xi_k^T \bar{\Xi}_{\hbar(k)}\xi_k - \sum_{l=k-\tau}^{k-1} \tilde{G}^T(e_l)P_3\tilde{G}(e_l) - \frac{1}{\rho}V_{1,k}\right\}+ \\ \mathrm{tr}\left\{\mathscr{R}^T B^T P_{1,\hbar(k+1)} B\mathscr{R}\right\} + \lambda_3\|v_k\|^2 \tag{2.17}$$

in which

$$\bar{\Xi}_{\hbar(k)} = \Xi_{\hbar(k+1)} - \mathrm{diag}\left\{\left(1-\frac{1}{\rho}\right)P_{1,\hbar(k)}, 0, -P_2 - \tau P_3, P_2, \lambda_3 I\right\}.$$

On the other hand, it follows from the inequality (2.3) that

$$\begin{bmatrix} e_k \\ \tilde{F}(e_k) \end{bmatrix}^T \begin{bmatrix} \hat{U} & -\bar{U}^T \\ * & 2I \end{bmatrix} \begin{bmatrix} e_k \\ \tilde{F}(e_k) \end{bmatrix} \leq 0, \begin{bmatrix} e_k \\ \tilde{G}(e_k) \end{bmatrix}^T \begin{bmatrix} \hat{L} & -\bar{L}^T \\ * & 2I \end{bmatrix} \begin{bmatrix} e_k \\ \tilde{G}(e_k) \end{bmatrix} \leq 0. \tag{2.18}$$

According to the results in (2.17), (2.18) and (2.14), one can infer that

$$\begin{aligned} \mathbb{E}\{\Delta V_k\} \leq & \mathbb{E}\left\{\xi_k^T \bar{\Xi}_{\hbar(k)}\xi_k - \frac{1}{\rho}\left(V_{1,k} + \sum_{l=k-\tau_M}^{k-1} \tilde{G}^T(e_l)(P_2+\tau P_3)\tilde{G}(e_l)\right)\right. \\ & - \lambda_1 \begin{bmatrix} e_k \\ \tilde{F}(e_k) \end{bmatrix}^T \begin{bmatrix} \hat{U} & -\bar{U}^T \\ * & 2I \end{bmatrix} \begin{bmatrix} e_k \\ \tilde{F}(e_k) \end{bmatrix} - \lambda_2 \begin{bmatrix} e_k \\ \tilde{G}(e_k) \end{bmatrix}^T \begin{bmatrix} \hat{L} & -\bar{L}^T \\ * & 2I \end{bmatrix} \\ & \left. \times \begin{bmatrix} e_k \\ \tilde{G}(e_k) \end{bmatrix} + \lambda_3\|v_k\|^2 + \mathrm{tr}\left\{\mathscr{R}^T B^T P_{1,\hbar(k+1)} B\mathscr{R}\right\}\right\} \\ = & \mathbb{E}\left\{\xi_k^T \hat{\Xi}_{\hbar(k)}\xi_k - \frac{1}{\rho}\left(V_{1,k} + \sum_{l=k-\tau_M}^{k-1} \tilde{G}^T(e_l)(P_2+\tau P_3)\tilde{G}(e_l)\right)\right. \\ & \left. + \lambda_3\|v_k\|^2 + \mathrm{tr}\left\{\mathscr{R}^T B^T P_{1,\hbar(k+1)} B\mathscr{R}\right\}\right\}. \end{aligned} \tag{2.19}$$

Since

$$V_{3,k} \leq \tau \sum_{l=k-\tau}^{k-1} \tilde{G}^T(e_l)P_3\tilde{G}(e_k),$$

it follows from (2.13) that

$$\mathbb{E}\{\Delta V_k\} \leq -\frac{1}{\rho}\mathbb{E}\{V_k\} + \theta \tag{2.20}$$

where $\theta = \text{tr}\{\mathscr{R}^T B^T P_{1,\hbar(k+1)} B\mathscr{R}\} + \lambda_3 \bar{\nu}^2$. Therefore, it can be derived that for any scalar $\sigma > 0$,

$$\begin{aligned}
&\mathbb{E}\{\sigma^{k+1} V_{k+1}\} - \mathbb{E}\{\sigma^k V_k\} \\
&= \sigma^{k+1} (\mathbb{E}\{V_{k+1}\} - \mathbb{E}\{V_k\}) + \sigma^k(\sigma - 1)\mathbb{E}\{V_k\} \\
&\leq \sigma^k \left(\sigma - \frac{\sigma}{\rho} - 1\right) \mathbb{E}\{V_k\} + \sigma^{k+1}\theta.
\end{aligned} \tag{2.21}$$

Letting $\sigma = \sigma_0 = \frac{\rho}{\rho-1}$ and summing up both sides of (2.21) from 0 to $\mu - 1$ with respect to k, we obtain

$$\mathbb{E}\{\sigma_0^\mu V_\mu\} - \mathbb{E}\{V_0\} \leq \frac{\sigma_0(1-\sigma_0^\mu)}{1-\sigma_0}\theta, \tag{2.22}$$

which implies that

$$\begin{aligned}
\mathbb{E}\{V_\mu\} &\leq \sigma_0^{-\mu} \left(\mathbb{E}\{V_0\} + \frac{\sigma_0}{1-\sigma_0}\theta\right) + \frac{\sigma_0}{\sigma_0 - 1}\theta \\
&= \left(1 - \frac{1}{\rho}\right)^\mu (\mathbb{E}\{V_0\} - \rho\theta) + \rho\theta.
\end{aligned} \tag{2.23}$$

Now, it follows easily from Definition 2.1 that the error dynamical system (2.11) is exponentially ultimately bounded in mean square. The asymptotic upper bound of the filtering error can be computed as $\rho\theta / \left(\min_{1\leq i\leq N, i\in\mathbb{N}}\{\lambda_{\min}\{P_{1,i}\}\}\right)$. The proof is complete.

Remark 2.5 Theorem 2.1 provides a sufficient condition under which the filtering error e_k is ultimately bounded subject to two kinds of noise (e.g. ω_k and ν_k) in mean square. According to the inequality (2.14), it is easy to find that $\rho > \tau$, which indicates that the decay rate of the filtering error (i.e. $1 - 1/\rho$) should be large enough to guarantee the ultimate boundedness of e_k. Obviously, the decay rate of the output filtering error can be regarded as an important performance index for the filtering performance. Next, we focus our attention on the design of the filter gain matrices $K_{1,i}$ and $K_{2,i}$ $(i = 1, 2, \ldots, N)$ based on the condition that the decay rate of the output filtering error is bounded by $\bar{\alpha}$.

Corollary 2.1 *For system (2.1), under Assumptions 2.1, 2.2 and the RR protocol, let a scalar $1 > \bar{\alpha} > 0$ be given. Suppose that there exist $N + 5$ positive definite matrices $P_{1,i}$ $(i = 1, 2, \ldots, N)$, P_2, P_3, $Q \in \mathbb{R}^{N(n+q)\times N(n+q)}$, $\bar{P}_x = \text{diag}\{\bar{P}_{1,1}, \bar{P}_{1,2}, \ldots, \bar{P}_{1,N}\} \in \mathbb{R}^{Nn\times Nn}$ $\bar{P}_y = \text{diag}\{\bar{P}_{2,1}, \bar{P}_{2,2}, \ldots, \bar{P}_{2,N}\} \in \mathbb{R}^{Nq\times Nq}$, three positive scalars λ_i $(i = 1, 2, 3)$, two matrices $\mathscr{K}_x = \text{diag}\{\mathscr{K}_{1,1}, \mathscr{K}_{1,2}, \ldots, \mathscr{K}_{1,N}\} \in \mathbb{R}^{Nn\times Nq}$ and $\mathscr{K}_y = \text{diag}\{\mathscr{K}_{2,1}, \mathscr{K}_{2,2}, \ldots, \mathscr{K}_{2,N}\} \in \mathbb{R}^{Nq\times Nq}$ satisfying*

$$\hat{\Xi}_i = \begin{bmatrix} \hat{\chi}_{11} & 0 & \hat{\chi}_{13} \\ * & \hat{\chi}_{22} & \hat{\chi}_{23} \\ * & * & \hat{\chi}_{33} \end{bmatrix} < 0 \tag{2.24}$$

$$P_2 + \tau P_3 \leq \frac{1}{1-\bar{\alpha}} P_3 \tag{2.25}$$

$$\begin{bmatrix} -Q & \bar{P} \\ * & -P_{1,i} \end{bmatrix} < 0 \tag{2.26}$$

$$2\bar{P} - P_{1,i} \geq \bar{M}^T \bar{M}, \tag{2.27}$$

where

$$\hat{\chi}_{11} = \begin{bmatrix} \hat{\chi}_{11}^{11} & \hat{\chi}_{11}^{12} & \hat{\chi}_{11}^{13} \\ * & \hat{\chi}_{11}^{22} & 0 \\ * & * & \hat{\chi}_{11}^{33} \end{bmatrix}, \quad \hat{\chi}_{13} = \begin{bmatrix} \hat{\chi}_{13}^{11} \\ \hat{\chi}_{13}^{21} \\ 0 \end{bmatrix}, \quad \hat{\chi}_{22} = \begin{bmatrix} \hat{\chi}_{22}^{11} & 0 \\ * & \hat{\chi}_{22}^{22} \end{bmatrix}, \quad \hat{\chi}_{23} = \begin{bmatrix} \hat{\chi}_{23}^{11} \\ \hat{\chi}_{23}^{21} \end{bmatrix},$$

$$\hat{\chi}_{33} = -P_{1,i+1}, \quad \bar{P} = \mathrm{diag}\{\bar{P}_x, \bar{P}_y\}, \quad \mathscr{K} = \begin{bmatrix} \mathscr{K}_x^T & \mathscr{K}_y^T \end{bmatrix}^T, \quad \hat{\chi}_{11}^{13} = \lambda_2 \bar{L}^T,$$
$$\hat{\chi}_{11}^{11} = -\bar{\alpha}(2\bar{P} - P_{1,i}) - \lambda_1 \hat{U} - \lambda_2 \hat{L}, \quad \hat{\chi}_{11}^{12} = \lambda_1 \bar{U}^T, \quad \hat{\chi}_{11}^{22} = -2\lambda_1 I,$$
$$\hat{\chi}_{11}^{33} = P_2 + \tau P_3 - 2\lambda_2 I, \quad \hat{\chi}_{13}^{11} = \bar{A}_i^T \bar{P} - \bar{C}_i^T \mathscr{K}^T, \quad \hat{\chi}_{13}^{21} = \bar{P},$$
$$\hat{\chi}_{22}^{11} = -P_2, \quad \hat{\chi}_{22}^{22} = -\lambda_3 I, \quad \hat{\chi}_{23}^{11} = \bar{P}, \quad \hat{\chi}_{23}^{21} = \bar{E}_i^T \bar{P} - \bar{D}_i^T \mathscr{K}^T$$

with $P_{1,N+1} = P_{1,1}$ *for all* $i \in \{1, 2, \ldots, N\}$. *Then, the dynamics of the filtering error is ultimately bounded in mean square subject to the process noise* ω_k *and exogenous disturbance* v_k. *Furthermore, the decay rate of the output filtering error* $\mathbb{E}\{\|\tilde{z}_k\|^2\}$ *is less than* $\bar{\alpha}$, *and the minimum of the asymptotic upper bound of* $\mathbb{E}\{\|\tilde{z}_k\|^2\}$ *can be derived by solving the following minimization problem*

$$\min \left\{ tr\{\mathscr{R}^T B^T Q B \mathscr{R}\} + \lambda_3 \bar{v}^2 \right\} \tag{2.28}$$

subject to the matrix inequality constraints (2.24)–(2.27). The desired filter gains can be obtained by

$$K_{1,i} = \bar{P}_{1,i}^{-1} \mathscr{K}_{1,i}, \quad K_{2,i} = \bar{P}_{2,i}^{-1} \mathscr{K}_{2,i}, \ (i = 1, 2, \ldots, N). \tag{2.29}$$

Proof Choose the Lyapunov-like functional

$$\bar{V}_k = \sum_{i=1}^{3} \bar{V}_{i,k}, \tag{2.30}$$

where

$$\bar{V}_{1,k} = e_k^T \bar{P} P_{1,\hbar(k)}^{-1} \bar{P} e_k, \quad \bar{V}_{2,k} = \sum_{l=k-\tau}^{k-1} \tilde{G}^T(e_l) P_2 \tilde{G}(e_l),$$

$$\bar{V}_{3,k} = \sum_{j=-\tau}^{-1} \sum_{l=k+j}^{k-1} \tilde{G}^T(e_l) P_3 \tilde{G}(e_l).$$

Along similar lines as the proof of Theorem 2.1, it can be obtained from (2.26) and (2.29) that

$$\mathbb{E}\{\Delta \bar{V}_k\} \leq \xi_k^T \big(\Theta_{\hbar(k)} + \zeta_{\hbar(k)} P_{1,\hbar(k+1)}^{-1} \zeta_{\hbar(k)}^T\big)\xi_k - (1-\bar{\alpha})\bar{V}_k + \hat{\theta}, \tag{2.31}$$

in which

$$\Theta_{\hbar(k)} = \begin{bmatrix} \overrightarrow{\chi} & \hat{\chi}_{11}^{12} & \hat{\chi}_{11}^{13} & 0 & 0 \\ * & \hat{\chi}_{11}^{22} & 0 & 0 & 0 \\ * & * & \hat{\chi}_{11}^{33} & 0 & 0 \\ * & * & * & \hat{\chi}_{22}^{11} & 0 \\ * & * & * & * & \hat{\chi}_{22}^{22} \end{bmatrix}, \quad \zeta_{\hbar(k)} = \begin{bmatrix} \bar{A}_{\hbar(k)}^T \bar{P} - \bar{C}_{\hbar(k)}^T K^T \bar{P} \\ \bar{P} \\ 0 \\ \bar{P} \\ \bar{E}_{\hbar(k)}^T \bar{P} - \bar{D}_{\hbar(k)}^T K^T \bar{P} \end{bmatrix},$$

$$\overrightarrow{\chi} = -\bar{\alpha} \bar{P} P_{1,\hbar(k)}^{-1} \bar{P} - \lambda_1 \hat{U} - \lambda_2 \hat{L}, \; \hat{\theta} = \mathrm{tr}\{\mathscr{R}^T B^T \bar{P} P_{1,\hbar(k+1)}^{-1} \bar{P} B \mathscr{R}\} + \lambda_3 \|v_k\|^2$$

and ξ_k is defined as the same one in Theorem 2.1.

Since

$$2\bar{P} - P_{1,\hbar(k)} - \bar{P} P_{1,\hbar(k)}^{-1} \bar{P} = -(P_{1,\hbar(k)} - \bar{P}) P_{1,\hbar(k)}^{-1} (P_{1,\hbar(k)} - \bar{P}) \leq 0,$$

it follows from (2.31) that

$$\overrightarrow{\chi} \leq -\bar{\alpha}(2\bar{P} - P_{1,\hbar(k)}) - \lambda_1 \hat{U} - \lambda_2 \hat{L}. \tag{2.32}$$

By using the Schur Complement Lemma, it can be seen from (2.24), (2.29) and (2.32) that

$$\Theta_{\hbar(k)} + \zeta_{\hbar(k)} P_{1,\hbar(k+1)}^{-1} \zeta_{\hbar(k)}^T < 0$$

which means that

$$\mathbb{E}\{\Delta \bar{V}_k\} \leq -(1-\bar{\alpha})\bar{V}_k + \hat{\theta}. \tag{2.33}$$

Taking (2.26) into consideration, we have

$$\hat{\theta} \leq \hat{\phi} \tag{2.34}$$

where $\hat{\phi} = \mathrm{tr}\{\mathscr{R}^T B^T Q B \mathscr{R}\} + \lambda_3 \bar{v}^2$. According to the inequality (2.33) and (2.34), it can be derived that

$$\mathbb{E}\{\bar{V}_k\} \le \bar{\alpha}^k \left(\mathbb{E}\{\bar{V}_0\} - \frac{\hat{\phi}}{1-\bar{\alpha}}\right) + \frac{\hat{\phi}}{1-\bar{\alpha}}. \tag{2.35}$$

Therefore, one can conclude that the filtering error is exponentially ultimately bounded in mean square.

Next, it can be derived from (2.27) that

$$\begin{aligned}\mathbb{E}\{\|\tilde{z}_k\|^2\} &\le \mathbb{E}\{e_k^T(2\bar{P}_1 - P_{1,\hbar(k)})e_k\} \le \mathbb{E}\{\bar{V}_{1,k}\} \\ &\le \bar{\alpha}^k \left(\mathbb{E}\{\bar{V}_0\} - \frac{\hat{\phi}}{1-\bar{\alpha}}\right) + \frac{\hat{\phi}}{1-\bar{\alpha}}.\end{aligned} \tag{2.36}$$

Accordingly, we can conclude that the asymptotic bound of $\mathbb{E}\{\|\tilde{z}_k\|^2\}$ is $\frac{\hat{\phi}}{1-\bar{\alpha}}$ and the minimum of this asymptotic bound can be derived by minimizing $\hat{\phi}$ which is equivalent to the condition (2.28). The proof is complete.

Remark 2.6 In Corollary 2.1, the design problem for complex networks with time-delays and RR protocol is considered, and the desired filter gains are derived by solving a minimization problem subject to some inequality constraints. The main result established in Theorem 2.1 and Corollary 2.1 contains all the information about the system parameters, the RR protocol, the value of the time-delay and the bound of the disturbance input. The main novelty is threefold: (1) the RR protocol is introduced to the filtering problem for the complex network with time-delay; (2) sufficient conditions are derived to guarantee the ultimate boundedness of the filtering error in mean square; and (3) the filter is constructed to minimize an asymptotic upper bound of the output filtering error in mean square.

Remark 2.7 Our main results are derived based on the LMI-based algorithm which has a polynomial time complexity. Specifically, the number $\mathscr{N}(\varepsilon)$ of flops needed to compute an ε-accurate solution is bounded by $O(\mathscr{M}\mathscr{N}^3 \log(\mathscr{V}/\varepsilon))$, where $\mathscr{M}$ is the total row size of the LMI system, $\mathscr{N}$ is the total number of scalar decision variables, $\mathscr{V}$ is a data-dependent scaling factor and ε is relative accuracy set for algorithm. For the investigated discrete-time complex network (2.1) with the RR protocol (2.5), the variable dimensions can be seen from $x_{i,k} \in \mathbb{R}^n$, $y_{i,k} \in \mathbb{R}^q$, $z_{i,k} \in \mathbb{R}^m$ and $v_k \in \mathbb{R}^{r2}$. Furthermore, according to Corollary 2.1, one has both $\mathscr{M} = 10N(n+q) + r_2$ and $\mathscr{N} = 0.5N(n+q)^2(N^2+4N+1) + 0.5N(n+q)(N+2q+5) + 3$. Therefore, the computational complexity of the established result can be represented as $O(N^{10}(n+q)^7)$. Note that the computation of the filter gains is executed offline for the subsequent filtering, which implies that the computation of the filter gains would not affect the implementation of the filter in practical applications.

Now, let us show that Corollary 2.1 can be specialized to the following two kinds of complex networks: (1) nonlinear complex networks without delays; and (2) linear complex networks with time-delays.

Case 1: We first consider a special class of nonlinear complex networks without delays. For this purpose, we assume that the nonlinear function $g(x) = 0$. Then, the complex dynamical network with a RR protocol can be rewritten as follows:

$$\begin{cases} \bar{x}_{k+1} = \bar{A}_{\hbar(k)}\bar{x}_k + \bar{F}(\bar{x}_k) + \bar{B}\omega_k + \bar{E}_{\hbar(k)}\nu_k, \\ \bar{y}_k = \bar{C}_{\hbar(k)}\bar{x}_k + \bar{D}_{\hbar(k)}\nu_k, \\ z_k = \bar{M}\bar{x}_k. \end{cases} \tag{2.37}$$

The corresponding filter for (2.37) can be constructed as follows:

$$\begin{cases} \hat{x}_{k+1} = \bar{A}_{\hbar(k)}\hat{x}_k + \bar{F}(\hat{x}_k) + K\big(\bar{y}_k - \bar{C}_{\hbar(k)}\hat{x}_k\big), \\ \hat{z}_k = \bar{M}\hat{x}_k. \end{cases} \tag{2.38}$$

In this case, the filtering error dynamics for the complex network (2.37) can be represented as follows:

$$\begin{cases} e_{k+1} = \big(\bar{A}_{\hbar(k)} - K\bar{C}_{\hbar(k)}\big)e_k + \tilde{F}(e_k) + \bar{B}\omega_k + \big(\bar{E}_{\hbar(k)} - K\bar{D}_{\hbar(k)}\big)\nu_k, \\ \tilde{z}_k = \bar{M}e(k). \end{cases} \tag{2.39}$$

The following corollary can be obtained immediately from Corollary 2.1.

Corollary 2.2 *For the complex network with RR protocol shown in (2.37), under Assumptions 2.1 and 2.2, let a scalar* $1 > \bar{\alpha} > 0$ *be given. Suppose that there exist two block diagonal matrices* $\mathscr{K}_x = \mathrm{diag}\{\mathscr{K}_{1,1}, \mathscr{K}_{1,2}, \ldots, \mathscr{K}_{1,N}\} \in \mathbb{R}^{Nn\times Nq}$ *and* $\mathscr{K}_y = \mathrm{diag}\{\mathscr{K}_{2,1}, \mathscr{K}_{2,2}, \ldots, \mathscr{K}_{2,N}\} \in \mathbb{R}^{Nq\times Nq}$, $N+3$ *positive definite matrices* $P_{1,i}$ $(i = 1, 2, \ldots, N)$, $Q \in \mathbb{R}^{N(n+q)\times N(n+q)}$, $\bar{P}_x = \mathrm{diag}\{\bar{P}_{1,1}, \bar{P}_{1,2}, \ldots, \bar{P}_{1,N}\} \in \mathbb{R}^{Nn\times Nn}$, $\bar{P}_y = \mathrm{diag}\{\bar{P}_{1,1}, \bar{P}_{1,2}, \ldots, \bar{P}_{1,N}\} \in \mathbb{R}^{Nq\times Nq}$, *two positive scalars* λ_i $(i = 1, 2)$ *satisfying*

$$\hat{\Xi}_i = \begin{bmatrix} \overrightarrow{\chi}_{11} & \overrightarrow{\chi}_{12} & 0 & \overrightarrow{\chi}_{14} \\ * & \overrightarrow{\chi}_{22} & 0 & \overrightarrow{\chi}_{24} \\ * & * & \overrightarrow{\chi}_{33} & \overrightarrow{\chi}_{34} \\ * & * & * & \overrightarrow{\chi}_{44} \end{bmatrix} < 0 \tag{2.40}$$

$$\begin{bmatrix} -Q & \bar{P} \\ * & -P_{1,i} \end{bmatrix} < 0 \tag{2.41}$$

$$2\bar{P} - P_{1,i} \geq \bar{M}^T\bar{M}, \tag{2.42}$$

where

$$\vec{\chi}_{11} = -\bar{\alpha}(2\bar{P} - P_{1,i}) - \lambda_1\hat{U}, \quad \vec{\chi}_{12} = \lambda_1\hat{U}^T, \quad \vec{\chi}_{14} = \bar{A}_i^T\bar{P} - \bar{C}_i^T\mathscr{K}^T,$$
$$\vec{\chi}_{22} = -2\lambda_1 I, \quad \vec{\chi}_{24} = \bar{P}, \quad \vec{\chi}_{33} = -\lambda_2 I, \quad \vec{\chi}_{34} = \bar{E}_i^T\bar{P} - \bar{D}_i^T\mathscr{K}^T,$$
$$\vec{\chi}_{44} = -P_{1,i+1}, \quad \bar{P} = \text{diag}\{\bar{P}_x, \bar{P}_y\}, \quad \mathscr{K} = \left[\mathscr{K}_x^T \; \mathscr{K}_y^T\right]^T$$

with $P_{1,N+1} = P_{1,1}$ for all $i \in \{1, 2, \ldots, N\}$. Then, the dynamics of the filtering error is ultimately bounded in mean square subject to the process noise ω_k and exogenous disturbance ν_k. Furthermore, the decay rate of the output filtering error $\mathbb{E}\{\|\tilde{z}_k\|^2\}$ is less than $\bar{\alpha}$, and the minimum of the asymptotic upper bound of $\mathbb{E}\{\|\tilde{z}_k\|^2\}$ can be derived by solving the following minimization problem

$$\min\left\{tr\{\mathscr{R}^T B^T Q B \mathscr{R}\} + \lambda_2\bar{\nu}^2\right\} \tag{2.43}$$

subject to the matrix inequality constraints (2.40)–(2.42). The desired filter gains can be obtained by

$$K_{1,i} = \bar{P}_{1,i}^{-1}\mathscr{K}_{1,i}, \; K_{2,i} = \bar{P}_{2,i}^{-1}\mathscr{K}_{2,i}, \; (i = 1, 2, \ldots, N). \tag{2.44}$$

Case 2: Let us now deal with the linear complex networks with time-delays where the nonlinear functions $f(\cdot)$ and $g(\cdot)$ take the linear form as $f(x_{i,k}) = \vec{F}x_{i,k}$ and $g(x_{i,k-\tau}) = \vec{G}x_{i,k-\tau}$. In this case, the complex network with a RR protocol can be described as follows:

$$\begin{cases} \bar{x}_{k+1} = \bar{\mathscr{A}}_{\hbar(k)}\bar{x}_k + \bar{\mathscr{G}}\bar{x}_{k-\tau} + \bar{B}\omega_k + \bar{E}_{\hbar(k)}\nu_k, \\ \bar{y}_k = \bar{C}_{\hbar(k)}\bar{x}_k + \bar{D}_{\hbar(k)}\nu_k, \\ z_k = \bar{M}\bar{x}_k, \end{cases} \tag{2.45}$$

where

$$\bar{\mathscr{A}}_{\hbar(k)} = \begin{bmatrix} (W \otimes \Gamma) + \mathscr{F} & 0 \\ \Phi_{\hbar(k)}C & I - \Phi_{\hbar(k)} \end{bmatrix}, \; \bar{\mathscr{G}} = \begin{bmatrix} \mathscr{G} & 0 \\ 0 & 0 \end{bmatrix}, \; \mathscr{F} = \text{diag}_N\{\vec{F}\}, \; \mathscr{G} = \text{diag}_N\{\vec{G}\}.$$

The filter for system (2.45) is described by

$$\begin{cases} \hat{x}_{k+1} = \bar{\mathscr{A}}_{\hbar(k)}\hat{x}_k + \bar{\mathscr{G}}\hat{x}_{k-\tau} + K\left(\bar{y}_k - \bar{C}_{\hbar(k)}\hat{x}_k\right), \\ \hat{z}_k = \bar{M}\hat{x}_k. \end{cases} \tag{2.46}$$

Along the similar line in Sect. 2.1, we arrive at the following filtering error dynamics for the system (2.45)

$$\begin{cases} e_{k+1} = \left(\bar{\mathscr{A}}_{\hbar(k)} - K\bar{C}_{\hbar(k)}\right)e_k + \bar{\mathscr{G}}e_{k-\tau} + \bar{B}\omega_k + \left(\bar{E}_{\hbar(k)} - K\bar{D}_{\hbar(k)}\right)\nu_k, \\ \tilde{z}_k = \bar{M}e_k. \end{cases} \tag{2.47}$$

Corollary 2.3 *For the linear complex networks with time-delays and RR protocol shown in (2.45), under Assumptions 2.1 and 2.2, let a scalar $1 > \bar{\alpha} > 0$ be given. Suppose that there exist $N+5$ positive definite matrices $P_{1,i}$ $(i = 1, 2, \ldots, N)$, P_2, P_3, $Q \in \mathbb{R}^{N(n+q)\times N(n+q)}$, $\bar{P}_x = \mathrm{diag}\{\bar{P}_{1,1}, \bar{P}_{1,2}, \ldots, \bar{P}_{1,N}\} \in \mathbb{R}^{Nn\times Nn}$, $\bar{P}_y = \mathrm{diag}\{\bar{P}_{1,1}, \bar{P}_{1,2}, \ldots, \bar{P}_{1,N}\} \in \mathbb{R}^{Nq\times Nq}$, a positive scalar λ_1 and two block diagonal matrices $\mathscr{K}_x = \mathrm{diag}\{\mathscr{K}_{1,1}, \mathscr{K}_{1,2}, \ldots, \mathscr{K}_{1,N}\} \in \mathbb{R}^{Nn\times Nq}$ and $\mathscr{K}_y = \mathrm{diag}\{\mathscr{K}_{2,1}, \mathscr{K}_{2,2}, \ldots, \mathscr{K}_{2,N}\} \in \mathbb{R}^{Nq\times Nq}$ satisfying*

$$\hat{\Xi}_i = \begin{bmatrix} \check{\chi}_{11} & 0 & 0 & \check{\chi}_{14} \\ * & \check{\chi}_{22} & 0 & \check{\chi}_{24} \\ * & * & \check{\chi}_{33} & \check{\chi}_{34} \\ * & * & * & \check{\chi}_{44} \end{bmatrix} < 0 \tag{2.48}$$

$$P_2 + \tau P_3 \leq \frac{1}{1-\bar{\alpha}} P_3 \tag{2.49}$$

$$\begin{bmatrix} -Q & \bar{P} \\ * & -P_{1,i} \end{bmatrix} < 0 \tag{2.50}$$

$$2\bar{P} - P_{1,i} \geq \bar{M}^T \bar{M}, \tag{2.51}$$

where

$$\begin{aligned}
\check{\chi}_{11} &= -\bar{\alpha}(2\bar{P} - P_{1,i}) + \mathscr{G}^T(P_2 + \tau P_3)\mathscr{G}, \quad \check{\chi}_{14} = \bar{\mathscr{A}}_i^T \bar{P} - \bar{C}_i^T \mathscr{K}^T, \\
\check{\chi}_{22} &= -P_2, \quad \check{\chi}_{24} = \bar{P}, \quad \check{\chi}_{33} = -\lambda_1 I, \quad \check{\chi}_{34} = \bar{E}_i^T \bar{P} - \bar{D}_i^T \mathscr{K}^T, \\
\check{\chi}_{44} &= -P_{1,i+1}, \quad \bar{P} = \mathrm{diag}\{\bar{P}_x, \bar{P}_y\}, \quad \mathscr{K} = \left[\mathscr{K}_x^T \ \mathscr{K}_y^T\right]^T
\end{aligned}$$

with $P_{1,N+1} = P_{1,1}$ for all $i \in \{1, 2, \ldots, N\}$. Then, the dynamics of the filtering error is ultimately bounded in mean square subject to the process noise $\omega(k)$ and exogenous disturbance $v(k)$. Furthermore, the decay rate of the output filtering error $\mathbb{E}\{\|\tilde{z}(k)\|^2\}$ is less than $\bar{\alpha}$, and the minimum of the asymptotic upper bound of $\mathbb{E}\{\|\tilde{z}(k)\|^2\}$ can be derived by solving the following minimization problem

$$\min \left\{ tr\{\mathscr{R}^T B^T Q B \mathscr{R}\} + \lambda_1 \bar{v}^2 \right\} \tag{2.52}$$

subject to the matrix inequality constraints (2.48)–(2.52). The desired filter gains can be obtained by

$$K_{1,i} = \bar{P}_{1,i}^{-1} \mathscr{K}_{1,i}, \; K_{2,i} = \bar{P}_{2,i}^{-1} \mathscr{K}_{2,i}, \; (i = 1, 2, \ldots, N). \tag{2.53}$$

Remark 2.8 Up until now, several criteria have been proposed for the filtering issue for complex networks with RR protocol. In Corollaries 2.2 and 2.3, the estimator design problems for two special complex networks have been considered, and some simplified conditions are obtained. It is worth mentioning that both the existence of time-delays and the forms of the nonlinear functions are reflected in the matrix inequality constraints of the minimization problem.

2.3 An Illustrative Example

In this section, a simulation example is presented to demonstrate the effectiveness of the theoretical results obtained above. Consider a three-node complex network of the form (2.1). The system parameters for each node are given as follows:

	B_i	C_i	D_i	E_i	M_i
Node 1 ($i = 1$)	$[0.5\ -0.6]^T$	$[0.5\ -0.4]$	0.5	$[0.6\ 0.3]^T$	$[0.2\ 0.8]$
Node 2 ($i = 2$)	$[0.6\ 0.4]^T$	$[0.3\ 0.2]$	0.7	$[0.4\ 0.5]^T$	$[0.2\ 0.8]$
Node 3 ($i = 3$)	$[0.5\ -0.6]^T$	$[0.4\ -0.5]$	0.4	$[0.8\ 0.5]^T$	$[0.2\ 0.8]$

The nonlinear functions are taken to be

$$f(x_{i,k}) = 0.5\left((U_1 + U_2)x_{i,k} + (U_2 - U_1)\sin(k)x_{i,k}\right), \quad (i = 1, 2, 3),$$
$$g(x_{i,k}) = 0.5\left((L_1 + L_2)x_{i,k} + (L_2 - L_1)\cos(k)x_{i,k}\right), \quad (i = 1, 2, 3),$$

where

$$U_1 = \text{diag}\{0.88, 0.79\}, \qquad U_2 = \text{diag}\{0.82, 0.72\},$$
$$L_1 = \text{diag}\{0.16, 0.18\}, \qquad L_2 = \text{diag}\{0.18, 0.16\}.$$

The coupling configuration matrix is assumed to be $W = \left[w_{i,j}\right]_{N\times N}$ with

$$w_{i,j} = \begin{cases} 0.1, & \text{if } i \neq j \\ -0.2, & \text{otherwise} \end{cases}$$

and the inner-coupling matrix is given as $\Gamma = \text{diag}\{2, 2\}$. The time-delay parameter is selected as $\tau = 4$. ω_k is assumed to be a one-dimensional Gaussian white noise with the covariance $R = I$. The decay rate is selected as 0.95. Note that the above complex network under consideration is set to be *unstable* in order to make our simulation non-trivial.

According to Corollary 2.1, the minimization problem (2.28) subject to the inequality constraints (2.24)–(2.27) can be solved via the LMI toolbox in MATLAB software. The filter gain matrices are derived as follows:

$$K_{1,1} = \begin{bmatrix} 2.583 \\ 1.233 \end{bmatrix}, \quad K_{1,2} = \begin{bmatrix} 1.848 \\ 1.482 \end{bmatrix}, \quad K_{1,3} = \begin{bmatrix} 2.561 \\ 0.342 \end{bmatrix},$$
$$K_{2,1} = 0.9831, \quad K_{2,2} = 0.9509, \quad K_{2,3} = 0.9168.$$

The asymptotic upper bound of the output filtering error is 16.345. Based on the derived filter gain matrices, numerical simulation results are given in Figs. 2.2, 2.3, 2.4, 2.5, 2.6 and 2.7 which plot the state responses and their estimations for

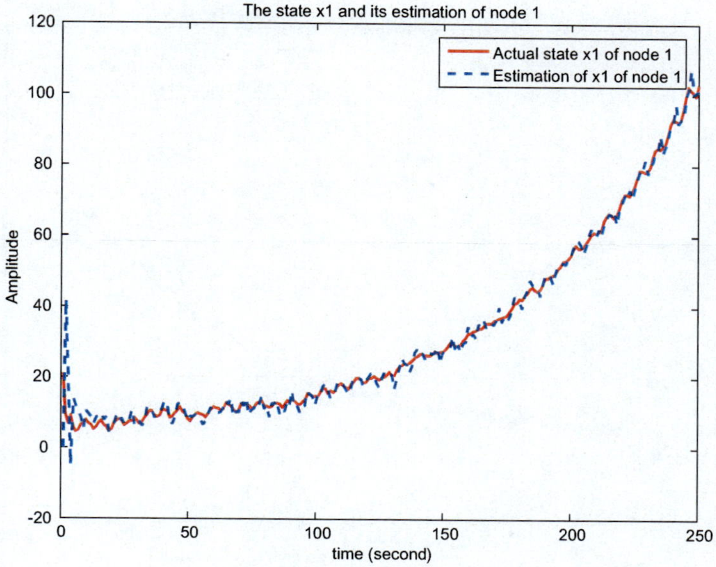

Fig. 2.2 State trajectories of x_1 and $\hat{x}_1$ for node 1

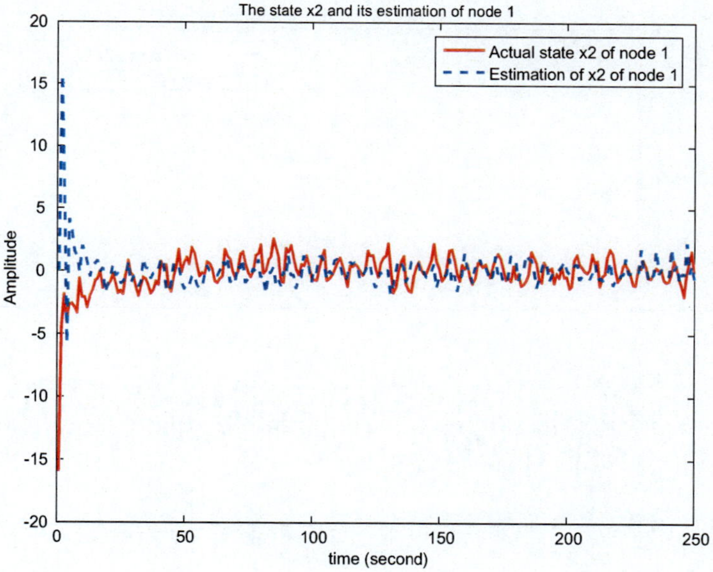

Fig. 2.3 State trajectories of x_2 and $\hat{x}_2$ for node 1

node i ($i = 1, 2, 3$). The simulation has verified that our theoretical analysis of the filtering problem is indeed effective for the addressed class of complex networks with time-varying delay and RR protocol.

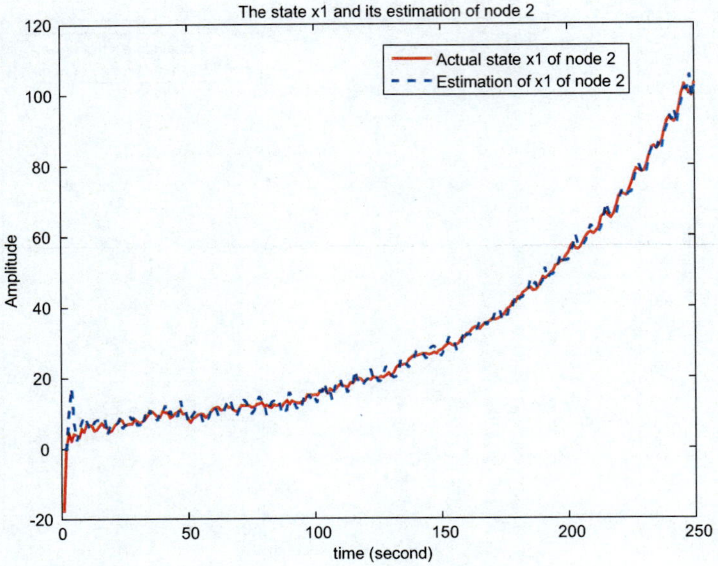

Fig. 2.4 State trajectories of x_1 and $\hat{x}_1$ for node 2

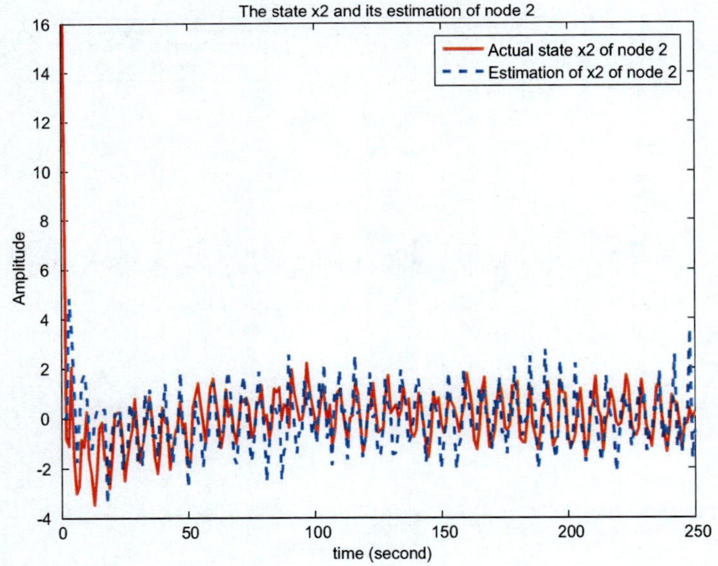

Fig. 2.5 State trajectories of x_2 and $\hat{x}_2$ for node 2

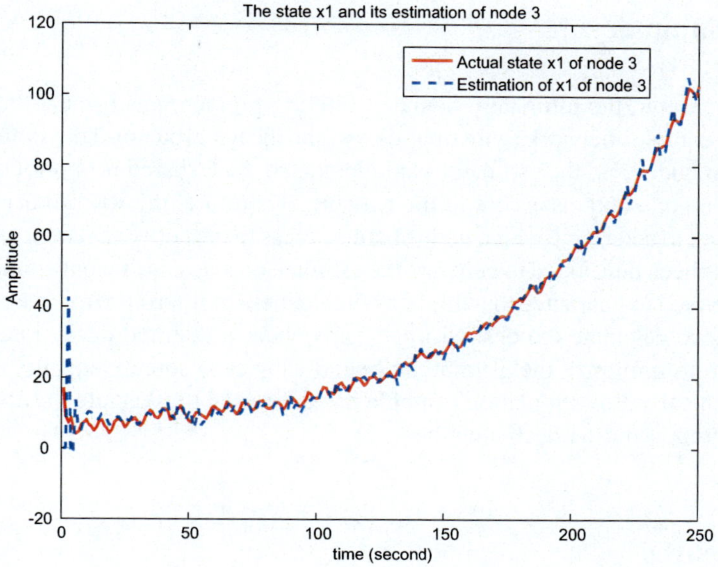

Fig. 2.6 State trajectories of x_1 and $\hat{x}_1$ for node 3

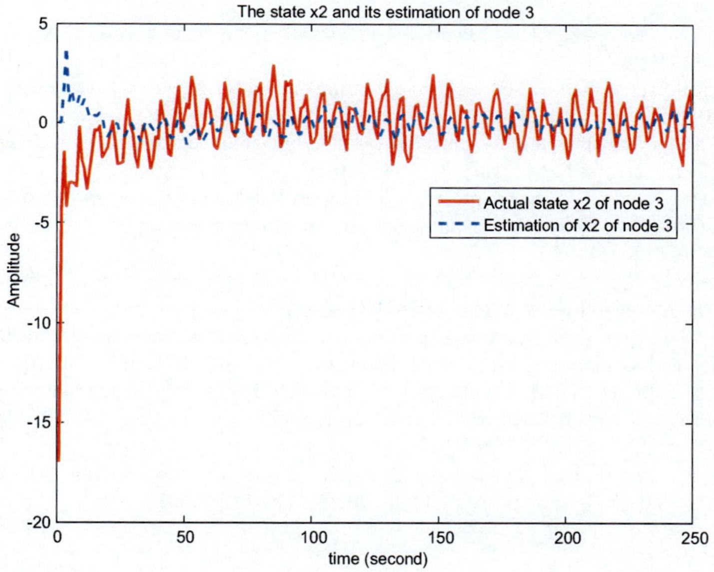

Fig. 2.7 State trajectories of x_2 and $\hat{x}_2$ for node 3

2.4 Summary

In this chapter, the ultimately bounded filtering problem has been addressed for a class of complex networks with time-delays and the RR protocol. The communication between nodes and the estimator is implemented via a shared network under which only one node obtains access to the network. Therefore, the RR protocol has been employed to determine which node obtains access to the network. The corresponding filter has been developed to generate the estimate of states. Sufficient conditions have been derived to guarantee the ultimate boundedness of filtering error in mean square. In order to calculate the desired filter gains under a required decay rate, a convex problem to minimize the asymptotic bound subject to some inequality constraints has been solved. A simulation example has been used to illustrate the effectiveness of the proposed filter design method.

References

1. Albert, R., Barabási, A.-L.: Statistical mechanics of complex networks. Rev. Mod. Phys. **74**(1), 47–97 (2002)
2. Watts, D.J., Strogatz, S.H.: Collective dynamics of small-world networks. Nature **393**(6684), 440–442 (1998)
3. Barabási, A.L., Albert, R.: Emergence of scaling in random networks. Science **286**(5439), 509–512 (1999)
4. He, W., Cao, J.: Exponential synchronization of hybrid coupled networks with delayed coupling. IEEE Trans. Neural Netw. **21**(4), 571–583 (2010)
5. Caballero-Aguila, R., Hermoso-Carazo, A., Linares-Perez, J.: Linear estimation based on covariances for networked systems featuring sensor correlated random delays. Int. J. Syst. Sci. **44**(7), 1233–1244 (2013)
6. He, Y., Wang, Q., Wu, M., Lin, C.: Delay-dependent state estimation for delayed neural networks. IEEE Trans. Neural Netw. **17**(4), 1077–1081 (2006)
7. Basin, M.V., Rodriguez-Ramirez, P.C.: Sliding mode controller design for stochastic polynomial systems with unmeasured states. IEEE Trans. Ind. Electron. **61**(1), 387–396 (2014)
8. Tang, Y., Fang, J., Xia, M., Gu, X.: Synchronization of Takagi-Sugeno fuzzy stochastic discrete-time complex networks with mixed time-varying delays. Appl. Math. Model. **34**(4), 843–855 (2010)
9. Wang, Z., Lam, J., Liu, X.: Filtering for a class of nonlinear discrete-time stochastic systems with state delays. J. Comput. Appl. Math. **201**(1), 153–163 (2007)

Chapter 3
Finite-Horizon $\mathscr{H}_\infty$ Filtering with Random Access Protocol and High-Rate Communication

$\mathscr{H}_\infty$ filtering is one of the most well-known filtering approaches. To date, the $\mathscr{H}_\infty$ filtering approach has found successful applications in a wide range of areas including the aerospace industry, process control, unmanned aerial vehicles and geological prospecting. The $\mathscr{H}_\infty$ filtering approach has gained a huge amount of research attention since it provides a guaranteed noise attenuation level against the external disturbance. Compared with the well-known Kalman filters, one of the primary advantages of the $\mathscr{H}_\infty$ filters lies in the fact that it is insensitive to the exact knowledge of the statistics of the external disturbance.

On the other hand, it should be pointed out that almost all practical systems are subject to time-varying parameters variations (e.g. the temperature fluctuation, operating point shifting, graduate aging of system components, etc) in reality. Hence, one would be naturally interested in studying their transient dynamics over a finite horizon. As such, considerable research attention has been devoted to the finite-horizon $\mathscr{H}_\infty$ filtering problem for time-varying systems due mainly to the insight of their engineering applications. By now, there are mainly two frameworks dealing with the finite-horizon $\mathscr{H}_\infty$ filtering and control problems, namely, the recursive linear matrix inequality-based framework and backward recursive Riccati difference equation-based framework.

With the rapid development of network technologies, the $\mathscr{H}_\infty$ filtering problem of networked systems has attracted an ever-growing research interest. The existence of networked-induced effects would largely complicate the filter design and performance analysis. For a networked systems with protocol scheduling, the use of the communication protocol gives rise to some specific protocol-induced constraints on the network communication, which, in turn, affect the corresponding filtering performance.

In networked systems, the signal transmission rate over a network channel is sometimes *much faster* than the sampling rate of the sensors and, in this chapter, such kind of network is customarily referred to as the *high-rate* communication network.

L. Zou et al., *Communication-Protocol-Based Filtering and Control of Networked Systems*, Studies in Systems, Decision and Control 430,
https://doi.org/10.1007/978-3-030-97512-8_3

For example, the well-known Control Area Network (CAN) is able to broadcast signals at speeds from 20 kb/s to 1 Mb/s. The data transmission rate of Process Field Bus—Process Automation (Profibus-PA) network is 31.25 kb/s. Nevertheless, the sampling period of the 13-bit LM95172 temperature sensor is only 35 ms (about 0.37 kb/s) and the maximum sampling rate of the 12-bit acceleration sensor DT138 is 100 Hz (1.2 kb/s). In this case, high-rate network will lead to the oversampling on transmitted signals (i.e. *multiple transmissions* will occur between two adjacent sampling instants of the sensors).

For a filtering system over a high-rate network with multi-sensors subject to certain protocol scheduling (e.g. the RA protocol), multiple transmissions would consume more network bandwidth and result in different transmission behavior compared with the single transmission between two adjacent sampling instants. In this chapter, we aim to deal with the finite-horizon $\mathscr{H}_\infty$ filtering problem for the time-delay networked system in the simultaneous presence of the high-rate communication network and the RA protocol scheduling effects. In this context, two challenges are identified as follows: (1) how to formulate a mathematical model accounting for both RA protocol scheduling effects and the high-rate communication behavior? and (2) how to develop appropriate methodology to analyze the transient dynamics of the filtering error due to the time-varying nature over a finite-horizon? We endeavor to answer the above two questions in this chapter by launching a major study on the finite-horizon $\mathscr{H}_\infty$ filtering problem subject to RA protocol scheduling effects and the high-rate communication behavior.

The rest of this chapter is organized as follows. In Sect. 3.1, the time-varying system with the high-rate communication network and communication constraints is introduced and the problem under consideration is formulated. In Sect. 3.2, by employing the stochastic analysis techniques, some sufficient conditions are established to guarantee the desired filtering performance and then the filter gains are obtained by solving a set of recursive matrix inequalities. Two simulation examples are given in Sect. 3.3 to demonstrate the main results obtained. Finally, we conclude the chapter in Sect. 3.4.

3.1 Problem Formulation

3.1.1 The Random Access Protocol

Consider a networked system with N transmission nodes labeled as $\{1, 2, \ldots, N\}$. The main idea of the RA protocol for *discrete-time systems* is that *only one node* is selected to transmit/receive data via the communication network at each transmission instant. Let $\alpha(t_k)$ denote the selected node obtaining access to the communication network at time t_k ($k \in \mathbb{N}^+$) where k represents the kth transmission instant. Then, as shown in [1], due to the scheduling of the RA protocol, $\alpha(t_k)$ can be regarded as a stochastic process, which could be modeled by a Markov chain. The occurrence probability of $\alpha(t_{k+1}) = j$ conditioned on $\alpha(t_k) = i$ is given by

$$\text{Prob}\{\alpha(t_{k+1}) = j|\alpha(t_k) = i\} = p_{ij}(t_k),$$

where $\forall i, j \in \{1, 2, \ldots, N\}$, $p_{ij}(t_k) \geq 0$ is the transition probability from i to j at time instant t_k and $\sum_{j=1}^{N} p_{ij}(t_k) = 1$ $(i \in \{1, 2, \ldots, N\}, k \in \mathbb{N}^+)$.

Remark 3.1 The so-called RA protocol has been first studied in [2] for continuous-time systems and in [1] for discrete-time systems. Such a protocol is also known as the Stochastic Communication Protocol. The well-known CSMA protocol is one good example of the RA protocols in use. In [1], the Markov chain has been employed to characterize the "random switch" behavior of the RA protocol.

3.1.2 The System Model

The plant under consideration is a nonlinear discrete time-varying system with time-delay defined on the finite horizon $\big([t_0, t_N]\big)$ of the form:

$$\begin{cases} x(t_{k+1}) = A(t_k)x(t_k) + A_d(t_k)x(t_{k-\tau}) + f(t_k, x(t_k)) + B(t_k)\omega(t_k) \\ y(t_k) = C(t_k)x(t_k) + C_d(t_k)x(t_{k-\tau}) + D(t_k)\omega(t_k) \\ z(t_k) = M(t_k)x(t_k) \\ x(t_i) = \varphi(i), \quad i = -\tau, -\tau+1, \ldots, 0 \end{cases} \tag{3.1}$$

where t_k $(k \in \mathbb{N}^+)$ represents the kth sampling instant, $x(t_k) \in \mathbb{R}^{n_x}$, $y(t_k) \in \mathbb{R}^{n_y}$ and $z(t_k) \in \mathbb{R}^{n_z}$ denote, respectively, the state vector, the measurement output before transmitted and the signal to be estimated. In this chapter, sampling instants are generated periodically with a certain interval. Denote $h_p \triangleq t_{k+1} - t_k$ $(\forall k \in \mathbb{N}^+)$ as the sampling period of the system (3.1). τ is a known constant positive scalar. $\omega(t_k) \in l_2([0, N], \mathbb{R}^{n_\omega})$ is the disturbance input where $l_2([0, N], \mathbb{R}^{n_\omega})$ is the space of square-summable n_ν-dimensional vector functions over the interval $[0, N]$. $A(t_k)$, $A_d(t_k)$, $B(t_k)$, $C(t_k)$, $C_d(t_k)$, $D(t_k)$ and $M(t_k)$ are known time-varying matrices with appropriate dimensions. $\varphi(i)$ $(i = -\tau, -\tau+1, \ldots, 0)$ are the initial conditions. The known nonlinear vector-valued function $f : [t_0, t_N] \times \mathbb{R}^{n_x} \to \mathbb{R}^{n_x}$ is continuous and satisfies $f(t_k, 0) = 0$ and the following sector bounded condition

$$\left[f(t_k, x_1) - f(t_k, x_2) - F_1(t_k)(x_1 - x_2)\right]^T \left[f(t_k, x_1) - f(t_k, x_2) - F_2(t_k)(x_1 - x_2)\right] \leq 0 \tag{3.2}$$

for all $x_1, x_2 \in \mathbb{R}^{n_x}$, where $F_1(t_k)$ and $F_2(t_k)$ are known time-varying matrices with appropriate dimensions.

3.1.3 Issues with the High-Rate Network

Let us now discuss the effect induced by high-rate communication network. Before proceeding further, we introduce the following assumptions.

Assumption 3.1 The signal transmissions are implemented periodically between the sensors and the filter with a constant period h_c satisfying $h_c = h_p/d$, where $d \in \mathbb{N}^+$ and $d \geq 1$.

Assumption 3.2 The initial transmission instant equals to the 0th sampling instant, i.e. $\bar{t}_0 = t_0$.

In this chapter, we refer to h_c as the communication period of the network. For simplicity, we define the $(\hat{k}+1)$th transmission instant as $\bar{t}_{\hat{k}+1} = \bar{t}_{\hat{k}} + h_c$ $(k \in \mathbb{N}^+)$. Let us now study the effect induced by the network. For technical analysis, we first write $y(t_k) = \left[y_1^T(t_k)\ y_2^T(t_k)\ \cdots\ y_{n_y}^T(t_k)\right]^T$, where $y_i(t_k)$ is the measurement of the ith sensor.

In this chapter, the RA protocol is employed to determine which sensor obtains access to the high-rate communication network. According to the RA protocol scheduling, only one sensor is allowed to get access to communication network at each transmission instant. For the sake of examining the influence of the communication constraints induced by the RA protocol, let $\theta(\bar{t}_{\hat{k}}) \in \{1, 2, \ldots, n_y\}$ be the selected sensor obtaining access to the communication network at transmission instant $\bar{t}_{\hat{k}}$. Under the RA protocol scheduling, $\theta(\bar{t}_{\hat{k}})$ can be governed by the Markov chain with the transition probability matrix $\mathscr{P}(\bar{t}_{\hat{k}})$. The transition probability matrix $\mathscr{P}(\bar{t}_{\hat{k}}) \triangleq \left[p_{ij}(\bar{t}_{\hat{k}})\right]_{n_y \times n_y}$ is defined as follows:

$$p_{ij}(\bar{t}_{\hat{k}}) \triangleq \text{Prob}\left(\theta\left(\bar{t}_{\hat{k}+1}\right) = j | \theta\left(\bar{t}_{\hat{k}}\right) = i\right). \tag{3.3}$$

Let $\bar{y}(\bar{t}_{\hat{k}}) \triangleq \left[\bar{y}_1^T(\bar{t}_{\hat{k}})\ \bar{y}_2^T(\bar{t}_{\hat{k}})\ \cdots\ \bar{y}_{n_y}^T(\bar{t}_{\hat{k}})\right]^T \in \mathbb{R}^{n_y}$ be the measurement output after transmitted through the communication network at time $t_{\hat{k}}$. Similarly, in this chapter, we choose the ZOH strategy to compensate the signals, under which the updating rule for $\bar{y}_i(\bar{t}_{\hat{k}})$ $(i \in \{1, 2, \ldots, n_y\}, t_k \leq \bar{t}_{\hat{k}} < t_{k+1})$ subject to the RA protocol scheduling is set to be

$$\bar{y}_i(\bar{t}_{\hat{k}}) = \begin{cases} y_i(t_k) + E_i(\bar{t}_{\hat{k}})v(\bar{t}_{\hat{k}}), & \text{if } i = \theta(\bar{t}_{\hat{k}}) \\ \bar{y}_i(\bar{t}_{\hat{k}-1}), & \text{otherwise} \end{cases} \tag{3.4}$$

where $v(\bar{t}_{\hat{k}}) \in l_2([0, N]; \mathbb{R}^{n_v})$ is the network-induced disturbance, and $E_i(\bar{t}_{\hat{k}}) \in \mathbb{R}^{1\times n_v}$ $(i = 1, 2, \ldots, n_y)$ are known, real, time-varying matrices. Defining

$$E(\bar{t}_{\hat{k}}) \triangleq \left[E_1^T(\bar{t}_{\hat{k}})\ E_2^T(\bar{t}_{\hat{k}})\ \cdots\ E_{n_y}^T(\bar{t}_{\hat{k}})\right]^T,$$
$$\Phi(\theta(\bar{t}_{\hat{k}})) \triangleq \text{diag}\left\{\delta_1(\theta(\bar{t}_{\hat{k}})), \delta_2(\theta(\bar{t}_{\hat{k}})), \cdots, \delta_{n_y}(\theta(\bar{t}_{\hat{k}}))\right\},$$

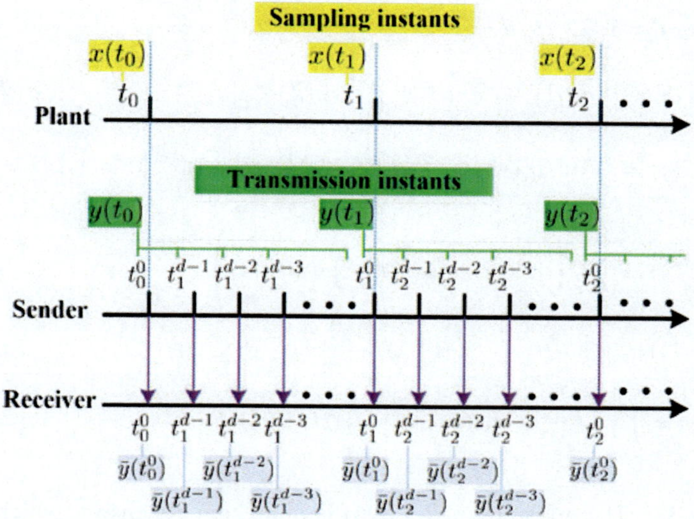

Fig. 3.1 Timelines of the sent and received sampled signals

where $\delta_i(j) \triangleq \delta(i-j) \in \{0, 1\}$ is the Kronecker delta function, we have

$$\bar{y}(\bar{t}_{\hat{k}}) = \Phi(\theta(\bar{t}_{\hat{k}}))\left(y(t_k) + E(\bar{t}_{\hat{k}})\nu(\bar{t}_{\hat{k}})\right) + \left(I - \Phi(\theta(\bar{t}_{\hat{k}}))\right)\bar{y}(\bar{t}_{\hat{k}-1}). \tag{3.5}$$

It is easy to see from Assumptions 3.1 and 3.2 that $t_k = t_0 + kh_p = \bar{t}_0 + kdh_c = \bar{t}_{kd}$, which implies that there are d signal transmissions occurring in the time interval $(t_{k-1,\, t_k}]$ (i.e. $t_{k-1} = \bar{t}_{(k-1)d} < \bar{t}_{(k-1)d+1} < \cdots < \bar{t}_{kd} = t_k$). Furthermore, it is assumed that the clocks are synchronized and there is no clock drifts and communication delays between the sender and the receiver. The timelines of the sent and received sampled signals are shown in Fig. 3.1.

Consider the updating models of $\bar{y}(t_k^l)$ in the time interval $(t_{k-1}, t_k]$ where $t_k^l = t_k - lh_c$ and $l \in \{0, 1, \ldots, d-1\}$. It follows from the updating rule (3.5) that:

$$\begin{aligned}
\bar{y}(t_k^{d-1}) &= \Phi(\theta(t_k^{d-1}))\big(y(t_{k-1}) + E(t_k^{d-1})\nu(t_k^{d-1})\big) + \big(I - \Phi(\theta(t_k^{d-1}))\big)\bar{y}(t_{k-1}),\\
\bar{y}(t_k^{d-2}) &= \Phi(\theta(t_k^{d-2}))\big(y(t_{k-1}) + E(t_k^{d-2})\nu(t_k^{d-2})\big) + \big(I - \Phi(\theta(t_k^{d-2}))\big)\bar{y}(t_k^{d-1}),\\
&\;\;\vdots\\
\bar{y}(t_k^{0}) &= \Phi(\theta(t_k^{0}))\big(y(t_{k}) + E(t_k^{0})\nu(t_k^{0})\big) + \big(I - \Phi(\theta(t_k^{0}))\big)\bar{y}(t_k^{1}),
\end{aligned}$$

which implies

$$\begin{aligned}
\bar{y}(t_k) = \bar{y}(t_k^0) &= \Phi(\theta(t_k^0))y(t_k) + \Theta_1(\bar{\theta}(t_k^{d-1}))y(t_{k-1}) + \Theta_2(\bar{\theta}(t_k^{d-1}))\bar{y}(t_{k-1})\\
&\quad + \Theta_3(\bar{\theta}(t_k^{d-1}))\bar{\nu}(t_k)
\end{aligned} \tag{3.6}$$

where, for $i = 1, 2, \ldots, d-1$

$$\bar{v}(t_k) = \left[v^T(t_k^0)\ v^T(t_k^1) \cdots v^T(t_k^{d-1})\right]^T, \quad \bar{\theta}(t_k^i) = \left(\theta(t_k^0), \theta(t_k^1), \ldots, \theta(t_k^i)\right),$$

$$\Theta_3(\bar{\theta}(t_k^{d-1})) = \left[\Phi(\theta(t_k^0))E(t_k^0)\ \Theta_4(\bar{\theta}(t_k^{d-1}))\right], \quad \Theta_2(\bar{\theta}(t_k^{d-1})) = \prod_{i=0}^{d-1}\left(I - \Phi(\theta(t_k^i))\right),$$

$$\Theta_1(\bar{\theta}(t_k^{d-1})) = \sum_{i=1}^{d-1}\left(\prod_{j=0}^{i-1}\left(I - \Phi(\theta(t_k^j))\right)\Phi(\theta(t_k^i))\right),$$

$$\Theta_4(\bar{\theta}(t_k^{d-1})) = \left[\bar{\Theta}_1(\bar{\theta}(t_k^1))\ \bar{\Theta}_2(\bar{\theta}(t_k^2)) \cdots \bar{\Theta}_{d-1}(\bar{\theta}(t_k^{d-1}))\right],$$

$$\bar{\Theta}_i(\hat{\theta}(t_k^i)) = \prod_{j=0}^{i-1}\left(I - \Phi(\theta(t_k^j))\right)\Phi(\theta(t_k^i))E(\theta(t_k^i)).$$

Remark 3.2 The updating model (3.6) is quite comprehensive, which covers the channel noise $\bar{v}(t_k)$, the communication constraints $\bar{\theta}(t_k^{d-1})$ and the scale parameter d. In the case that the communication period h_c is equal to the sampling period h_p (i.e. $d = 1$), the update model (3.6) reduces to the following one where $\Phi_{\theta(t_k)} \triangleq \Phi(\theta(t_k))$:

$$\bar{y}(t_k) = \Phi_{\theta(t_k)}y(t_k) + \Theta_2(\theta(t_k))\bar{y}(t_{k-1}) + \Theta_3(\theta(t_k))v(t_k).$$

3.1.4 System Augmentation

Denoting $\bar{y}(k) \triangleq \bar{y}(t_k)$, $\tilde{z}(k) \triangleq z(t_k)$ and $\tilde{x}(k) \triangleq \left[x^T(t_k)\ y^T(t_{k-1})\ \bar{y}^T(t_{k-1})\right]^T$, the time-varying delayed system (3.1) with the communication network can be reformulated as follows:

$$\begin{cases} \tilde{x}(k+1) = \bar{A}(k, \bar{\theta}(t_k^{d-1}))\tilde{x}(k) + F(k, \tilde{x}(k)) + \bar{A}_d(k, \bar{\theta}(t_k^{d-1}))\tilde{x}(k-\tau) \\ \qquad\qquad + \bar{B}(k, \bar{\theta}(t_k^{d-1}))\tilde{\omega}(k) \\ \bar{y}(k) = \bar{C}(k, \bar{\theta}(t_k^{d-1}))\tilde{x}(k) + \bar{C}_d(k, \bar{\theta}(t_k^{d-1}))\tilde{x}(k-\tau) \\ \qquad\qquad + \bar{D}(k, \bar{\theta}(t_k^{d-1}))\tilde{\omega}(k) \\ \tilde{z}(k) = \tilde{M}(k)\tilde{x}(k) \end{cases} \tag{3.7}$$

where

$$\bar{A}(k,\bar{\theta}(t_k^{d-1})) = \begin{bmatrix} A(t_k) & 0 & 0 \\ C(t_k) & 0 & 0 \\ \Phi(\theta(t_k^0))C(t_k) & \Theta_1(\bar{\theta}(t_k^{d-1})) & \Theta_2(\bar{\theta}(t_k^{d-1})) \end{bmatrix},$$

$$\bar{A}_d(k,\bar{\theta}(t_k^{d-1})) = \begin{bmatrix} A_d(t_k) & 0\,0 \\ C_d(t_k) & 0\,0 \\ \Phi(\theta(t_k^0))C_d(t_k) & 0\,0 \end{bmatrix},$$

$$\bar{B}(k,\bar{\theta}(t_k^{d-1})) = \begin{bmatrix} B(t_k) & 0 \\ D(t_k) & 0 \\ \Phi(\theta(t_k^0))D(t_k) & \Theta_3(\bar{\theta}(t_k^{d-1})) \end{bmatrix},$$

$$\bar{D}(k,\bar{\theta}(t_k^{d-1})) = \left[\Phi(\theta(t_k^0))D(t_k)\ \Theta_3(\bar{\theta}(t_k^{d-1}))\right],\ \tilde{M}(k) = \left[M(t_k)\ 0\ 0\right],$$

$$\bar{C}(k,\bar{\theta}(t_k^{d-1})) = \left[\Phi(\theta(t_k^0))C(t_k)\ \overrightarrow{\Theta}(\bar{\theta}(t_k^{d-1}))\right],\ \tilde{\omega}(k) = \left[\omega^T(t_k)\ \bar{v}^T(t_k)\right]^T,$$

$$\overrightarrow{\Theta}(\bar{\theta}(t_k^{d-1})) = \left[\Theta_1(\bar{\theta}(t_k^{d-1}))\ \Theta_2(\bar{\theta}(t_k^{d-1}))\right],$$

$$\bar{C}_d(k,\bar{\theta}(t_k^{d-1})) = \left[\Phi(\theta(t_k^0))C_d(t_k)\ 0\ 0\right],\quad F(k,\tilde{x}(k)) = \left[f^T(t_k,x(t_k))\ 0\ 0\right]^T.$$

For analysis convenience, we now reformulate the system (3.7) via a mapping method. The following proposition can be easily accessible from Lemma 1 of [3].

Proposition 3.1 *The communication constraints governed by $\bar{\theta}(t_k^{d-1})$ of the augmented system (3.7) can be mapped to the sequence $r(k) \in \mathscr{R} \triangleq \{1, 2, \ldots, (n_y)^d\}$ by the mapping $\mathfrak{R}(\cdot)$:*

$$r(k) = \mathfrak{R}\big(\bar{\theta}(t_k^{d-1})\big) \triangleq \sum_{i=0}^{d-1}(n_y)^i\big(\theta(t_k^i) - 1\big) + 1. \tag{3.8}$$

Moreover, if $r(k)$ is given, the values of $\theta(t_k^i)$ can be derived by $\phi_i(r(k))$ ($i = 0, 1, \ldots, d-1$):

$$\theta(t_k^i) = \phi_i(r(k)) \triangleq \mathrm{mod}\left(\left\lfloor \frac{r(k)-1}{(n_y)^i} \right\rfloor, n_y\right) + 1. \tag{3.9}$$

Proof First, it is easy to see that the value of $r(k)$ obtained by (3.8) satisfies $r(k) \in \mathscr{R}$. Next, we shall prove that $\theta(t_k^i)$ can be derived by (3.9). For any given $r(k) \in \mathscr{R}$, we have

$$\begin{aligned} \phi_i(r(k)) &= \mathrm{mod}\left(\left\lfloor \frac{r(k)-1}{(n_y)^i} \right\rfloor, n_y\right) + 1 \\ &= \mathrm{mod}\left(\sum_{j=i}^{d-1}(n_y)^{j-i}\big(\theta(t_k^j) - 1\big), n_y\right) + 1 \\ &= \theta(t_k^i) - 1 + 1 = \theta(t_k^i). \end{aligned}$$

The proof is complete.

The characteristics of the new stochastic process $r(k)$ are studied in Proposition 3.2 whose proof is omitted for brevity.

Proposition 3.2 *The stochastic process $r(k) \in \mathscr{R}$ is a Markov chain, and the transition probability matrix $\bar{\mathscr{P}}(k) \triangleq [\bar{p}_{ij}(k)]_{(n_y)^d \times (n_y)^d}$ is obtained as follows:*

$$\begin{aligned}\bar{p}_{ij}(k) &= \text{Prob}\,(r(k+1) = j | r(k) = i) \\ &= p_{\phi_0(i)\phi_{d-1}(j)}(t_k^0) \prod_{s=1}^{d-1} p_{\phi_s(j)\phi_{s-1}(j)}(t_{k+1}^s)\end{aligned} \tag{3.10}$$

where $p_{ij}(\cdot)$ has been defined in (3.3).

Based on Propositions 3.1 and 3.2, the augmented system (3.7) can be rewritten as follows:

$$\begin{cases}\tilde{x}(k+1) = \tilde{A}_{r(k)}(k)\tilde{x}(k) + F(k, \tilde{x}(k)) + \tilde{E}_{1,r(k)}(k)\tilde{x}(k-\tau) + \tilde{B}_{r(k)}(k)\tilde{\omega}(k) \\ \bar{y}(k) = \tilde{C}_{r(k)}(k)\tilde{x}(k) + \tilde{D}_{r(k)}(k)\tilde{\omega}(k) + \tilde{E}_{2,r(k)}(k)\tilde{x}(k-\tau) \\ \tilde{z}(k) = \tilde{M}(k)\tilde{x}(k)\end{cases} \tag{3.11}$$

where

$$\tilde{A}_{r(k)}(k) = \bar{A}(k, \bar{\theta}(t_k^{d-1})),\ \tilde{B}_{r(k)}(k) = \bar{B}(k, \bar{\theta}(t_k^{d-1})),\ \tilde{E}_{1,r(k)}(k) = \bar{A}_d(k, \bar{\theta}(t_k^{d-1})),$$
$$\tilde{D}_{r(k)}(k) = \bar{D}(k, \bar{\theta}(t_k^{d-1})),\ \tilde{C}_{r(k)}(k) = \bar{C}(k, \bar{\theta}(t_k^{d-1})),\ \tilde{E}_{2,r(k)}(k) = \bar{C}_d(k, \bar{\theta}(t_k^{d-1})),$$

and $r(k)$ is a Markov chain with the probability matrix $\bar{\mathscr{P}}(k)$ defined in Proposition 3.2.

3.1.5 Time-Varying Filter

The filter design problem is illustrated in Fig. 3.2 for system (3.1) with the measurement output (3.5) transmitted from the given high-rate communication network under the communication constraints.

Consider the following nonlinear filter for the discrete-time-varying stochastic system (3.11)

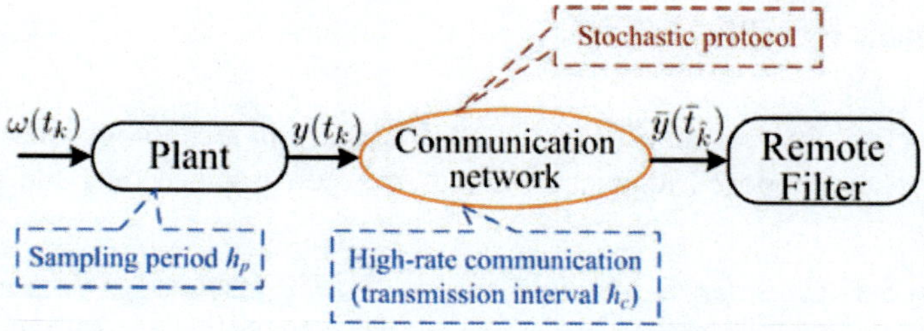

Fig. 3.2 Filtering over high-rate communication network with RA protocol

$$\begin{cases} \hat{x}(k+1) = \tilde{A}_{r(k)}(k)\hat{x}(k) + F(k,\hat{x}(k)) + \tilde{E}_{1,r(k)}(k)\hat{x}(k-\tau) + K_{r(k)}(k)\Big(\bar{y}(k) \\ \qquad - \tilde{C}_{r(k)}(k)\hat{x}(k) - \tilde{E}_{2,r(k)}(k)\hat{x}(k-\tau)\Big) \\ \hat{z}(k) = \tilde{M}(k)\hat{x}(k) \\ \hat{x}(i) = 0, \quad i = -\tau, -\tau+1, \ldots, 0 \end{cases} \tag{3.12}$$

where $\hat{x}(k) \in \mathbb{R}^n$ is the state estimate of the system (3.11) with $n = n_x + 2n_y$, $\hat{z}(k)$ denotes the estimate of $z(t_k)$, and $K_{r(k)}(k)$ is the filter parameter to be determined.

Let the filtering error be $e(k) \triangleq \tilde{x}(k) - \hat{x}(k)$ and the output filtering error be $z_e(k) \triangleq \tilde{z}(k) - \hat{z}(k)$. Then the filtering error dynamics for the augmented system (3.11) can be obtained as follows:

$$\begin{cases} e(k+1) = \Big(\tilde{A}_{r(k)}(k) - K_{r(k)}(k)\tilde{C}_{r(k)}(k)\Big)e(k) + \big(\tilde{E}_{1,r(k)}(k) \\ \qquad - K_{r(k)}(k)\tilde{E}_{2,r(k)}(k)\big)e(k-\tau) + \tilde{F}(k,e(k)) + \Big(\tilde{B}_{r(k)}(k) \\ \qquad - K_{r(k)}(k)\tilde{D}_{r(k)}(k)\Big)\tilde{\omega}(k) \\ z_e(k) = \tilde{M}(k)e(k) \end{cases} \tag{3.13}$$

where $\tilde{F}(k,e(k)) = F(k,\tilde{x}(k)) - F(k,\hat{x}(k))$.

Our aim in this chapter is to design a finite-horizon filter in the form of (3.12) such that, for the given disturbance attenuation level $\gamma > 0$, the following $\mathscr{H}_\infty$ performance is satisfied:

$$J \triangleq \mathbb{E}\left\{\sum_{k=0}^{N}\left(\|z_e(k)\|^2 - \gamma^2\|\tilde{\omega}(k)\|^2\right)\right\} - \gamma^2\sum_{i=-\tau}^{0} e^T(i)Q_i e(i) \le 0, \quad \forall\tilde{\omega}(k) \neq 0 \tag{3.14}$$

where $Q_i \in \mathbb{R}^{n\times n}$ $(i = -\tau, -\tau+1, \ldots, 0)$ are some weighting matrices satisfying $Q_i > 0$.

3.2 Main Results

In this section, by resorting to the stochastic analysis technology, some sufficient conditions are proposed to guarantee the $\mathscr{H}_\infty$ performance over the given finite horizon.

Theorem 3.1 *Consider the dynamic system (3.13) with the high-rate communication network and the communication constraints governed by the Markov chain $r(k)$ whose transition probability matrix is $\bar{\mathscr{P}}$ defined in (3.3). Let the disturbance attenuation level $\gamma > 0$, the positive definite matrices Q_i $(i = -\tau, -\tau+1, \ldots, 0)$ and the filter gain matrices $K_i(k)$ $(i \in \mathscr{R}, 0 \le k \le N)$ be given. The $\mathscr{H}_\infty$ performance index defined in (3.14) is achieved for all nonzero $\tilde{\omega}(k)$ if there exist families of positive scalars $\{\lambda_i(k)\}$ $(i \in \mathscr{R}, 0 \le k \le N)$, positive definite matrices $P_i(k)$ $(i \in \mathscr{R}, 0 \le k \le N+1)$ and $S(k)$ $(-\tau \le k \le N)$ satisfying the following recursive matrix inequalities*

$$\Pi_i(k) = \begin{bmatrix} \bar{\Pi}_i(11) & \Pi_i(12) & \bar{\Pi}_i(13) & \Pi_i(14) \\ * & \Pi_i(22) & \Pi_i(23) & \Pi_i(24) \\ * & * & \bar{\Pi}_i(33) & \Pi_i(34) \\ * & * & * & \Pi_i(44) \end{bmatrix} < 0 \tag{3.15}$$

and

$$P_i(0) \le \gamma^2 Q_0, \quad S(j) \le \gamma^2 Q_j, \quad (-\tau \le j \le -1) \tag{3.16}$$

for all $i \in \mathscr{R}$, where $\bar{P}_i(k+1) = \sum_{j \in \mathscr{R}} \bar{p}_{ij}(k) P_j(k+1)$ and

$$\begin{aligned}
&\mathscr{A}_i(k) = \tilde{A}_i(k) - K_i(k)\tilde{C}_i(k), \ \Pi_i(23) = \mathscr{E}_i^T(k)\bar{P}_i(k+1), \ \mathscr{B}_i(k) = \tilde{B}_i(k) - K_i(k)\tilde{D}_i(k), \\
&\Pi_i(34) = \bar{P}_i(k+1)\mathscr{B}_i(k), \ \mathscr{E}_i(k) = \tilde{E}_{1,i}(k) - K_i(k)\tilde{E}_{2,i}(k), \ \hat{F}(k) = \bar{F}_1^T(k) + \bar{F}_2^T(k), \\
&\bar{\Pi}_i(11) = \mathscr{A}_i^T(k)\bar{P}_i(k+1)\mathscr{A}_i(k) + S(k) - P_i(k) + \tilde{M}^T(k)\tilde{M}(k) - \lambda_i(k)\overrightarrow{F}(k), \\
&\Pi_i(11) = \bar{\Pi}_i(11) + \lambda_i(k)\overrightarrow{F}(k), \ \bar{\Pi}_i(33) = \Pi_i(33) - 2\lambda_i(k)I, \ \Pi_i(33) = \bar{P}_i(k+1), \\
&\Pi_i(12) = \mathscr{A}_i^T(k)\bar{P}_i(k+1)\mathscr{E}_i(k), \ \bar{\Pi}_i(13) = \Pi_i(13) + \lambda_i(k)\hat{F}(k), \\
&\Pi_i(13) = \mathscr{A}_i^T(k)\bar{P}_i(k+1), \Pi_i(14) = \mathscr{A}_i^T(k)\bar{P}_i(k+1)\mathscr{B}_i(k), \\
&\Pi_i(22) = \mathscr{E}_i^T(k)\bar{P}_i(k+1)\mathscr{E}_i(k) - S(k-\tau), \ \Pi_i(24) = \mathscr{E}_i^T(k)\bar{P}_i(k+1)\mathscr{B}_i(k), \\
&\Pi_i(44) = \mathscr{B}_i^T(k)\bar{P}_i(k+1)\mathscr{B}_i(k) - \gamma^2 I, \ \bar{F}_1(k) = \mathrm{diag}\{F_1(t_k), 0, 0\}, \\
&\bar{F}_2(k) = \mathrm{diag}\{F_2(t_k), 0, 0\}, \ \overrightarrow{F}(k) = \bar{F}_1^T(k)\bar{F}_2(k) + \bar{F}_2^T(k)\bar{F}_1(k).
\end{aligned}$$

Proof Define the following Lyapunov-like function

$$V(k) = e^T(k)P_{r(k)}(k)e(k) + \sum_{j=k-\tau}^{k-1} e^T(j)S(j)e(j). \tag{3.17}$$

Then, it can be obtained from (3.13) that

$$\begin{aligned}&\mathbb{E}\{V(k+1)-V(k)\big|r(k)=i\}\\ =&\mathbb{E}\left\{\alpha^T(k)\bar{\Pi}_i(k)\alpha(k)-\|z_e(k)\|^2+\gamma^2\|\tilde{\omega}(k)\|^2\big|r(k)=i\right\}\end{aligned} \tag{3.18}$$

where

$$\alpha(k)=\left[e^T(k)\ e^T(k-\tau)\ \tilde{F}^T(k,e(k))\ \tilde{\omega}^T(k)\right]^T,$$
$$\bar{\Pi}_i(k)=\begin{bmatrix}\Pi_i(11) & \Pi_i(12) & \Pi_i(13) & \Pi_i(14)\\ * & \Pi_i(22) & \Pi_i(23) & \Pi_i(24)\\ * & * & \Pi_i(33) & \Pi_i(34)\\ * & * & * & \Pi_i(44)\end{bmatrix}.$$

Since the nonlinear function $f(t_k, x(t_k))$ satisfies the section bounded condition (3.2), it is easy to see that

$$\begin{bmatrix}e(k)\\ \tilde{F}(k,e(k))\end{bmatrix}^T\begin{bmatrix}\overrightarrow{F}(k) & -\hat{F}(k)\\ * & 2I\end{bmatrix}\begin{bmatrix}e(k)\\ \tilde{F}(k,e(k))\end{bmatrix}\leq 0. \tag{3.19}$$

Hence, it can be derived from (3.15), (3.18) and (3.19) that

$$\begin{aligned}&\mathbb{E}\{V(k+1)-V(k)\big|r(k)=i\}\\ \leq&\mathbb{E}\left\{\alpha^T(k)\Pi_i(k)\alpha(k)-\|z_e(k)\|^2+\gamma^2\|\tilde{\omega}(k)\|^2\big|r(k)=i\right\}\\ \leq&-\mathbb{E}\left\{\|z_e(k)\|^2-\gamma^2\|\tilde{\omega}(k)\|^2\right\}.\end{aligned} \tag{3.20}$$

Taking the sum on both sides of (3.20) from 0 to N, we obtain

$$\mathbb{E}\{V(N+1)-V(0)\}\leq\mathbb{E}\Big\{\sum_{k=0}^{N}\left(\gamma^2\|\tilde{\omega}(k)\|^2-\|z_e(k)\|^2\right)\Big\} \tag{3.21}$$

which implies from (3.14) that

$$\begin{aligned}J\leq&-\mathbb{E}\{V(e(N+1),r(N+1))-V(e(0),r(0))\}-\gamma^2\sum_{i=-\tau}^{0}e^T(i)Q_ie(i)\\ \leq&\mathbb{E}\{V(e(0),r(0))\}-\gamma^2\sum_{i=-\tau}^{0}e^T(i)Q_ie(i).\end{aligned} \tag{3.22}$$

Finally, it can be easily concluded from (3.16) to (3.22) that

$$J \le \mathbb{E}\{V(e(0), r(0))\} - \gamma^2 \sum_{i=-\tau}^{0} e^T(i) Q_i e(i) \le 0$$

which means the pre-specified $\mathscr{H}_\infty$ performance is satisfied. The proof is complete.

Theorem 3.2 *Consider the dynamic system (3.13) with the high-rate communication network and the communication constraints governed by the Markov chain $r(k)$ whose transition probability matrix is $\bar{\mathscr{P}}$ defined in (3.3). Let the disturbance attenuation level $\gamma > 0$ and the positive definite matrices Q_i $(i = -\tau, -\tau+1, \ldots, 0)$ be given. The $\mathscr{H}_\infty$ performance index defined in (3.14) is achieved for all nonzero $\tilde{\omega}(k)$ if there exist families of positive scalars $\{\lambda_i(k)\}$ $(i \in \mathscr{R}, 0 \le k \le N)$, positive definite matrices $P_i(k)$ $(i \in \mathscr{R}, 0 \le k \le N+1)$, $S(k)$ $(-\tau \le k \le N)$, and real-valued matrices $\mathscr{K}_i(k)$ $(i \in \mathscr{R}, 0 \le k \le N)$ satisfying the following recursive matrix inequalities*

$$\hat{\Pi}_i(k) = \begin{bmatrix} \hat{\Pi}_i(11) & \hat{\Pi}_i(12) & \hat{\Pi}_i(13) & \hat{\Pi}_i(14) & \hat{\Pi}_i(15) \\ * & \hat{\Pi}_i(22) & \hat{\Pi}_i(23) & \hat{\Pi}_i(24) & \hat{\Pi}_i(25) \\ * & * & \hat{\Pi}_i(33) & \hat{\Pi}_i(34) & 0 \\ * & * & * & \hat{\Pi}_i(44) & \hat{\Pi}_i(45) \\ * & * & * & * & \hat{\Pi}_i(55) \end{bmatrix} < 0 \tag{3.23}$$

and

$$P_i(0) \le \gamma^2 Q_0, \; S(j) \le \gamma^2 Q_j, \; (-\tau \le j \le -1) \tag{3.24}$$

for all $i \in \mathscr{R}$, where

$$\begin{aligned}
\hat{\Pi}_i(11) =& \tilde{A}_i^T(k)\bar{P}_i(k+1)\tilde{A}_i(k) - \tilde{C}_i^T(k)\mathscr{K}_i^T(k)\tilde{A}_i(k) - \tilde{A}_i^T(k)\mathscr{K}_i(k)\tilde{C}_i(k) \\
&+ S(k) - P_i(k) + \tilde{M}^T(k)\tilde{M}(k) - \lambda_i(k)\overrightarrow{F}(k), \\
\hat{\Pi}_i(12) =& \tilde{A}_i^T(k)\bar{P}_i(k+1)\tilde{E}_{1,i}(k) - \tilde{A}_i^T(k)\mathscr{K}_i(k)\tilde{E}_{2,i}(k) - \tilde{C}_i^T(k)\mathscr{K}_i^T(k)\tilde{E}_{1,i}(k), \\
\hat{\Pi}_i(45) =& \tilde{D}_i^T(k)\mathscr{K}_i^T(k), \;\; \hat{\Pi}_i(13) = \tilde{A}_i^T(k)\bar{P}_i(k+1) - \tilde{C}_i^T(k)\mathscr{K}_i^T(k) + \lambda_i(k)\hat{F}(k), \\
\hat{\Pi}_i(14) =& \tilde{A}_i^T(k)\bar{P}_i(k+1)\tilde{B}_i(k) - \tilde{A}_i^T(k)\mathscr{K}_i(k)\tilde{D}_i(k) - \tilde{C}_i^T(k)\mathscr{K}_i^T(k)\tilde{B}_i(k), \\
\hat{\Pi}_i(22) =& \tilde{E}_{1,i}^T(k)\bar{P}_i(k+1)\tilde{E}_{1,i}(k) - \tilde{E}_{1,i}^T(k)\mathscr{K}_i(k)\tilde{E}_{2,i}(k) - \tilde{E}_{2,i}^T(k)\mathscr{K}_i^T(k)\tilde{E}_{1,i}(k) \\
&- S(k-\tau), \;\; \hat{\Pi}_i(34) = \bar{P}_i(k+1)\tilde{B}_i(k) - \mathscr{K}_i(k)\tilde{D}_i(k), \\
\hat{\Pi}_i(23) =& \tilde{E}_{1,i}^T(k)\bar{P}_i(k+1) - \tilde{E}_{2,i}^T(k)\mathscr{K}_i^T(k), \;\; \hat{\Pi}_i(15) = \tilde{C}_i^T(k)\mathscr{K}_i^T(k), \\
\hat{\Pi}_i(24) =& \tilde{E}_{1,i}^T(k)\bar{P}_i(k+1)\tilde{B}_i(k) - \tilde{E}_{1,i}^T(k)\mathscr{K}_i(k)\tilde{D}_i(k) - \tilde{E}_{2,i}^T(k)\mathscr{K}_i^T(k)\tilde{B}_i(k), \\
\hat{\Pi}_i(55) =& -\bar{P}_i(k+1), \;\; \hat{\Pi}_i(25) = \tilde{E}_{2,i}^T(k)\mathscr{K}_i^T(k), \;\; \hat{\Pi}_i(33) = \bar{P}_i(k+1) - 2\lambda_i(k)I, \\
\hat{\Pi}_i(44) =& \tilde{B}_i^T(k)\bar{P}_i(k+1)\tilde{B}_i(k) - \tilde{B}_i^T(k)\mathscr{K}_i(k)\tilde{D}_i(k) - \tilde{D}_i^T(k)\mathscr{K}_i^T(k)\tilde{B}_i(k) - \gamma^2 I
\end{aligned}$$

and other matrices are defined previously. Furthermore, if $(P_i(k), S(k), \mathscr{K}_i(k))$ is the feasible solution of (3.23) and (3.24), then the desired filter parameters in the *form of (3.12) are given as follows:*

$$K_i(k) = \bar{P}_i^{-1}(k+1)\mathscr{K}_i(k). \tag{3.25}$$

Proof The proof follows directly from that of Theorem 3.1 by using the Schur Complement Lemma.

By means of Theorem 3.2, we can summarize the Finite-Horizon $\mathscr{H}_\infty$ Filter Design (FHFD) algorithm as follows:

Algorithm FHFD:
Step 1. Give the $\mathscr{H}_\infty$ performance index γ, positive definite matrices Q_i $(i = -\tau, -\tau+1, \ldots, 0)$ and the error initial condition $e(i)$ $(i = -\tau, -\tau+1, \ldots, 0)$. Select the initial values for matrices $P_i(0)$ $(i \in \mathscr{R})$ and $S(i)$ $(-\tau \le i \le -1)$ which satisfy the condition (3.24) and set $k = 0$.
Step 2. For the time instant k, solve the recursive matrix inequality (3.23) to obtain the positive matrices $P_i(k+1)$ $(i \in \mathscr{S})$ and $S(k)$.
Step 3. Derive the filter gain matrices $K_i(k)$ $(i \in \mathscr{S})$ according to (3.25), and set $k = k+1$.
Step 4. If $k \le N$, then go to *Step 2*, else go to *Step 5*.
Step 5. Stop.

Remark 3.3 The main result established in Theorem 3.2 contains all the information about the $\mathscr{H}_\infty$ index, the transition probabilities of the RA protocol, the ratio of the sampling period and the communication period, the system parameters and the delay size. The main novelty of this chapter is twofold: (1) this research represents one of the first attempts to consider the filtering problem for time-varying systems with high-rate communication network and scheduled RA protocol; and (2) the mapping technology is employed to characterize the effect induced by the high-rate communication and the RA protocol scheduling. It can be observed from Algorithm FHFD that the recursive filtering process is particularly suitable for real-time implementation such as online process filtering.

Remark 3.4 The main results of in this chapter are derived based on the LMI-based algorithm, which has a polynomial time complexity. Specifically, the number $\mathscr{N}(\varepsilon)$ of flops needed to compute an ε-accurate solution is bounded by $O(\mathscr{M}\mathscr{N}^3\log(\mathscr{V}/\varepsilon))$, where $\mathscr{M}$ is the total row size of the LMI system, $\mathscr{N}$ is the total number of scalar decision variables, $\mathscr{V}$ is a data-dependent scaling factor, and ε is relative accuracy set for the algorithm. For the investigated filtering error dynamics (13), the computational complexity of the established result in Theorem 3.2 can be represented as $O(\mathscr{M}\mathscr{N}^3)$ where $\mathscr{M} = (4n + n_\omega + dn_y)(n_y)^d$ and $\mathscr{N} = (1 + 0.5n(n+1) + nn_y)(n_y)^d + 0.5n(n+1)$. Such a computational complexity depends polynomially on the variable dimensions. Obviously, the value of d has an important influence on the algorithm complexity. With the increase of d, the computational complexity becomes bigger.

3.3 Illustrative Examples

In this section, two numerical examples are presented to illustrate the effectiveness of the proposed time-varying $\mathscr{H}_\infty$ filter design scheme for the time-varying nonlinear delayed system (3.1) with the high-rate communication network and RA protocol scheduling.

Example 3.1 Consider the time-varying delayed system (3.1) with the following parameters:

$$A(t_k) = \begin{bmatrix} 0.73 + 0.4\sin(1.5t_k) & -0.57\sin(0.7t_k) \\ 0.25\cos(2.8t_k) & -0.24 - 0.36(-1)^{t_k}e^{-0.01t_k} \end{bmatrix},\ B(t_k) = \begin{bmatrix} 0.15 \\ -0.26 \end{bmatrix},$$

$$A_d(t_k) = \begin{bmatrix} 0.07 & -0.1\sin(1.5t_k) \\ 0.1e^{-0.04t_k} & -0.08 \end{bmatrix},\ C(t_k) = \begin{bmatrix} 0.65 & -0.30 \\ 0.55 & 0.40 \end{bmatrix},$$

$$M(t_k) = \mathrm{diag}\{0.20, 0.20\},\ C_d(t_k) = \begin{bmatrix} -0.1 & 0.2 \\ 0.2 & -0.2 \end{bmatrix},\ D(t_k) = \begin{bmatrix} 0.3 & 0.2 \end{bmatrix}^T,\ \tau = 2,$$

$$f(t_k, x(t_k)) = 0.5\big((F_1(t_k) + F_2(t_k))x(t_k) + (F_2(t_k) - F_1(t_k))\sin(t_k)x(t_k)\big),$$
$$F_1(t_k) = \mathrm{diag}\{0.08, 0.12\},\ F_2(t_k) = \mathrm{diag}\{0.10, 0.15\}.$$

Let the transition probability matrix $\mathscr{P}(\bar{t}_{\hat{k}})$ and the matrix $E(\bar{t}_{\hat{k}})$ be

$$\mathscr{P}(\bar{t}_{\hat{k}}) = \begin{bmatrix} 0.1 & 0.9 \\ 0.9 & 0.1 \end{bmatrix}$$

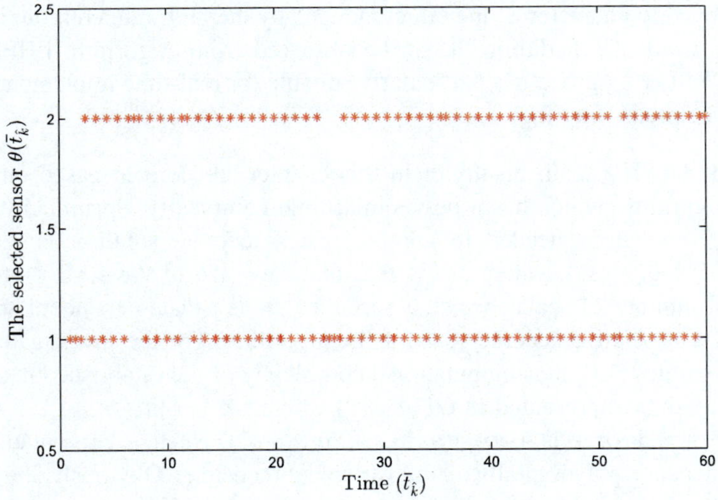

Fig. 3.3 The selected sensor obtaining access to the network

Table 3.1 The filter parameters $K_i(k)$ $(i \in \mathscr{R})$

k	0	1	2	3	...
$K_1(k)$	$\begin{bmatrix} 0.2787 & 0.0089 \\ 0.1467 & -0.0007 \\ 0.4087 & 0.0048 \\ 0.0323 & 0.0094 \\ 0.5180 & 0.0028 \\ -0.0006 & 1.0753 \end{bmatrix}$	$\begin{bmatrix} 0.2790 & -0.0042 \\ -0.1738 & 0.0061 \\ 0.2745 & -0.0004 \\ 0.0052 & 0.0116 \\ 0.3756 & -0.0140 \\ -0.0492 & 0.9931 \end{bmatrix}$	$\begin{bmatrix} 0.0355 & 0.0781 \\ -0.0257 & 0.1296 \\ 0.1127 & -0.1892 \\ -0.0621 & 0.2681 \\ 0.1274 & -0.0225 \\ -0.0194 & 1.4102 \end{bmatrix}$	$\begin{bmatrix} 0.0831 & 0.0326 \\ -0.0742 & 0.0195 \\ 0.1443 & 0.1629 \\ -0.0263 & 0.3289 \\ 0.1986 & 0.0172 \\ -0.0104 & 1.5332 \end{bmatrix}$	...
$K_2(k)$	$\begin{bmatrix} -0.0819 & 0.8813 \\ -0.0439 & 0.1073 \\ -0.1002 & 0.7529 \\ -0.0176 & 0.6016 \\ 0.8898 & -0.0157 \\ -0.0149 & 0.6511 \end{bmatrix}$	$\begin{bmatrix} 0.5092 & 0.7619 \\ -0.0943 & -0.1414 \\ 0.3109 & 0.4433 \\ 0.2631 & 0.5440 \\ 1.0170 & -0.1248 \\ 0.2607 & 0.5968 \end{bmatrix}$	$\begin{bmatrix} 1.6256 & 0.0484 \\ 0.4151 & 0.4013 \\ 1.4234 & -0.8651 \\ 0.3533 & 0.8857 \\ 0.8969 & -0.0053 \\ 0.3473 & 0.8801 \end{bmatrix}$	$\begin{bmatrix} 0.2566 & 0.1762 \\ -0.0729 & -0.0401 \\ 0.3025 & 0.6303 \\ 0.2208 & 0.8515 \\ 0.9309 & -0.0138 \\ 0.2272 & 0.8780 \end{bmatrix}$	...
$K_3(k)$	$\begin{bmatrix} 0.9336 & -0.0581 \\ 0.3049 & 0.0079 \\ 0.6818 & -0.0161 \\ 0.7294 & -0.0785 \\ 0.7451 & -0.0115 \\ -0.0156 & 0.8931 \end{bmatrix}$	$\begin{bmatrix} 1.0369 & 0.5471 \\ -0.2466 & -0.0567 \\ 0.5462 & 0.2086 \\ 0.4964 & 0.2743 \\ 0.6241 & 0.1937 \\ -0.0849 & 0.9597 \end{bmatrix}$	$\begin{bmatrix} 0.1102 & 0.6782 \\ -0.2110 & 0.1351 \\ 0.7056 & 0.5370 \\ -0.6050 & 0.3877 \\ 0.7207 & 0.5348 \\ -0.0174 & 0.9265 \end{bmatrix}$	$\begin{bmatrix} 0.4611 & 0.0209 \\ -0.0686 & 0.0066 \\ 0.6523 & 0.2207 \\ 0.9904 & 0.3203 \\ 0.6923 & 0.2271 \\ -0.0344 & 0.9300 \end{bmatrix}$	...
$K_4(k)$	$\begin{bmatrix} 0.0093 & 0.2437 \\ 0.0039 & -0.1269 \\ 0.0094 & 0.0206 \\ 0.0044 & 0.3831 \\ 1.0737 & 0.0001 \\ 0.0030 & 0.4295 \end{bmatrix}$	$\begin{bmatrix} 0.0225 & 0.1521 \\ -0.0127 & 0.0331 \\ 0.0043 & 0.0163 \\ -0.0284 & 0.3218 \\ 0.9978 & -0.0599 \\ -0.0385 & 0.3742 \end{bmatrix}$	$\begin{bmatrix} 0.1065 & 0.0703 \\ -0.0344 & 0.0558 \\ 0.2534 & -0.0550 \\ -0.1476 & 0.1600 \\ 1.3944 & -0.0176 \\ -0.0117 & 0.1385 \end{bmatrix}$	$\begin{bmatrix} 0.1186 & 0.0535 \\ -0.0350 & 0.0229 \\ 0.1528 & -0.0429 \\ 0.1408 & 0.1272 \\ 1.3945 & -0.0074 \\ 0.0166 & 0.1453 \end{bmatrix}$	...

and $E(\bar{t}_{\hat{k}}) = \begin{bmatrix}0.3 & -0.2\end{bmatrix}^T$, respectively. Under the scheduling of RA protocol, the selected sensor obtaining access to the network is plotted in the following Fig. 3.3.

The sampling period h_p and the communication period h_c are selectively as 1 and 0.5, respectively. In this simulation, the $\mathscr{H}_\infty$ performance level γ, the positive definite matrix Q_i $(i = -\tau, -\tau + 1, \ldots, 0)$ and time-horizon t_N are selected as $\gamma = 0.98$, $Q_i = 0.8I$ and $t_N = 60$, respectively. The state initial value is set to be $x(t_i) = \begin{bmatrix}0.5 & 0.5\end{bmatrix}^T$ $(i = -\tau, -\tau + 1, \ldots, 0)$. The exogenous disturbance inputs are

$$\nu(\bar{t}_{\hat{k}}) = 0.1e^{-0.02\bar{t}_{\hat{k}}} \sin(0.4\bar{t}_{\hat{k}}), \quad \omega(t_k) = 0.3e^{-0.01t_k} \cos(0.3t_k).$$

Using the given algorithm and MATLAB software, the set of solutions to recursive linear matrix inequalities in Theorem 3.2 are obtained, and the filter gain matrices are shown in Table 3.1.

The simulation results are shown in Figs. 3.4 and 3.5, which depict the actual states $x_1(t_k)$, $x_2(t_k)$ and their estimate $\hat{x}_1(t_k)$, $\hat{x}_2(t_k)$. The simulation results confirm that the desired finite-horizon $\mathscr{H}_\infty$ filtering performance is well achieved, and the proposed FHFD algorithm is indeed effective.

Example 3.2 Consider a time-varying nonlinear delayed system with the form (3.1). The corresponding parameters are given as follows:

$$A(t_k) = \begin{bmatrix} 0.73 + 0.1\sin(0.5t_k) & 0.17 \\ 0.18 & 0.74 \end{bmatrix}, \quad A_d(t_k) = \begin{bmatrix} 0.07 & -0.02 \\ 0.1 & -0.08 - 0.01\sin(0.5t_k) \end{bmatrix},$$

$$C(t_k) = \begin{bmatrix} 0.65 & -0.30 \\ 0.55 & 0.40 \end{bmatrix}, \quad B(t_k) = \begin{bmatrix} 0.15 \\ -0.26 \end{bmatrix}, \quad M(t_k) = \text{diag}\{0.59, 0.58\},$$

$$C_d(t_k) = \begin{bmatrix} -0.1 & 0.2 \\ 0.2 & -0.2 \end{bmatrix}, \quad D(t_k) = \begin{bmatrix}0.3 & 0.2\end{bmatrix}^T, \quad \tau = 2,$$

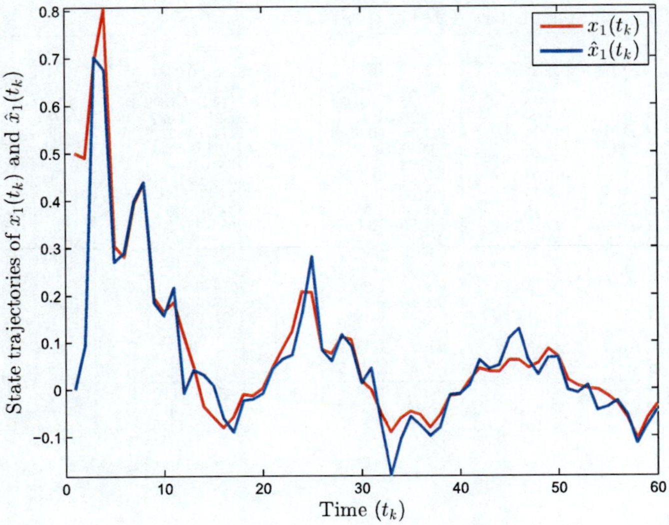

Fig. 3.4 The state trajectories of $x_1(t_k)$ and $\hat{x}_1(t_k)$

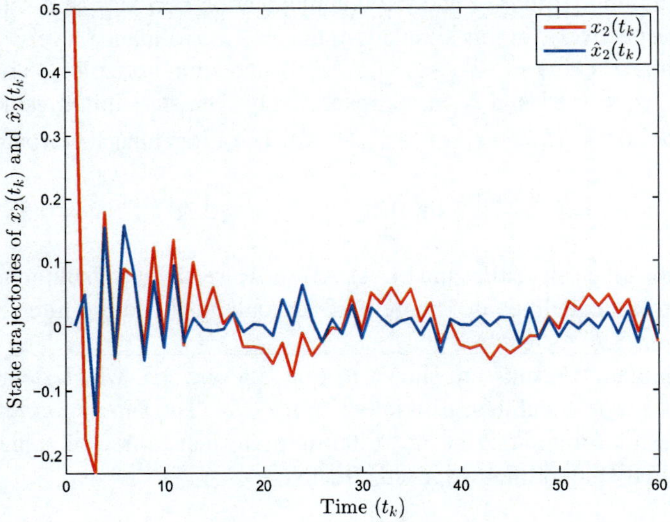

Fig. 3.5 The state trajectories of $x_2(t_k)$ and $\hat{x}_2(t_k)$

$$f(t_k, x(t_k)) = 0.5\big((F_1(t_k) + F_2(t_k))x(t_k) + (F_2(t_k) - F_1(t_k))\sin(t_k)x(t_k)\big),$$
$$F_1(t_k) = \text{diag}\{0.08, 0.12\}, \; F_2(t_k) = \text{diag}\{0.10, 0.15\}.$$

The transition probability matrix $\mathscr{P}(\bar{t_{\hat{k}}})$ is taken to be $\mathscr{P}(\bar{t_{\hat{k}}}) = \begin{bmatrix} 0.1 & 0.9 \\ 0.9 & 0.1 \end{bmatrix}$. The matrix $E(\bar{t_{\hat{k}}})$ is defined as $\begin{bmatrix} 0.3 & -0.2 \end{bmatrix}^T$. The sampling period h_p is chosen as 1.

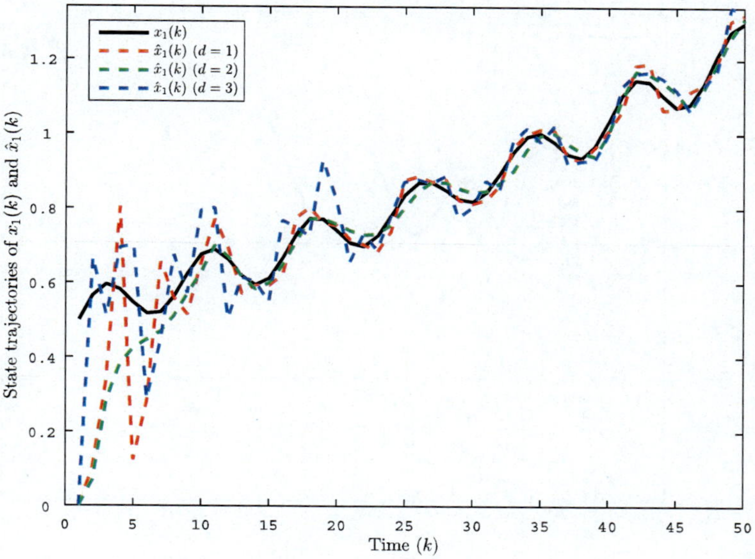

Fig. 3.6 The state trajectories of $x_1(t_k)$ and $\hat{x}_1(t_k)$

In this simulation, the $\mathscr{H}_\infty$ performance level γ, the positive definite matrix Q_i $(i = -\tau, -\tau + 1, \ldots, 0)$ and time-horizon t_N are selected as $\gamma = 0.98$, $Q_i = 6.5I$ and $t_N = 50$, respectively. The state initial value is set to be $x(t_i) = \begin{bmatrix} 0.5 \; 0.5 \end{bmatrix}^T$ $(i = -\tau, -\tau + 1, \ldots, 0)$. The disturbance inputs are

$$\nu(\bar{t}_{\hat{k}}) = 0.1 \sin(0.7\bar{t}_{\hat{k}}), \quad \omega(t_k) = 0.3 \cos(0.8 t_k).$$

Using the given algorithm and MATLAB software, the set of solutions to recursive linear matrix inequalities in Theorem 3.2 are obtained. The simulation results are shown in Figs. 3.6, 3.7 and 3.8. Figures 3.6 and 3.7 depict the actual states $x_1(t_k)$, $x_2(t_k)$ and their estimates $\hat{x}_1(t_k), \hat{x}_2(t_k)$ subject to different values of d. The trajectories of $\|z_e(k)\|$ are shown in Fig. 3.8. All the simulation results confirm that the desired finite-horizon $\mathscr{H}_\infty$ filtering performance is well achieved, and the proposed FHFD algorithm is indeed effective.

3.4 Summary

In this chapter, the $\mathscr{H}_\infty$ filtering problem has been investigated for a class of time-varying nonlinear delayed system subject to the scheduled RA protocol and high-rate communication network over a given finite horizon. The mapping technology has been employed to formulate the transmission model subject to the effect induced by the high-rate communication and RA protocol scheduling. Based on the stochastic

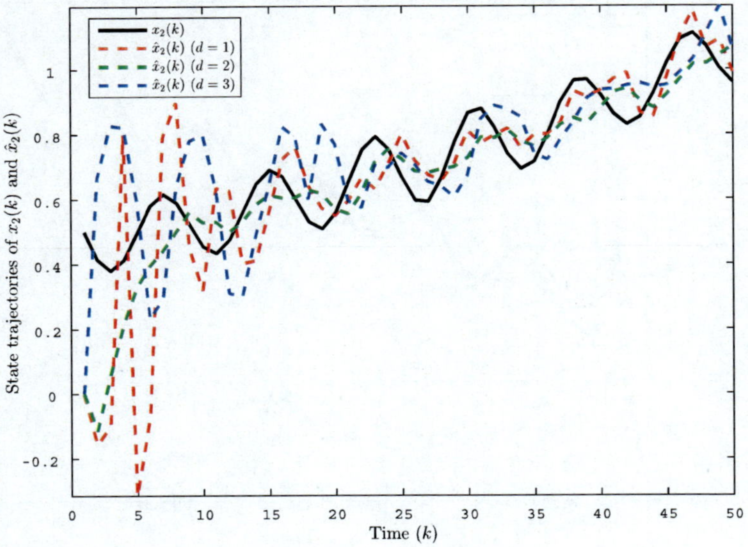

Fig. 3.7 The state trajectories of $x_2(t_k)$ and $\hat{x}_2(t_k)$

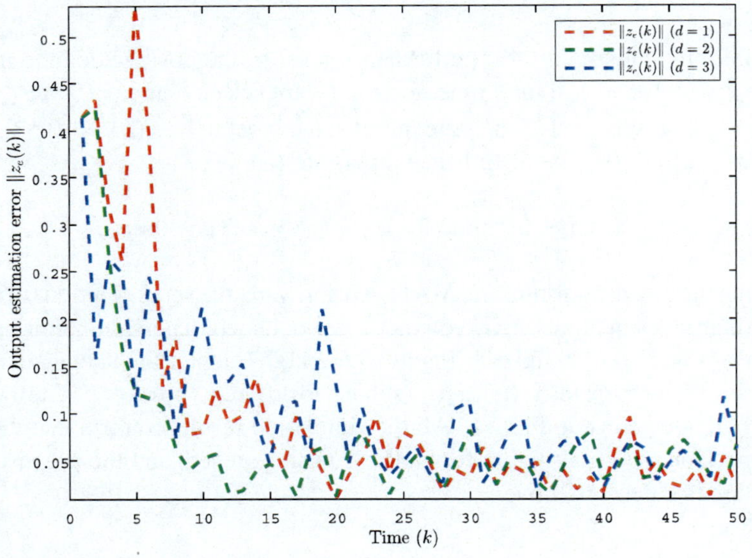

Fig. 3.8 The trajectories of $\|z_e(k)\|$ subject to different d

analysis techniques, the desired filter parameters have been obtained by solving a set of recursive matrix inequalities to guarantee the $\mathscr{H}_\infty$ performance of the filtering error dynamics. Finally, an illustrative example has been given to highlight the effectiveness of the proposed design strategy.

References

1. Donkers, M.C.F., Heemels, W.P.M.H., Bernardini, D., Bemporad, A., Shneer, V.: Stability analysis of stochastic networked control systems. Automatica **48**(5), 917–925 (2012)
2. Tabbara, M., Nešić, D.: Input-output stability of networked control systems with stochastic protocols and channels. IEEE Trans. Autom. Control **53**(5), 1160–1175 (2008)
3. Xu, Y., Su, H., Pan, Y., Wu, Z., Xu, W.: Stability analysis of network control systems with round-robin scheduling and packet dropouts. J. Frankl. Inst. **350**(8), 2013–2027 (2013)

Chapter 4
Finite-Horizon $\mathscr{H}_\infty$ Fault Estimation of Time-Varying Systems with Random Access Protocol

For decades, fault diagnosis and fault-tolerant control have been active research topics due mainly to the growing demand for high safety in various industrial systems, and a large body of research results have been reported in the literature. One of the foundational problems of fault diagnosis and fault-tolerant control is the fault estimation, which could generate the accurate sizes and shapes of the faults based on the received measurement signals. In fact, fault estimation can be regarded as a special filtering technique whose aim is to estimate the fault signals in a system. By now, the fault estimation problem has stirred considerable research interests, and a number of fault estimation approaches have been developed in the literature.

As we have discussed in Chap. 3, time-varying nature is an indispensable factor reflecting the fast changes in system dynamics since the system parameters might be changeable in time due to various reasons (e.g. temperature fluctuation, operating point shifting, graduate aging of system components). When it comes to the fault estimation issue, it is of vital practical significance to study the fault estimation problem for time-varying systems over a given finite horizon, which is referred to as the finite-horizon fault estimation problem. A reasonable way tackling the finite-horizon fault estimation issue is to generate the fault estimates based on the finite-horizon filtering framework.

On the other hand, owing to the rapid development of network technologies, network-induced constraints have become an emerging research topic, which has drawn a great deal of research attention. These network-induced constraints would largely affect the system performance, which motivates researchers to study the fault estimation problems subject to various network-induced constraints. As an important yet complex network-induced effects, the scheduling behavior of communication protocol would also give rise to some specific protocol-induced effects, which, in turn, has a significant impact on the fault estimation performance, and this motivates us to investigate the communication protocol-based fault estimation problem for

L. Zou et al., *Communication-Protocol-Based Filtering and Control of Networked Systems*, Studies in Systems, Decision and Control 430,
https://doi.org/10.1007/978-3-030-97512-8_4

time-varying systems. In this context, two challenges are identified as follows: (1) how to develop the suitable fault estimator for time-varying systems subject to the RA protocol scheduling? (2) how to derive the estimator gains recursively? It is, therefore, the main purpose of this chapter is to offer satisfactory answers to the two questions.

The rest of this chapter is organized as follows. In Sect. 4.1, the networked time-varying system with the RA protocol scheduling is introduced, and the corresponding fault estimator is proposed. In Sect. 4.2, the finite-horizon $\mathscr{H}_\infty$ fault estimation issue of the time-varying system is investigated based on the stochastic analysis approach. Furthermore, the required estimator gain matrices are obtained by solving a family of recursive linear matrix inequalities. In Sect. 4.3, the effectiveness of the proposed estimation scheme is demonstrated by two numerical simulation examples. Finally, conclusions are drawn in Sect. 4.4.

4.1 Problem Formulation

4.1.1 The Random Access Protocol

In this chapter, let us address another modeling method of the RA protocol scheduling (different from the discrete-time Markov chain introduced in Sect. 3.1.3). Consider a networked system with M nodes (labeled as $\{1, 2, \ldots, M\}$) where nodes transmit their data via a *shared communication network*. In order to prevent the data from collisions during transmission, *only one sensor node* is permitted to send data via the communication network at each transmission instant. According to such a *network access constraint*, various scheduling protocols have been utilized to orchestrate the transmission order of nodes. In this chapter, the RA protocol is employed to scheduling the data transmission of nodes.

Let $\alpha(k) \in \{1, 2, \ldots, M\}$ be the selected node obtaining access to the communication network at time instant k. As shown in [1], $\{\alpha(k)\}_{k\geq 0}$ could be regarded as a sequence of random variables due to the scheduling behavior of the RA protocol in which all the random variables are assumed to be *mutually independent*. The occurrence probability of $\alpha(k) = i$ ($\forall i \in \{1, 2, \ldots, M\}$) is given by

$$\text{Prob}\{\alpha(k) = i\} = p_i \tag{4.1}$$

where $p_i > 0$ ($i \in \{1, 2, \ldots, M\}$) is the occurrence probability that the node i is selected to transmit data via the communication network and $\sum_{i=1}^{M} p_i = 1$.

Remark 4.1 The RA protocol is a frequently used scheduling protocol to resolve the network access constraint, especially in wireless communications. Generally speaking, there are two different stochastic processes characterizing the scheduling behavior of the RA protocol. The first one is the independent and identically distributed (i.i.d.) sequence of random variables [1], and the other one is the discrete-time Markov chain [2]. In Chap. 3, we have considered the discrete-time Markov chain case. In this chapter, we consider the case that the RA protocol scheduling behavior is modeled by the i.i.d. sequence of random variables.

4.1.2 The System Model and Communication Model

Consider the following discrete time-varying systems defined on the finite horizon ($k \in [0, N]$):

$$\begin{cases} x_{k+1} = A_k x_k + B_k \omega_k + E_k f_k \\ y_k = C_k x_k + D_k \omega_k + F_k f_k \end{cases} \tag{4.2}$$

where $x_k \in \mathbb{R}^{n_x}$ and $y_k \in \mathbb{R}^{n_y}$ are, respectively, the state vector and the sensor output of the plant. $\omega_k \in l_2([0, N-1], \mathbb{R}^{n_\omega})$ is the disturbance input. $f_k \in \mathbb{R}^{n_f}$ is the fault vector. A_k, B_k, C_k, D_k, E_k and F_k are known time-varying matrices with appropriate dimensions. In practical applications, the faults in a system are always classified into two categories, namely, the sensor faults (which can be represented by $f_{s,k}$) and actuator faults (which can be represented by $f_{a,k}$). Nevertheless, by denoting $f_k = \left[f_{s,k}^T \; f_{a,k}^T\right]^T$, a system with such two kinds of faults can be reformulated by our proposed model (4.2). In other words, the model proposed in this chapter is quite general.

In this chapter, the faults concerned are assumed to be either incipient faults or abrupt faults, which are the most common faults in industrial processes [3]. In [3], the fault estimation problem has been studied for a class of continuous-time systems with such kind of faults, in which the second-order derivative of the fault signal should be zero. In this chapter, we consider the fault estimation problem for discrete- time-varying systems. Hence, the faults concerned are assumed to satisfy the condition $\Delta(\Delta(f_k)) = 0$ (i.e. the second order difference of the fault signal should be zero). The operator $\Delta(\cdot)$ is defined by the following expression: $\Delta(f_k) = f_{k+1} - f_k$. In fact, the fault model utilized in this chapter can be easily extended to some other complex fault signals by replacing the condition $\Delta(\Delta(f_k)) = 0$ by another condition (such as $\Delta(\Delta(\cdots(\Delta(f_k)))) = 0$).

Define an augmented state vector as follows: $\bar{x}_k = \left[x_k^T \; f_k^T \; \Delta^T(f_k)\right]^T \in \mathbb{R}^{\bar{n}}$ where $\bar{n} = n_x + 2n_f$. Then, we have

$$\begin{cases} \bar{x}_{k+1} = \bar{A}_k \bar{x}_k + \bar{B}_k \omega_k \\ y_k = \bar{C}_k \bar{x}_k + \bar{D}_k \omega_k \\ f_k = \bar{M} \bar{x}_k \end{cases} \tag{4.3}$$

where

$$\bar{A}_k = \begin{bmatrix} A_k & E_k & 0 \\ 0 & I & I \\ 0 & 0 & I \end{bmatrix}, \quad \bar{B}_k = \begin{bmatrix} B_k \\ 0 \\ 0 \end{bmatrix}, \quad \bar{C}_k = \begin{bmatrix} C_k & F_k & 0 \end{bmatrix}, \quad \bar{D}_k = D_k, \quad \bar{M} = \begin{bmatrix} 0 & I & 0 \end{bmatrix}.$$

We are now ready to introduce the signal transmission over the communication network. The signal transmitted through the network would be scheduled by the communication network subject to the communication protocol. In this chapter, the RA protocol scheduling is considered. Without loss of generality, we first assume that the sensors of the plant are grouped into M ($M > 1$) sensor nodes according to their spatial distribution. As such, for technical analysis, the measurement output y_k can be rewritten as follows:

$$y_k = \begin{bmatrix} y_{1,k}^T & y_{2,k}^T & \cdots & y_{M,k}^T \end{bmatrix}^T \tag{4.4}$$

where $y_{i,k}$ ($i \in \{1, 2, \ldots, M\}$) is the measurement of the ith sensor node before transmitted. For the sake of examining the influence of communication protocols, let $\alpha(k) \in \{1, 2, \ldots, M\}$ be the selected sensor node obtaining access to the communication network at time instant k. As shown in Sect. 3.1.1, $\alpha(k) \in \{1, 2, \ldots, M\}$ could be modeled as a sequence of random variables. Suppose that $\alpha(k)$ ($k \in \mathbb{N}^+$) is independent of all noise signals. Denote the measurement output after transmitted through the network by

$$\bar{y}_k \triangleq \begin{bmatrix} \bar{y}_{1,k}^T & \bar{y}_{2,k}^T & \cdots & \bar{y}_{M,k}^T \end{bmatrix}^T \in \mathbb{R}^{n_y}.$$

In this chapter, we choose the ZI strategy to compensate the signals. Then, the updating rule of $\bar{y}_{i,k}$ ($k \in \mathbb{N}^+, i \in \{1, 2, \ldots, N\}$) subject to the RA protocol scheduling is set to be

$$\bar{y}_{i,k} = \begin{cases} y_{i,k}, & \text{if } i = \alpha(k) \\ 0, & \text{otherwise.} \end{cases} \tag{4.5}$$

According to the updating rule (4.5), by defining

$$\Phi_{\alpha(k)} = \text{diag}\{\delta(\alpha(k) - 1)I, \delta(\alpha(k) - 2)I, \ldots, \delta(\alpha(k) - N)I\}, \quad 1 \le i \le N$$

where $\delta(\cdot) \in \{0, 1\}$ is the Kronecker delta function, it can be seen that

$$\bar{y}_k = \Phi_{\alpha(k)} y_k. \tag{4.6}$$

4.1.3 Time-Varying Fault Estimator

In this chapter, we adopt the following time-varying fault estimator for system (4.2):

$$\begin{cases} \hat{x}_{k+1} = \bar{A}_k \hat{x}_k + K_k \big(\bar{y}_k - \Phi_{\alpha(k)} \bar{C}_k \hat{x}_k\big) \\ \hat{f}_k = \bar{M} \hat{x}_k \end{cases} \tag{4.7}$$

where $\hat{x}_k$ is the estimate of the state $\bar{x}_k$ and $\hat{f}_k$ is the estimate of the fault f_k. The matrix K_k is the estimator gain to be designed.

Letting $e_k \triangleq \bar{x}_k - \hat{x}_k$ and $\tilde{f}_k \triangleq f_k - \hat{f}_k$, we have

$$\begin{cases} e_{k+1} = (\bar{A}_k - K_k \Phi_{\alpha(k)} \bar{C}_k) e_k + (\bar{B}_k - K_k \Phi_{\alpha(k)} \bar{D}_k) \omega_k \\ \tilde{f}_k = \bar{M} e_k. \end{cases} \tag{4.8}$$

Our objective of the addressed problem is to find a fault estimate $\hat{f}_k$ $(0 \leq N-1)$, such that for the given positive scalar γ, the estimation error system (4.8) satisfies the following fault estimation performance requirement:

$$J \triangleq \mathbb{E}\left\{ \sum_{k=0}^{N-1} \left(\|\tilde{f}_k\|^2 - \gamma^2 \|\omega_k\|^2 \right) \right\} - \gamma^2 e_0^T Q e_0 < 0, \tag{4.9}$$

where Q is known positive definite weighted matrix.

Remark 4.2 It can be easily seen from the estimation error dynamics (4.8) that the fault estimation error (i.e. $\tilde{f}_k$) is largely affected by the scheduling behavior described by $\Phi_{\alpha(k)}$. Due to the existence of such a parameter-dependent matrix, the resultant estimation error dynamics is described by a linear time-varying system with a stochastic parameter. In a typical filtering problem, the filtering performance is mainly dependent on the observability of the plant (4.2), which is largely affected by the protocol scheduling behavior. In this chapter, the fault estimation performance is only partial dependent on the observability of the plant. In other words, the inaccurate estimates on the system states might lead to the accurate estimates on the fault signal. In this chapter, we intend to analyze the impact of the protocol scheduling behavior on the fault estimation performance by using the recursive linear matrix inequality-based approach.

4.2 Main Results

In this section, let us investigate both the fault estimator performance analysis and design problems for system (4.8). Firstly, we shall propose some sufficient conditions ensuring the required $\mathscr{H}_\infty$ performance over the given finite horizon.

Theorem 4.1 *Consider the time-varying system (4.8) with the stochastic parameter matrix $\Phi_{\alpha(k)}$ whose statistical properties are given in (4.1). Let the disturbance attenuation level $\gamma > 0$ and the weighted matrix Q be given. For any disturbance sequence $\{\omega_k\}_{0\le k\le N-1}$, the time-varying system (4.8) satisfies the $\mathscr{H}_\infty$ performance requirement (4.9) if there exist family of positive definite matrices $\{P_k\}_{0\le k\le N-1}$ satisfying*

$$\bar{\Psi}_k = \begin{bmatrix} \bar{\Psi}_{11}^k & \bar{\Psi}_{12}^k \\ * & \bar{\Psi}_{22}^k \end{bmatrix} < 0 \tag{4.10}$$

and the initial condition

$$P_0 \le \gamma^2 Q \tag{4.11}$$

where

$$\begin{aligned}
&\tilde{\Phi}_i = \Phi_i - \bar{\Phi}, \quad \Theta_k = \sum_{i=1}^{N} p_i \tilde{\Phi}_i K_k^T P_{k+1} K_k \tilde{\Phi}_i, \quad \bar{\Phi} = \sum_{i=1}^{M} p_i \Phi_i, \\
&\bar{\Psi}_{11}^k = (\bar{A}_k - K_k \bar{\Phi} \bar{C}_k)^T P_{k+1} (\bar{A}_k - K_k \bar{\Phi} \bar{C}_k) - P_k + \bar{C}_k^T \Theta_k \bar{C}_k + \bar{M}^T \bar{M}, \\
&\bar{\Psi}_{12}^k = (\bar{A}_k - K_k \bar{\Phi} \bar{C}_k)^T P_{k+1} (\bar{B}_k - K_k \bar{\Phi} \bar{D}_k) + \bar{C}_k^T \Theta_k \bar{D}_k, \\
&\bar{\Psi}_{22}^k = (\bar{B}_k - K_k \bar{\Phi} \bar{D}_k)^T P_{k+1} (\bar{B}_k - K_k \bar{\Phi} \bar{D}_k) + \bar{D}_k^T \Theta_k \bar{D}_k - \gamma^2 I.
\end{aligned}$$

Proof Consider the following Lyapunov function candidate:

$$V_k = e_k^T P_k e_k. \tag{4.12}$$

Calculate the difference of V_k along the solution of system (4.8) and take expectation conditional on e_k. Then, we have

$$\begin{aligned}
&\mathbb{E}\{\Delta V_k | e_k\} = \mathbb{E}\{V_{k+1} - V_k | e_k\} \\
&= \mathbb{E}\Big\{\Big((\bar{A}_k - K_k \Phi_{\alpha(k)} \bar{C}_k) e_k + (\bar{B}_k - K_k \Phi_{\alpha(k)} \bar{D}_k)\omega_k\Big)^T P_{k+1} \Big((\bar{A}_k - K_k \Phi_{\alpha(k)} \bar{C}_k) e_k \\
&\quad + (\bar{B}_k - K_k \Phi_{\alpha(k)} \bar{D}_k)\omega_k\Big) - e_k^T P_k e_k \Big| e_k\Big\} \\
&= \mathbb{E}\Big\{\Big((\bar{A}_k - K_k \bar{\Phi} \bar{C}_k) e_k + (\bar{B}_k - K_k \bar{\Phi} \bar{D}_k)\omega_k\Big)^T P_{k+1} \Big((\bar{A}_k - K_k \bar{\Phi} \bar{C}_k) e_k + (\bar{B}_k \\
&\quad - K_k \bar{\Phi} \bar{D}_k)\omega_k\Big) - e_k^T P_k e_k + \Big(K_k \tilde{\Phi}_{\alpha(k)} (\bar{C} e_k + \bar{D}_k \omega_k)\Big)^T P_{k+1} K_k \tilde{\Phi}_{\alpha(k)} (\bar{C} e_k \\
&\quad + \bar{D}_k \omega_k) \Big| e_k\Big\}.
\end{aligned} \tag{4.13}$$

In view of the statistical properties of $\tilde{\Phi}_{\alpha(k)}$, it is easy to see that

$$\mathbb{E}\{\tilde{\Phi}_{\alpha(k)} K_k^T P_{k+1} K_k \tilde{\Phi}_{\alpha(k)} | e_k\} = \Theta_k, \tag{4.14}$$

which implies that

$$\mathbb{E}\{\Delta V_k | e_k\} = \begin{bmatrix} e_k \\ \omega_k \end{bmatrix}^T \begin{bmatrix} \Psi_{11}^k & \Psi_{12}^k \\ * & \Psi_{22}^k \end{bmatrix} \begin{bmatrix} e_k \\ \omega_k \end{bmatrix} \tag{4.15}$$

where

$$\begin{aligned}
\Psi_{11}^k &= (\bar{A}_k - K_k \bar{\Phi} \bar{C}_k)^T P_{k+1} (\bar{A}_k - K_k \bar{\Phi} \bar{C}_k) - P_k + \bar{C}_k^T \Theta_k \bar{C}_k, \\
\Psi_{12}^k &= (\bar{A}_k - K_k \bar{\Phi} \bar{C}_k)^T P_{k+1} (\bar{B}_k - K_k \bar{\Phi} \bar{D}_k) + \bar{C}_k^T \Theta_k \bar{D}_k, \\
\Psi_{22}^k &= (\bar{B}_k - K_k \bar{\Phi} \bar{D}_k)^T P_{k+1} (\bar{B}_k - K_k \bar{\Phi} \bar{D}_k) + \bar{D}_k^T \Theta_k \bar{D}_k.
\end{aligned}$$

Adding the following zero term

$$\|\tilde{f}_k\|^2 - \gamma^2 \|\omega_k\|^2 - (\|\tilde{f}_k\|^2 - \gamma^2 \|\omega_k\|^2) \tag{4.16}$$

to the right-hand side of (4.15), we have

$$\begin{aligned}
\mathbb{E}\{\Delta V_k | e_k\} &= \begin{bmatrix} e_k \\ \omega_k \end{bmatrix}^T \bar{\Psi}_k \begin{bmatrix} e_k \\ \omega_k \end{bmatrix} - (\|\tilde{f}_k\|^2 - \gamma^2 \|\omega_k\|^2) \\
&\leq -(\|\tilde{f}_k\|^2 - \gamma^2 \|\omega_k\|^2).
\end{aligned} \tag{4.17}$$

Taking the sum on both sides of (4.17) from 0 to $N-1$, we obtain

$$\mathbb{E}\{V_N\} - \mathbb{E}\{V_0\} \leq -\mathbb{E}\left\{ \sum_{k=0}^{N-1} (\|\tilde{f}_k\|^2 - \gamma^2 \|\omega_k\|^2) \right\}. \tag{4.18}$$

Then, it can be concluded from (4.18) that

$$\mathbb{E}\left\{ \sum_{k=0}^{N-1} (\|\tilde{f}_k\|^2 - \gamma^2 \|\omega_k\|^2) \right\} \leq \mathbb{E}\{V_0\} - \mathbb{E}\{V_N\} \leq \mathbb{E}\{V_0\}. \tag{4.19}$$

According to the condition (4.11), it is easy to see that the fault estimation performance requirement (4.9) is satisfied.

So far, we have analyzed the $\mathscr{H}_\infty$ performance in terms of the solvability of a set of matrix inequalities. In the next stage, we would like to deal with the design problem of the estimator parameters $\{K_k\}_{0 \leq k \leq N-1}$.

Theorem 4.2 *Consider the discrete time-varying system (4.8). Let the disturbance attenuation level $\gamma > 0$ and the weighted matrix Q be given. The fault estimate $\hat{f}_k$ satisfies the $\mathscr{H}_\infty$ performance criterion (4.9) if there exist families of positive definite matrices $\{P_k\}_{0\le k\le N-1}$ and $\{S_k\}_{0\le k\le N-1}$, a set of matrices $\{\bar{K}_k\}_{0\le k\le N-1}$ satisfying the following matrix inequalities:*

$$\alpha_k = \begin{bmatrix} \alpha_{11}^k & \alpha_{12}^k & \alpha_{13}^k \\ * & \alpha_{22}^k & \alpha_{23}^k \\ * & * & \alpha_{33}^k \end{bmatrix} < 0 \tag{4.20}$$

$$\begin{bmatrix} -S_k & \tilde{\Phi}_1 \bar{K}_k^T & \tilde{\Phi}_2 \bar{K}_k^T & \cdots & \tilde{\Phi}_N \bar{K}_k^T \\ * & -\frac{P_{k+1}}{p_1} & 0 & \cdots & 0 \\ * & * & -\frac{P_{k+1}}{p_2} & \cdots & 0 \\ \vdots & \vdots & \vdots & \ddots & \vdots \\ * & * & * & \cdots & -\frac{P_{k+1}}{p_N} \end{bmatrix} < 0 \tag{4.21}$$

and the initial condition

$$P_0 \le \gamma^2 Q \tag{4.22}$$

where

$$\tilde{\Phi}_i = \Phi_i - \bar{\Phi},\ \ \bar{\Phi} = \sum_{i=1}^{M} p_i \Phi_i,\ \ \alpha_{11}^k = -P_k + \bar{M}^T \bar{M} + \bar{C}_k^T S_k \bar{C}_k,\ \ \alpha_{12}^k = \bar{C}_k^T S_k \bar{D}_k,$$
$$\alpha_{13}^k = \bar{A}_k^T P_{k+1} - \bar{C}_k^T \bar{\Phi} \bar{K}_k^T,\ \ \alpha_{22}^k = -\gamma^2 I + \bar{D}_k^T S_k \bar{D}_k,\ \ \alpha_{23}^k = \bar{B}_k^T P_{k+1} - \bar{D}_k^T \bar{\Phi} \bar{K}_k^T,$$
$$\alpha_{33}^k = -\ P_{k+1}.$$

Proof According to (4.21), it is easy to see that

$$\Theta_k \triangleq \sum_{i=1}^{N} p_i \tilde{\Phi}_i K_k^T P_{k+1} K_k \tilde{\Phi}_i \le S_k. \tag{4.23}$$

Then, by apply Schur Complement lemma, it follows from (4.23) that the condition (4.10) could be guaranteed by the inequality (4.20). The proof is complete.

By means of Theorem 4.2, we can summarize the Finite-Horizon Fault Estimator Design (FHFED) algorithm as follows:

Algorithm FHFED:

Step 1. Give the $\mathscr{H}_\infty$ performance index γ, positive definite matrix Q and the error initial condition $e(0)$. Select the initial values for matrix P_0, which satisfy the condition (4.22) and set $k = 0$

Step 2. For the time instant k, solve the recursive matrix inequality (4.20) and (4.21) to obtain the positive matrices P_{k+1} and $\bar{K}_k$

Step 3. Derive the filter parameters K_k according to (4.22), and set $k = k + 1$

Step 4. If $k \leq N - 1$, then go to *Step 2*, else go to *Step 5*

Step 5. Stop

Remark 4.3 By now, we have completed the design of the finite-horizon $\mathscr{H}_\infty$ filter by solving a set of recursive linear matrix inequalities. It can be easily observed from the conditions proposed in Theorems 4.1 and 4.2 that the filtering performance and the filter design procedure are affected by the occurrence probabilities about the protocol scheduling behavior. The results of this chapter can be extended to the case that the occurrence probabilities about the protocol scheduling behavior are partially known by combining the robust analysis approach and stochastic analysis method.

4.3 Illustrative Examples

In this section, two numerical examples are presented to illustrate the effectiveness of the proposed fault estimation technology for time-varying system (4.2) with RA protocol scheduling.

Example 4.1 Consider a time-varying system (4.2) with the following parameters:

$$A_k = \begin{bmatrix} 0.99 + 0.02\sin(0.3k) & 0.01\cos(0.5k) \\ -0.02\sin(k) & 1.01 \end{bmatrix}, \quad B_k = \begin{bmatrix} 0.3 \\ 0.2 + 0.1(-1)^k \end{bmatrix},$$

$$C_k = \begin{bmatrix} 0.7 & 0.3 \\ 0.4 & 0.6 \end{bmatrix}, \quad D_k = \begin{bmatrix} 0.3 \\ -0.3 \end{bmatrix}, \quad E_k = \begin{bmatrix} 0.85 \\ 0.72 \end{bmatrix}, \quad F_k = \begin{bmatrix} 0.85 \\ 0.76 \end{bmatrix}.$$

Suppose that the sensors could be grouped into two nodes. The occurrence probabilities are given as follows:

$$p_1 = 0.68, \quad p_2 = 0.32.$$

The disturbance attenuation level is set to be $\gamma = 0.98$. The weighted matrix Q and the initial values P_0 are selected as follows:

$$Q = 4.82I, \quad P_0 = 0.92\gamma^2 Q.$$

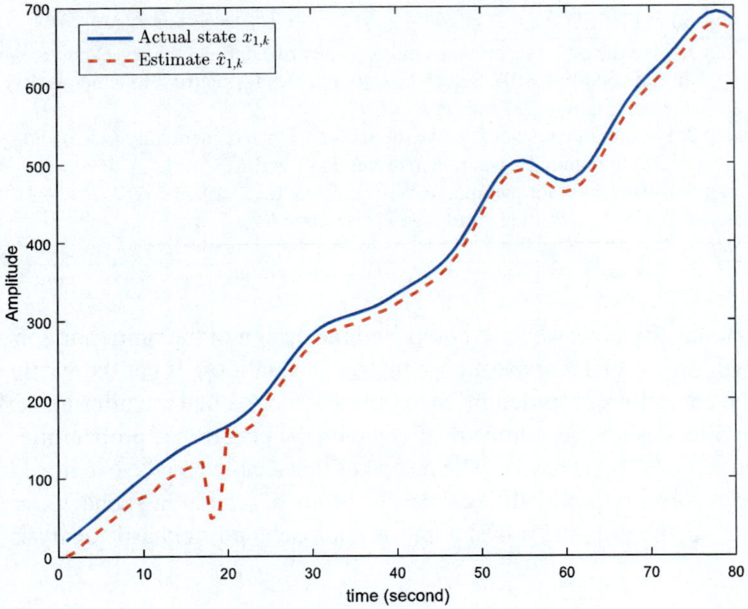

Fig. 4.1 The trajectories of $x_{1,k}$ and $\hat{x}_{1,k}$

According to Theorem 4.2, the fault estimator gains could be calculated recursively. Set the simulation run length to be 80 and let the initial state be $x(0) = \left[20\ -20\right]^T$. The fault signal is selected by $f_k = 10 + 0.1k$. Based on the derived estimator parameters, numerical simulation results are given in Figs. 4.1, 4.2 and 4.3. Figures 4.1 and 4.2 plot the state trajectories and the corresponding estimate trajectories. Figure 4.3 shows the fault signal and its estimate trajectory. The simulation results have confirmed that the designed fault estimator performs very well.

Example 4.2 Inspired by the three-tank system in [4], we consider the system (4.2) with the following parameters:

$$A_k = \begin{bmatrix} 0.9908 & 0 & 0.0091 \\ 0 & 0.9856 & 0.0072 \\ 0.0091 & 0.0072 & 0.9836 \end{bmatrix}, \quad B_k = \begin{bmatrix} 0.1 & 0.2 & 0 \\ 0 & 0.1 & -0.1 \\ 0 & 0 & 0.2 \end{bmatrix}, \quad E_k = \begin{bmatrix} 1 & 0 \\ 0 & 1 \\ 0 & 0 \end{bmatrix},$$

$$C_k = \begin{bmatrix} 1 & 0 & 0 \\ 0 & 1 & 0 \end{bmatrix}, \quad D_k = \begin{bmatrix} 0.1 & 0.1 & 0 \\ 0 & -0.2 & 0 \end{bmatrix}, \quad F_k = I.$$

Let the occurrence probabilities about the RA protocol scheduling behavior be

$$p_1 = 0.6, \quad p_2 = 0.4.$$

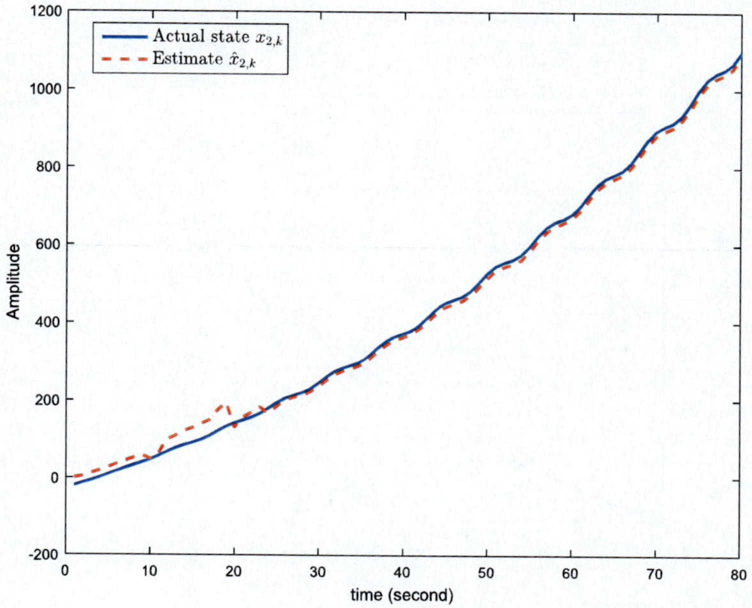

Fig. 4.2 The trajectories of $x_{2,k}$ and $\hat{x}_{2,k}$

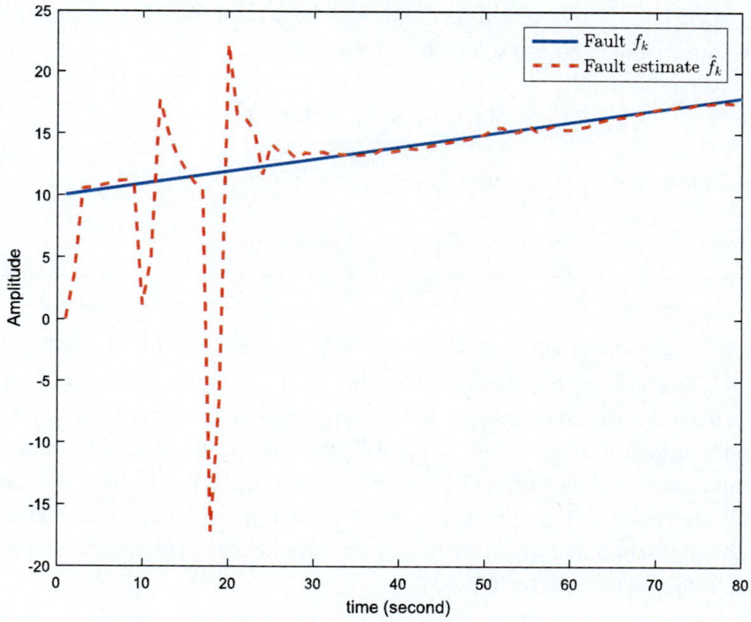

Fig. 4.3 The trajectories of f_k and $\hat{f}_k$

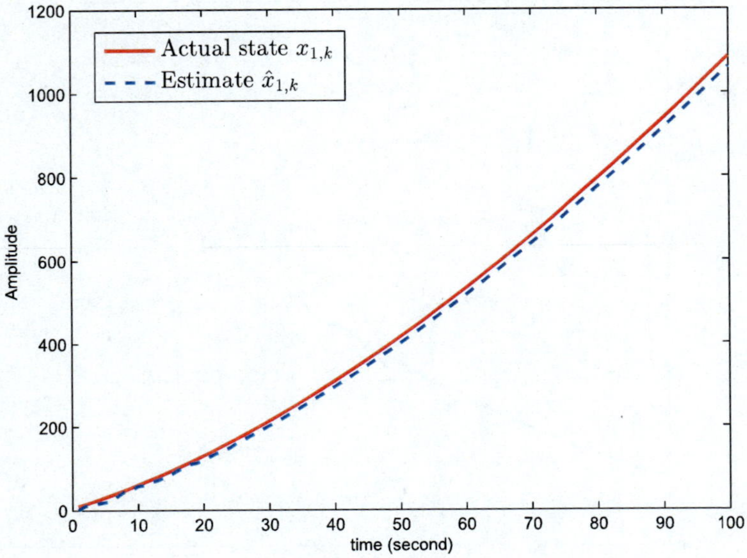

Fig. 4.4 The trajectories of $x_{1,k}$ and $\hat{x}_{1,k}$

The disturbance attenuation level is set to be $\gamma = 0.98$. The weighted matrix Q and the initial values P_0 are selected as follows:

$$Q = 4I, \quad P_0 = 0.9\gamma^2 Q.$$

The fault signal f_k is selected as follows:

$$f_k = \begin{bmatrix} f_{1,k} \\ f_{2,k} \end{bmatrix} = \begin{bmatrix} 5 + 0.2k \\ -3 - 0.3k \end{bmatrix}.$$

Set the simulation run length to be 100 and let the initial state be $x(0) = \begin{bmatrix} 8 & 10 & 5 \end{bmatrix}^T$.According to Theorem 4.2, the fault estimator gains could be calculated recursively. Based on these estimator parameters, the corresponding simulation results are shown in Figs. 4.4, 4.5, 4.6, 4.7 and 4.8. Figures 4.4, 4.5 and 4.6 plot the state trajectories and the corresponding estimate trajectories. Figures 4.7 and 4.8 plot the fault signals (i.e. $f_{i,k}$, $i = 1, 2$) and their estimate trajectories. The simulation results have verified that the fault estimation scheme developed in this chapter could achieve a satisfactory performance.

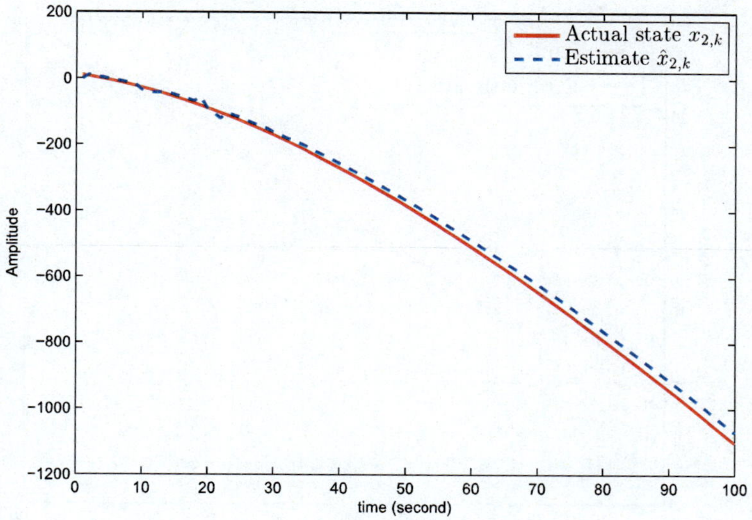

Fig. 4.5 The trajectories of $x_{2,k}$ and $\hat{x}_{2,k}$

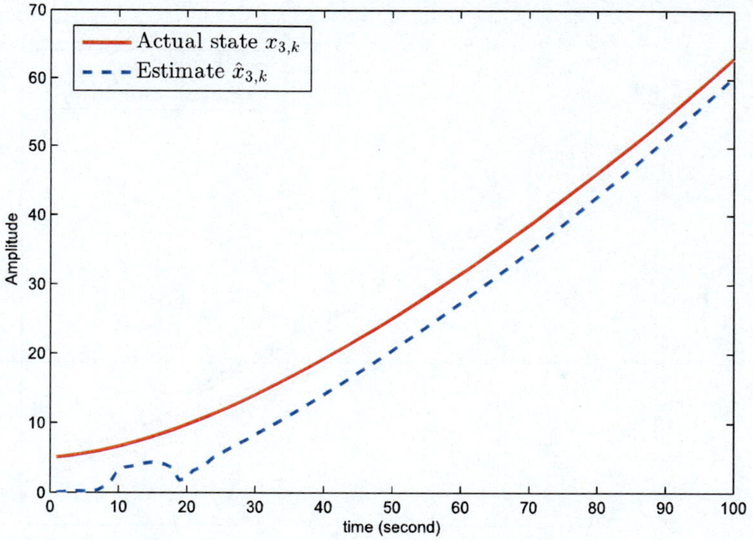

Fig. 4.6 The trajectories of $x_{3,k}$ and $\hat{x}_{3,k}$

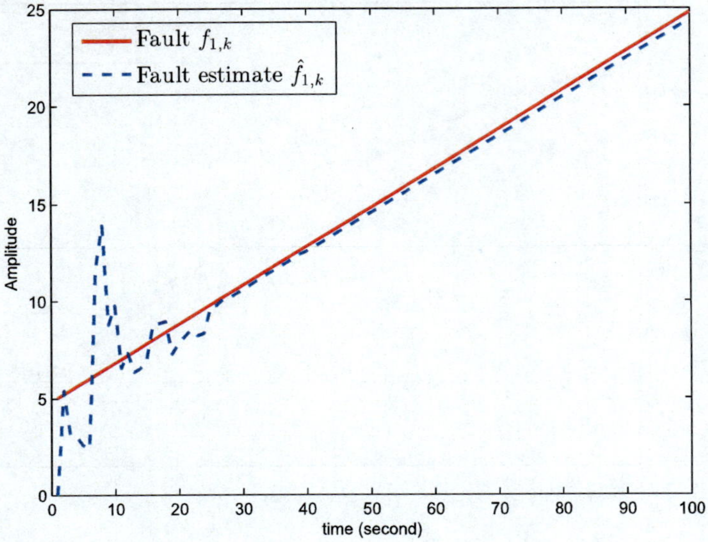

Fig. 4.7 The trajectories of $f_{1,k}$ and $\hat{f}_{1,k}$

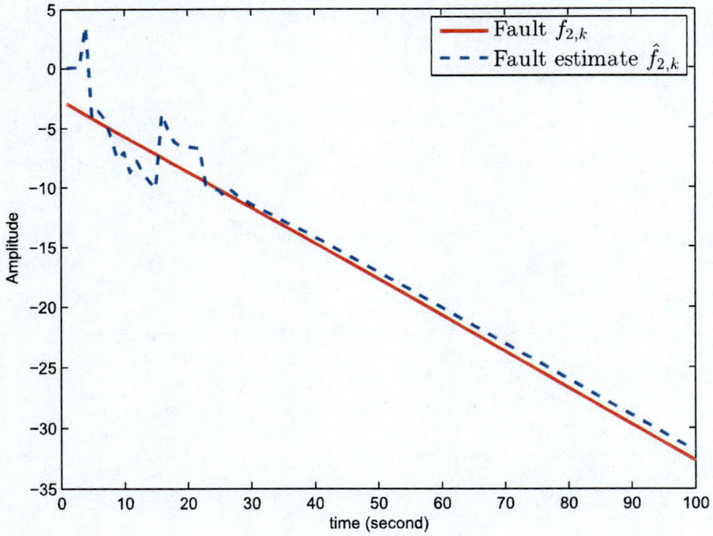

Fig. 4.8 The trajectories of $f_{2,k}$ and $\hat{f}_{2,k}$

4.4 Summary

In this chapter, we have made an attempt to investigate the fault estimation problem for time-varying systems subject to the RA protocol scheduling. The scheduling behavior of the RA protocol has been modeled by a sequence of random variables with certain known occurrence probabilities. The faults concerned have been assumed to be either incipient faults or abrupt faults, which are the most common faults in industrial processes. The corresponding fault estimator has been developed to generate the estimate of states as well as the faults. By employing the Lyapunov-based method and the stochastic analysis technique, sufficient conditions have been established to guarantee the prescribed $\mathscr{H}_\infty$ performance constraint. Furthermore, the estimator gains have been explicitly characterized by means of the solutions of a family of recursive linear matrix inequalities. Finally, two illustrative examples have been provided to highlight the effectiveness of the developed design approach.

References

1. Tabbara, M., Nešić, D.: Input-output stability of networked control systems with stochastic protocols and channels. IEEE Trans. Autom. Control **53**(5), 1160–1175 (2008)
2. Donkers, M.C.F., Heemels, W.P.M.H., Bernardini, D., Bemporad, A., Shneer, V.: Stability analysis of stochastic networked control systems. Automatica **48**(5), 917–925 (2012)
3. Gao, Z., Liu, X., Chen, M.Z.Q.: Unknown input observer-based robust fault estimation for systems corrupted by partially decoupled disturbances. IEEE Trans. Ind. Electron. **63**(4), 2537–2547 (2016)
4. He, X., Wang, Z., Liu, Y., Zhou, D.H.: Least-squares fault detection and diagnosis for networked sensing systems using a direct state estimation approach. IEEE Trans. Ind. Inf. **9**(3), 1670–1679 (2013)

Chapter 5
Set-Membership Filtering under Round-Robin Protocol and Try-Once-Discard Protocol

In practical applications, a system might suffer from various kinds of noises (or disturbances). For linear (or nonlinear) systems with Gaussian noises, the Kalman filtering (or extended Kalman filter) method is always recognized as a reliable filtering approach. Nevertheless, the Kalman filter may lead to unsatisfactory performance when the disturbances are non-Gaussian noises. Another well-known filtering method is the $\mathscr{H}_\infty$ filtering approach whose aim is to guarantee a given disturbance attenuation level on the estimation error subject to energy-bounded noises. For conventional $\mathscr{H}_\infty$ filtering method, there does not seem to be any provision to ensure the boundedness of the estimation error. In this regard, the so-called set-membership filtering method does well as it generates a satisfactory state estimate guaranteeing that the estimation error is confined to a bounded region in the state-space subject to the unknown-but-bounded noises.

The origination of the set-membership filtering problem dated back to the 1960s and such a problem has gained recurring research interest in the past decade. Compared with the widely studied filtering schemes (such as the minimum mean-square-error filtering, $\mathscr{H}_\infty$ filtering, energy-to-peak filtering, and peak-to-peak filtering), the set-membership filtering approach is developed according to an interval-based idea, which aims to generate a reasonable bounded region confining the actual state at each time instant. Such a filtering technique is particularly suitable to deal with time-varying systems with unknown-bounded disturbances. A remarkable scheme has been introduced in [1], where the recursive linear matrix inequality method has been developed for the design issue of the set-membership filter by minimizing the size of the bounded region.

In this chapter, we focus our attention on the set-membership filtering problem for a class of time-varying systems with mixed time-delays and communication protocols. Two kinds of well-known protocols (RR protocol and TOD protocol) are considered, with which the data transmission between the sensor nodes and the filter

L. Zou et al., *Communication-Protocol-Based Filtering and Control of Networked Systems*, Studies in Systems, Decision and Control 430,
https://doi.org/10.1007/978-3-030-97512-8_5

is implemented via a shared communication network that allows only one sensor node to send its measurement data at each transmission instant in order to prevent the data from collisions. The transmission order of sensor nodes is orchestrated by the underlying protocol of the network. The aim of the problem addressed is to design a set-membership filter capable of confining the state estimate of the system to certain ellipsoidal region subject to the bounded non-Gaussian noises.

The rest of this chapter is organized as follows. In Sect. 5.1, the time-varying system with the mixed time-delays and different communication protocols is introduced and the problem under consideration is formulated. In Sect. 5.2, for the given communication protocol (the RR protocol or TOD protocol), some sufficient conditions are established to guarantee the desired filtering performance and then the filter gains are obtained by solving the corresponding optimization problem subject to a recursive matrix inequality constraint. Two simulation examples are given in Sect. 5.3 to demonstrate the main results obtained. Finally, we conclude this chapter in Sect. 5.4.

5.1 Problem Formulation

5.1.1 The RR Protocol and TOD Protocol

First, let us recall the RR protocol scheduling (which has been proposed in Sect. 2.1.1) and introduce the TOD protocol scheduling. Consider a networked system with N nodes labeled as $\{1, 2, \ldots, N\}$. In this system, all the nodes transmit their data via a shared communication network in which *only one node* is allowed to get access to the network at each transmission instant for the purpose of preventing the data from collisions. Let $\theta(k) \in \{1, 2, \ldots, N\}$ be the selected node obtaining access to the communication network at time instant k. The value of $\theta(k)$ is determined by the communication protocol of the network.

(1) The Round-Robin (RR) Protocol

Under the scheduling of the RR protocol, the value of $\theta(k)$ satisfies $\theta(k + N) = \theta(k)$ for all $k \in \mathbb{N}^+$ and $\theta(k) = k$ for $k \in \{1, 2, \ldots, N\}$. In other words, $\theta(k)$ can be calculated as:

$$\theta(k) = \mathrm{mod}(k - 1, N) + 1. \tag{5.1}$$

In such a protocol, the number of nodes N can be regarded as the period of the RR protocol. During each period of the RR protocol, each node has access to the communication network exactly once.

(2) The Try-Once-Discard (TOD) Protocol

TOD protocol is a dynamical protocol in which the value of $\theta(k)$ is determined by the following selection principle:

$$\theta(k) \triangleq \arg \max_{1\leq i\leq N} \left(y_i(k) - y_i^*(k)\right)^T Q_i \left(y_i(k) - y_i^*(k)\right) \tag{5.2}$$

where $y_i^*(k)$ represents the last transmitted signal before time instant k of node i and Q_i ($i \in \{1, 2, \ldots, N\}$) is a known positive definite matrix denoting the weight matrix of the ith node.

Define

$$y(k) \triangleq [y_1^T(k)\ y_2^T(k) \cdots y_N^T(k)]^T,$$
$$y^*(k) \triangleq [\left(y_1^*(k)\right)^T\ \left(y_2^*(k)\right)^T \cdots \left(y_N^*(k)\right)^T]^T.$$

The selection principle (5.2) could be rewritten as

$$\theta(k) = \arg \max_{1\leq i\leq N} \left(y(k) - y^*(k)\right)^T \bar{Q}_i \left(y(k) - y^*(k)\right) \tag{5.3}$$

where $\bar{Q} = \text{diag}\{Q_1, Q_2, \ldots, Q_N\}$, $\Phi_i = \text{diag}\{\delta(i-1)I, \delta(i-2)I, \ldots, \delta(i-N)I\}$ $(1 \leq i \leq N)$, $\bar{Q}_i = \bar{Q}\Phi_i$ and $\delta(\cdot) \in \{0, 1\}$ is the Kronecker delta function.

5.1.2 The Plant and Filter

Consider the filtering problem for the following networked time-varying system:

$$\begin{cases} x(k+1) = A(k)x(k) + E(k)x(k-\tau_1) + F(k)\sum_{i=1}^{\tau_2} \mu_i x(k-i) + B(k)\omega(k) \\ y(k) = C(k)x(k) + D(k)\nu(k) \\ x(i) = \psi_1(i), \quad -\max\{\tau_1, \tau_2\} \leq i \leq 0 \end{cases} \tag{5.4}$$

where $x(k) \in \mathbb{R}^{n_x}$ and $y(k) \in \mathbb{R}^{n_y}$ denote, respectively, the state vector and the measurement output before transmitted through the communication network. The parameters $A(k)$, $E(k)$, $F(k)$, $B(k)$, $C(k)$ and $D(k)$ are real-valued time-varying matrices of appropriate dimensions. μ_i $(1 \leq i \leq \tau_2)$, τ_1 and τ_2 are known positive integers. $\omega(k) \in \mathbb{R}^{n_\omega}$ and $\nu(k) \in \mathbb{R}^{n_\nu}$ represent the process and measurement noises, respectively. $\psi_1(i)$ $(i = -\max\{\tau_1, \tau_2\}, -\max\{\tau_1, \tau_2\} + 1, \ldots, 0)$ are the initial conditions.

Assumption 5.1 The noise sequences $\omega(k)$ and $\nu(k)$ are confined to the following ellipsoidal sets:

$$\begin{cases} \mathcal{W}_k \triangleq \left\{\omega(k) : \omega^T(k)S^{-1}(k)\omega(k) \leq 1\right\} \\ \mathcal{V}_k \triangleq \left\{\nu(k) : \nu^T(k)R^{-1}(k)\nu(k) \leq 1\right\} \end{cases} \tag{5.5}$$

where $S(k)$ and $R(k)$ are known positive definite matrices with compatible dimensions.

Remark 5.1 As is well known, in practical engineering, the disturbance noises are known to be non-Gaussian due to man-made electromagnetic interference and other natural sources, see, e.g. [2]. As such, most conventional filtering approaches, such as the well-known Kalman filtering, are no longer applicable to the filtering problems subject to non-Gaussian noises. In this chapter, the process and measurement noises are assumed to be unknown, bounded, deterministic but reside within certain ellipsoidal sets. The time-varying matrices $S(k)$ and $R(k)$ can be regarded as the “special” upper bounds on the amplitude of the noises. On the other hand, the simultaneous presence of the discrete time-delay τ_1 and the distributed time-delay τ_2 gives rise to the so-called mixed time-delays that have been thoroughly investigated in [3]. Such mixed time-delays would affect the system dynamics tremendously, which poses some challenges in the corresponding analysis and synthesis issues.

We are now ready to introduce the signal transmission of the measurement output. Without loss of generality, it is assumed that sensors of the system are grouped into N sensor nodes according to their spatial distribution. As such, for technical analysis, the measurement output before transmitted can be rewritten as

$$y(k) = \left[y_1^T(k)\ y_2^T(k)\ \cdots\ y_N^T(k)\right]^T$$

where $y_i(k)$ $(i \in \{1, 2, \ldots, N\})$ is the measurement of the ith sensor node before transmitted.

The communication between the sensors and the remote filter is scheduled by certain communication protocol. In this chapter, we aim to study the set-membership filtering problem subject to two different kinds of communication protocols, namely, the RR protocol and the TOD protocol. For the sake of examining the influence of communication protocols, let $\theta(k) \in \{1, 2, \ldots, N\}$ be the selected sensor node obtaining access to the communication network at time instant k. As shown in Sect. 5.1.1, $\theta(k) \in \{1, 2, \ldots, N\}$ is a switching function dependent on the underlying protocol scheduling. The value of $\theta(k)$ is determined according to (5.1) in the case of RR scheduling and (5.3) in the case of TOD scheduling.

Next, let us denote the measurement output after transmitted through the network by

$$\bar{y}(k) \triangleq \left[\,\bar{y}_1^T(k)\ \bar{y}_2^T(k)\ \cdots\ \bar{y}_N^T(k)\,\right]^T \in \mathbb{R}^{n_y}.$$

The initial state of $\bar{y}(k)$ is assumed to be $\bar{y}(j) = \psi_2$ for $j < 0$ where ψ_2 is a known vector. In this chapter, the ZOH scheme is adopted in the signal update side. The updating rule of $\bar{y}_i(k)$ $(k \in \mathbb{N}^+, i \in \{1, 2, \ldots, N\})$ is set to be

$$\bar{y}_i(k) = \begin{cases} y_i(k), & \text{if } i = \theta(k) \\ \bar{y}_i(k-1), & \text{otherwise.} \end{cases} \tag{5.6}$$

According to the updating rule (5.6), it can be seen that

$$\begin{cases} \bar{y}(k) = \Phi_{\theta(k)} y(k) + \left(I_{n_y} - \Phi_{\theta(k)}\right) \bar{y}(k-1) \\ \bar{y}(j) = \psi_2, \quad j < 0. \end{cases} \tag{5.7}$$

Denoting $\bar{x}(k) \triangleq \left[x^T(k)\ \bar{y}^T(k-1)\right]^T$ and $\bar{\omega}(k) \triangleq \left[\omega^T(k)\ \nu^T(k)\right]^T$, the time-varying system (5.4) with a protocol scheduling can be reformulated as follows:

$$\begin{cases} \bar{x}(k+1) = \mathscr{A}(k)\bar{x}(k) + \mathscr{E}(k)\bar{x}(k-\tau_1) + \mathscr{F}(k)\displaystyle\sum_{i=1}^{\tau_2} \mu_i \bar{x}(k-i) + \mathscr{B}(k)\bar{\omega}(k) \\ \bar{y}(k) = \mathscr{C}(k)\bar{x}(k) + \mathscr{D}(k)\bar{\omega}(k) \\ \bar{x}(i) = \bar{\psi}(i), \quad -\max\{\tau_1, \tau_2\} \le i \le 0 \end{cases} \tag{5.8}$$

where

$$\mathscr{A}(k) = \begin{bmatrix} A(k) & 0 \\ \Phi_{\theta(k)} C(k) & I - \Phi_{\theta(k)} \end{bmatrix}, \quad \mathscr{B}(k) = \begin{bmatrix} B(k) & 0 \\ 0 & \Phi_{\theta(k)} D(k) \end{bmatrix}, \quad \mathscr{E}(k) = \begin{bmatrix} E(k) & 0 \\ 0 & 0 \end{bmatrix},$$

$$\mathscr{F}(k) = \begin{bmatrix} F(k) & 0 \\ 0 & 0 \end{bmatrix}, \quad \mathscr{C}(k) = \left[\Phi_{\theta(k)} C(k)\ \ I - \Phi_{\theta(k)}\right], \quad \mathscr{D}(k) = \left[0\ \ \Phi_{\theta(k)} D(k)\right],$$

$$\bar{\psi}(i) = \left[\psi_1^T(i)\ \psi_2^T(i)\right]^T.$$

5.1.3 *Time-Varying Filter*

In this chapter, a filter based on the signal $\bar{y}(k)$ is considered for the augmented system (5.8) with mixed time-delays, which is of the form:

$$\begin{cases} \hat{x}(k+1) = \mathscr{A}(k)\hat{x}(k) + \mathscr{E}(k)x(k-\tau_1) + K(k)\bar{y}(k) - K(k)\mathscr{C}(k)\hat{x}(k) \\ \qquad\qquad + \mathscr{F}(k)\displaystyle\sum_{i=1}^{\tau_2} \mu_i \hat{x}(k-i) \\ \hat{x}(i) = 0, \quad -\max\{\tau_1, \tau_2\} \le i \le 0 \end{cases} \tag{5.9}$$

where $\hat{x}(k) \in \mathbb{R}^{n_x+n_y}$ is the state estimate of $\bar{x}(k)$ and $K(k) \in \mathbb{R}^{(n_x+n_y)\times n_y}$ is the filter gain to be designed.

Letting the estimation error be $e(k) \triangleq \bar{x}(k) - \hat{x}(k)$, the estimation error dynamics for the augmented system (5.8) can be obtained as follows:

$$
\begin{aligned}
e(k+1) = &\big(\mathscr{A}(k) - K(k)\mathscr{C}(k)\big)e(k) + \mathscr{E}(k)e(k-\tau_1) + \mathscr{F}(k)\sum_{i=1}^{\tau_2}\mu_i e(k-i) \\
&+ \big(\mathscr{B}(k) - K(k)\mathscr{D}(k)\big)\bar{\omega}(k).
\end{aligned} \tag{5.10}
$$

Before proceeding further, we introduce the following definition and assumption.

Definition 5.1 For the time-varying system (5.8) with filter (5.9), let the sequence of the constraint matrices (ellipsoid matrices) $P(k) \in \mathbb{R}^{(n_x+n_y)\times(n_x+n_y)}$ $(k \in \mathbb{N}^+)$ be given. The filtering error $e(k)$ is said to satisfy the $P(k)$-dependent constraint if the following set of inequalities

$$
e^T(k)P^{-1}(k)e(k) \le 1 \tag{5.11}
$$

hold for $k \in \mathbb{N}^+$.

Assumption 5.2 The initial conditions $\bar{\psi}(i)$ $(i = -\max\{\tau_1, \tau_2\}, -\max\{\tau_1, \tau_2\} + 1, \ldots, 0)$ satisfy

$$
\bar{\psi}^T(i)P^{-1}(i)\bar{\psi}(i) \le 1 \tag{5.12}
$$

where $P(i)$ $(i = -\max\{\tau_1, \tau_2\}, -\max\{\tau_1, \tau_2\} + 1, \ldots, 0)$ are a known positive definite matrices.

We are now in the position to state the problem addressed in this chapter as follows. The objective of this chapter is to design appropriate filter parameter $K(k)$ subject to different protocols (the RR protocol and the TOD protocol). To be more specific, we are interested in dealing with the following analysis and design problems:

Problem 5.1 For a given sequence of constraint matrices $P(k)$, establish sufficient conditions for the existence of the filter gains under which the dynamics of the filtering error $e(k)$ satisfies the $P(k)$-dependent constraint (5.11) subject to the process and measurement noises;

Problem 5.2 Minimize the trace of the matrix $P(k)$ by appropriately choosing the filter parameter $K(k)$ satisfying the conditions established in *Problem* 5.1.

5.2 Main Results

In this section, we shall present some sufficient conditions guaranteeing that the dynamics of the filtering error $e(k)$ satisfies the $P(k)$-dependent constraint (5.11). Furthermore, two recursive algorithms are proposed to obtain the time-varying filter parameter $K(k)$ subject to the RR protocol and the TOD protocol, respectively. Before proceeding further, we introduce the following lemma which will be needed for the derivation of our main results.

Lemma 5.1 (S-procedure [4]) *Let $\phi_0(\cdot), \phi_1(\cdot), \cdots, \phi_p(\cdot)$ be quadratic functions of the variable $\varsigma \in \mathbb{R}^n : \phi_i(\varsigma) \triangleq \varsigma^T T_i \varsigma$ $(i = 0, 1, \ldots, p)$ with $T_i^T = T_i$. If there exist $\alpha_1, \alpha_2, \cdots, \alpha_p \geq 0$ such that*

$$T_0 - \sum_{i=1}^{p} \alpha_i T_i \leq 0, \tag{5.13}$$

then the following is true

$$\phi_1(\varsigma) \leq 0, \phi_2(\varsigma) \leq 0, \ldots, \phi_p(\varsigma) \leq 0 \Longrightarrow \phi_0(\varsigma) \leq 0. \tag{5.14}$$

Lemma 5.2 (Schur Complement Lemma) *Given constant matrices $\mathscr{S}_1$, $\mathscr{S}_2$ and $\mathscr{S}_3$ where $\mathscr{S}_1 = \mathscr{S}_1^T$ and $0 < \mathscr{S}_2 = \mathscr{S}_2^T$, then $\mathscr{S}_1 + \mathscr{S}_3^T \mathscr{S}_2^{-1} \mathscr{S}_3 < 0$ if and only if*

$$\begin{bmatrix} \mathscr{S}_1 & \mathscr{S}_3^T \\ * & -\mathscr{S}_2 \end{bmatrix} < 0, \quad \text{or} \quad \begin{bmatrix} -\mathscr{S}_2 & \mathscr{S}_3 \\ * & \mathscr{S}_1 \end{bmatrix} < 0. \tag{5.15}$$

5.2.1 *P(k)-Dependent Constraint Analysis*

In the following theorems, sufficient conditions are established to solve *Problem* 5.1 that aims to ensure the $P(k)$-dependent constraint to be satisfied for the dynamics of the filtering error subject to the TOD protocol and the RR protocol. A recursive matrix inequality approach is developed to cater for the effects from the ellipsoidal constraints, the mixed time-delays and communication protocols.

(1) The TOD Protocol

Theorem 5.1 *Consider the dynamic system (5.4), the TOD protocol given by (5.3) and the time-varying filter (5.9). Let the sequence of the constraint matrices $P(k) > 0$ $(k \in \mathbb{N}^+)$ be given. Assume that there exist real-valued matrices $K(k)$, positive scalars $\lambda_i(k)$ $(k \in \mathbb{N}^+, i \in \{1, 2, \ldots, N\})$ and $\varepsilon_j(k)$ $(k \in \mathbb{N}^+, j \in \{1, 2, 3, 4\})$ satisfying the following recursive matrix inequality*

$$\begin{bmatrix} -\Theta(k) & \Pi^T(k) \\ * & -P(k+1) \end{bmatrix} \leq 0 \tag{5.16}$$

where

$$\begin{aligned}\Theta(k) = {}& \Gamma^T(k) \sum_{i=1}^{N} \lambda_i(k) \Big(\bar{Q}(\Phi_i - \Phi_{\theta(k)}) \Big) \Gamma(k) \\ & + \mathrm{diag}\big\{\bar{\varepsilon}(k), \varepsilon_1(k)I, \varepsilon_2(k)I, \varepsilon_3(k)I, \varepsilon_4(k)\Omega^{-1}(k)\big\}, \\ \Pi(k) = {}& \big[0 \ \big(\mathscr{A}(k) - K(k)\mathscr{C}(k)\big)L(k) \ \mathscr{E}(k)L(k-\tau_1) \ \bar{\mathscr{F}}_\tau(k) \ \mathscr{B}(k) - K(k)\mathscr{D}(k)\big],\end{aligned}$$

$$\bar{\mathscr{F}}_\tau(k) = \left[\mu_1\mathscr{F}(k)L(k-1)\ \mu_2\mathscr{F}(k)L(k-2)\ \cdots\ \mu_{\tau_2}\mathscr{F}(k)L(k-\tau_2)\right],$$
$$\Omega(k) = \mathrm{diag}\{S(k), R(k)\},\ \Gamma(k) = \left[\vec{C}(k)\hat{x}(k)\ \vec{C}(k)L(k)\ 0\ 0\ \vec{D}(k)\right],$$
$$\vec{C}(k) = \left[C(k)\ \Phi_{\theta(k)} - I\right],\quad \bar{\varepsilon}(k) = 1 - \varepsilon_1(k) - \varepsilon_2(k) - \tau_2\varepsilon_3(k) - 2\varepsilon_4(k),$$
$$\vec{D}(k) = \left[0\ D(k)\right],$$

and $L(k)$ is the factorization of $P(k) = L(k)L^T(k)$. Then, the dynamics of the time-varying system (5.10) satisfies the $P(k)$-dependent constraint with the estimator parameters $K(k)$ ($k \in \mathbb{N}^+$).

Proof The proof of this theorem is performed by *mathematical induction*. In what follows, let us first prove the following assertion is true.

Assertion: The solution $e(k)$ of the time-varying system (5.10) satisfies

$$e^T(k)P^{-1}(k)e(k) \le 1 \tag{5.17}$$

for $k \in \mathbb{N}^+$.

The proof of the above assertion is divided into two steps, namely, *the initial step* and *the inductive step*.

Initial step. For $t = 0$, it can be immediately known from Assumption 5.2 that

$$e^T(0)P^{-1}(0)e(0) = \bar{x}^T(0)P^{-1}(0)\bar{x}(0) = \psi^T(0)P^{-1}(0)\psi(0) \le 1. \tag{5.18}$$

Inductive step. Now that the assertion is true for $t = 0$. Next, given that the assertion is true for $t \le k$, we aim to show that the same assertion is true for $t = k+1$.

Since the assertion is true for $t \le k$, if follows from [5] that there exists a vector $z(i)$ satisfying $e(i) = L(i)z(i)$ with $\|z(i)\| \le 1$ where $L(i)$ is a factorization of $P(i) = L(i)L^T(i)$ for $i \in \{0, 1, \ldots, k\}$. It remains to show that, for $t = k+1$, the solution $P(k+1)$ of the recursive matrix inequalities (5.16) guarantees $e^T(k+1)P^{-1}(k+1)e(k+1) \le 1$.

For notational simplicity, we first denote

$$\bar{z}_\tau(k) \triangleq \left[z^T(k-1)\ z^T(k-2)\ \cdots\ z^T(k-\tau_2)\right]^T,$$
$$\eta(k) \triangleq \left[1\ z^T(k)\ z^T(k-\tau_1)\ \bar{z}_\tau^T(k)\ \bar{\omega}^T(k)\right]^T.$$

Then, it follows from (5.10) that

$$e(k+1) = \Pi(k)\eta(k) \tag{5.19}$$

and therefore

$$\begin{aligned} &e^T(k+1)P^{-1}(k+1)e(k+1) - 1 \\ &= \eta^T(k)\left(\Pi^T(k)P^{-1}(k+1)\Pi(k) - \mathrm{diag}\{1, 0, 0, 0, 0\}\right)\eta(k). \end{aligned} \tag{5.20}$$

On the other hand, it can be obtained from Assumption 5.1 and the selection principle of the TOD that

$$\begin{cases} \|z(k)\|^2 \leq 1, \\ \|z(k-\tau_1)\|^2 \leq 1, \\ \|\bar{z}_\tau(k)\|^2 \leq \tau_2, \\ \bar{\omega}^T(k)\text{diag}\{S^{-1}(k), R^{-1}(k)\}\bar{\omega}(k) \leq 2, \\ \big(y(k)-\bar{y}(k-1)\big)^T \bar{Q}\big(\Phi_i - \Phi_{\theta(k)}\big)\big(y(k)-\bar{y}(k-1)\big) \leq 0, \quad i = 1, 2, \ldots, N, \end{cases} \tag{5.21}$$

which can be rearranged by means of $\eta(k)$ as follows:

$$\begin{cases} \eta^T(k)\text{diag}\{-1, I, 0, 0, 0\}\eta(k) \leq 0, \\ \eta^T(k)\text{diag}\{-1, 0, I, 0, 0\}\eta(k) \leq 0, \\ \eta^T(k)\text{diag}\{-\tau_2, 0, 0, I, 0\}\eta(k) \leq 0, \\ \eta^T(k)\text{diag}\big\{-2, 0, 0, 0, \Omega^{-1}(k)\big\}\eta(k) \leq 0, \\ \eta^T(k)\Gamma^T(k)\bar{Q}\big(\Phi_i - \Phi_{\theta(k)}\big)\Gamma(k)\eta(k) \leq 0, \quad i = 1, 2, \ldots, N. \end{cases} \tag{5.22}$$

On the other hand, it can be seen from the S-procedure (Lemma 5.1) that if there exist positive scalars $\varepsilon_1(k)$, $\varepsilon_2(k)$, $\varepsilon_3(k)$, $\varepsilon_4(k)$ and $\lambda_i(k)$ $(i = 1, 2, \ldots, N)$ such that

$$\begin{aligned} &\Pi^T(k)P^{-1}(k+1)\Pi(k) - \text{diag}\{1, 0, 0, 0, 0\} - \Gamma^T(k)\sum_{i=1}^{N}\lambda_i(k)\Big(\bar{Q}\big(\Phi_i - \Phi_{\theta(k)}\big)\Big)\Gamma(k) \\ &- \varepsilon_1(k)\text{diag}\{-1, I, 0, 0, 0\} - \varepsilon_2(k)\text{diag}\{-1, 0, I, 0, 0\} - \varepsilon_3(k)\text{diag}\{-\tau_2, 0, 0, I, 0\} \\ &- \varepsilon_4(k)\text{diag}\big\{-2, 0, 0, 0, \Omega^{-1}(k)\big\} \leq 0, \end{aligned} \tag{5.23}$$

then the inequality $e^T(k+1)P^{-1}(k+1)e(k+1) - 1 \leq 0$ can be guaranteed by the inequalities (5.22). The inequality (5.23) can be written in the following compact form:

$$\begin{aligned} &\Pi^T(k)P^{-1}(k+1)\Pi(k) - \text{diag}\big\{\bar{\varepsilon}(k), \varepsilon_1(k)I, \varepsilon_2(k)I, \varepsilon_3(k)I, \varepsilon_4(k)\Omega^{-1}(k)\big\} \\ &- \Gamma^T(k)\sum_{i=1}^{N}\lambda_i(k)\Big(\bar{Q}\big(\Phi_i - \Phi_{\theta(k)}\big)\Big)\Gamma(k) \leq 0. \end{aligned} \tag{5.24}$$

By denoting

$$\begin{aligned} \Theta(k) = {} & \Gamma^T(k)\sum_{i=1}^{N}\lambda_i(k)\Big(\bar{Q}\big(\Phi_i - \Phi_{\theta(k)}\big)\Big)\Gamma(k) \\ & + \text{diag}\big\{\bar{\varepsilon}(k), \varepsilon_1(k)I, \varepsilon_2(k)I, \varepsilon_3(k)I, \varepsilon_4(k)\Omega^{-1}(k)\big\}, \end{aligned}$$

(5.24) is rewritten as

$$\Pi^T(k)P^{-1}(k+1)\Pi(k) - \Theta(k) \leq 0. \tag{5.25}$$

By using the Schur complements (Lemma 5.2), (5.25) is equivalent to

$$\begin{bmatrix} -\Theta(k) & \Pi^T(k) \\ * & -P(k+1) \end{bmatrix} \leq 0. \tag{5.26}$$

Thus, it can be obtained from (5.16) that the inequality $e^T(k+1)P^{-1}(k+1)e(k+1) - 1 \leq 0$ is satisfied. Hence, by the induction, it can be concluded that the assertion is true for $k \in \mathbb{N}^+$, which means that the dynamics of the time-varying system (5.10) satisfies the $P(k)$-dependent constraint with the filter gain $K(k)$.

(2) The RR Protocol

Theorem 5.2 *Consider the dynamic system (5.4), the RR protocol given by (5.1) and the time-varying filter (5.9). Let the sequence of the constraint matrices $P(k) > 0$ ($k \in \mathbb{N}^+$) be given. If there exist real-valued matrices $K(k)$, positive scalars $\varepsilon_j(k)$ ($k \in \mathbb{N}^+$, $j \in \{1, 2, 3, 4\}$) satisfying the following recursive matrix inequality*

$$\begin{bmatrix} -\bar{\Theta}(k) & \Pi^T(k) \\ * & -P(k+1) \end{bmatrix} \leq 0 \tag{5.27}$$

where

$$\bar{\Theta}(k) = \text{diag}\{\bar{\varepsilon}(k), \varepsilon_1(k)I, \varepsilon_2(k)I, \varepsilon_3(k)I, \varepsilon_4(k)\Omega^{-1}(k)\}$$

with other corresponding matrices defined in Theorem 5.1, then the dynamics of the time-varying system (5.10) satisfies the $P(k)$-dependent constraint with the estimator parameters $K(k)$ ($k \in \mathbb{N}^+$).

Proof The proof is similar to that of Theorem 5.1 and is therefore omitted for the conciseness.

Remark 5.2 Sufficient conditions are provided in Theorem 5.1 and Theorem 5.2, respectively, under which the estimation error dynamics (5.10) satisfies the required $P(k)$-dependent constraint subject to the well-known TOD protocol and RR protocol. The main idea of the set-membership filter is to design the time-varying filter parameter $K(k)$ via the transmitted measurement output such that the estimation error is confined to the given ellipsoidal set at each time step. Obviously, the $P(k)$-dependent constraint (5.11) is equivalent to the *quadratic error-bounded constraint* $e(k)e^T(k) \leq P(k)$.

5.2.2 Design of Filter Gains

In this subsection, two optimization problems are solved that would help the design of the locally optimal filter gains.

Corollary 5.1 *Consider the dynamic system (5.4), the TOD protocol given by (5.3) and the time-varying filter (5.9). The ellipsoid constraint $P(k)$ on the estimation error is minimized (in the sense of the matrix trace) if there exist real-valued matrix $K(k)$, positive scalars $\lambda_i(k)$ ($k \in \mathbb{N}^+$, $i \in \{1, 2, \ldots, N\}$) and $\varepsilon_j(k)$ ($k \in \mathbb{N}^+$, $j \in \{1, 2, 3, 4\}$) solving the following optimization problem:*

$$\min_{P(k+1), K(k), \overrightarrow{\lambda}(k), \overrightarrow{\varepsilon}(k)} \quad \operatorname{tr}\{P(k+1)\} \tag{5.28}$$

subject to (5.16), where

$$\overrightarrow{\lambda}(k) = \{\lambda_i(k)\}_{i \in \{1,2,\ldots,N\}}, \quad \overrightarrow{\varepsilon}(k) = \{\varepsilon_i(k)\}_{i \in \{1,2,3,4\}}.$$

Corollary 5.2 *For the dynamic system (5.4) subject to the RR protocol (5.1), the ellipsoid constraint $P(k)$ on the estimation error is minimized (in the sense of the matrix trace) if there exist real-valued matrix $K(k)$ and positive scalars $\varepsilon_j(k)$ ($k \in \mathbb{N}^+$, $j \in \{1, 2, 3, 4\}$) solving the following optimization problem:*

$$\min_{P(k+1), K(k), \overrightarrow{\varepsilon}(k)} \quad \operatorname{tr}\{P(k+1)\} \tag{5.29}$$

subject to (5.27), where $\overrightarrow{\varepsilon}(k) := \{\varepsilon_i(k)\}_{i \in \{1,2,3,4\}}$.

Remark 5.3 Corollaries 5.1 and 5.2, whose proofs are straightforward and omitted, offer an optimization-based method to design an ellipsoid of the smallest size in order to constrain the state estimate by minimizing the trace of $P(k+1)$ at each time step. It can be seen that the inequalities (5.16) and (5.27) are linear to all the variables (e.g. $P(k+1)$, $K(k)$, $\varepsilon_i(k)$), and therefore the optimization problems (5.28) and (5.29) can be solved by the existing semi-definite programming algorithms such as the subgradient method, the subgradient projection method and the interior-point method.

By means of Corollaries 5.1 and 5.2, we can summarize the Protocol-Based Set-Membership Filter Design (PBSMFD) algorithm as follows.

Algorithm PBSMFD:

Step 1. Initialization: Set $k = 0$ and give initial constraint matrices $P(t)$ satisfying Assumption 5.2 for $t \in \{-\max\{\tau_1, \tau_2\}, -\max\{\tau_1, \tau_2\} + 1, \ldots, 0\}$.

Step 2. Calculate matrix factorizations $L(t)$ according to $P(t) = L(t)L^T(t)$ for $k - \tau_2 \leq t \leq k$ and $t = k - \tau_1$.

Step 3. Solve the corresponding optimization problem according to the utilized scheduling protocol of the network. If the network is scheduled under the TOD protocol, solve the optimization problem (5.28) subject to (5.16). Otherwise, solve the optimization problem (5.29) subject to (5.27). Then, the filter parameter $K(k)$ and positive matrices $P(k+1)$ are obtained according to the solution of the optimization problem.

Step 4. Set $k = k + 1$ and go to *Step 2*.

Remark 5.4 So far, we have discussed the set-membership filtering problems for a class of discrete time-varying systems with mixed time-delays subject to two kinds of communication protocols (the RR protocol and the TOD protocol). It can be observed from Algorithm PBSMFD that, in the filter design procedure, all the important factors contributing to the system complexity are reflected in the main results, which include: (1) the time-varying system parameters; (2) the delay sizes; (3) the estimation accuracy (characterized by $P(k)$); (4) the noise information (characterized by $\mathcal{W}_k$ and $\mathcal{V}_k$); and (5) the information about the utilized communication protocol (determined by (5.1) or (5.3)).

Remark 5.5 Our main results are derived in terms of recursive linear matrix inequalities, which has a polynomial time complexity. Specifically, the number $\mathscr{N}(\varepsilon)$ of flops needed to compute an ε-accurate solution is bounded by $O(\mathscr{M}\mathscr{N}^3 \log(\mathscr{V}/\varepsilon))$, where $\mathscr{M}$ is the total row size of the LMI system, $\mathscr{N}$ is the total number of scalar decision variables, $\mathscr{V}$ is a data-dependent scaling factor, and ε is relative accuracy set for algorithm. For the investigated discrete time-varying system (4) with certain communication protocol (RR protocol or TOD protocol), the variable dimensions can be seen from $x(k) \in \mathbb{R}^{n_x}$, $y(k) \in \mathbb{R}^{n_y}$, $\omega(k) \in \mathbb{R}^{n_\omega}$ and $\nu(k) \in \mathbb{R}^{n_\nu}$. Furthermore, according to Corollaries 1 and 2, one has $\mathscr{M} = (\tau_2 + 3)(n_x + n_y) + n_\omega + n_\nu$ and $\mathscr{N} = 0.5(n_x + n_y + 1)(n_x + n_y) + (n_x + n_y)n_y + N + 4$ for the TOD case, and $\mathscr{M} = (\tau_2 + 3)(n_x + n_y) + n_\omega + n_\nu$ and $\mathscr{N} = 0.5(n_x + n_y + 1)(n_x + n_y) + (n_x + n_y)n_y + 4$ for the RR case. Obviously, such a computational complexity depends polynomially on the variable dimensions.

5.3 Illustrative Examples

In this section, two numerical examples are presented to illustrate the effectiveness of the proposed set-membership filter design scheme for system (5.4) subject to the RR protocol and TOD protocol, respectively.

Example 5.1 Consider a discrete-time delayed system (5.4) with the following system parameters:

$$A(k)=\begin{bmatrix}0.75+0.35\sin(0.3k) & 0.02+0.01\cos(0.1k)\\ 0.03 & 0.75+0.39\sin(0.3k)\end{bmatrix},\ B(k)=\begin{bmatrix}0.64\\ 0.52+0.2(-1)^k\end{bmatrix},$$
$$E(k)=\begin{bmatrix}0.2 & 0\\ 0 & 0.2\end{bmatrix},\ D(k)=\begin{bmatrix}0.6\\ -0.4\end{bmatrix},\ C(k)=\begin{bmatrix}0.65 & -0.3\\ -0.55 & 0.40\end{bmatrix},$$
$$F(k)=\begin{bmatrix}0.1 & 0\\ 0 & 0.1\end{bmatrix},\ \tau_1=1,\ \tau_2=2,\ \mu_1=\mu_2=0.2.$$

The sensors of this system are grouped into two sensor nodes. The weight matrices of the TOD protocol is taken to be $Q_1=1.2$ and $Q_2=0.9$, and the bounded noises are assumed to be $\omega(k)=0.8^k\sin(2k)$ and $\nu(k)=0.8^k\cos(2k)$, respectively. Obviously, the matrices $S(k)$ and $R(k)$ could be chosen as $S(k)=R(k)=0.8^k$. Also, select the initial state and the initial constraint matrices as follows: $\psi(l)=\begin{bmatrix}4 & 3 & 1 & 1\end{bmatrix}^T$, $P(l)=\text{diag}\{64,36,4,4\}$ $(l=-\max\{\tau_1,\tau_2\},-\max\{\tau_1,\tau_2\}+1,\ldots,0)$.

Using the given algorithm and MATLAB software (YALMIP 3.0), the set of solutions to the optimization problem (5.28) subject to (5.16) in Corollary 5.1 and the optimization problem (5.29) subject to (5.27) in Corollary 5.2 are obtained, respectively. The simulation results are shown in Figs. 5.1, 5.2, 5.3, 5.4, 5.5 and 5.6. Figures 5.1 and 5.2 depict the state responses and their estimations under the TOD scheduling. Figure 5.3 plots the responses of the estimation error $e(k)$ with respect to the TOD protocol. Figures 5.4 and 5.5 depict the state responses and their estimations under the RR scheduling. Figure 5.6 plots the responses of the estimation

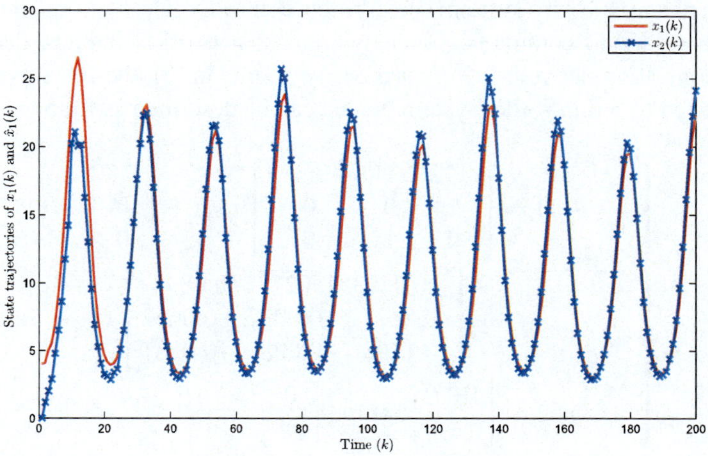

Fig. 5.1 The state trajectories of $x_1(k)$ and $\hat{x}_1(k)$ under the TOD scheduling

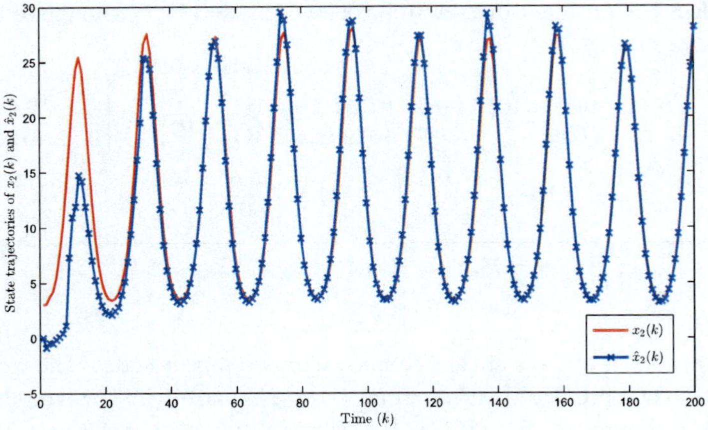

Fig. 5.2 The state trajectories of $x_2(k)$ and $\hat{x}_2(k)$ under the TOD scheduling

error $e(k)$ with respect to the RR protocol. All the simulation results confirm that the estimation performance is well achieved, and the proposed PBSMFD algorithm is indeed effective. Furthermore, it can be seen from the simulation that the set-membership estimation performs better under the TOD scheduling than that under the RR scheduling (at the cost of higher computational complexity), which is mainly due to the advantages of scheduling principle for the TOD protocol. According to the TOD scheduling, the node with the greatest changed measurement obtains access to the network.

Example 5.2 In this example, we consider a networked-based test rig, which consists of a plant (DC servo system) and a remote controller. The plant and the controller are connected via a communication network. The networked induced delay of such a communication network is 3. Based on the results in [6], the DC servo system is identified to be a third-order system, which can be described as follows:

$$
\begin{cases}
x(k+1) = \begin{bmatrix} 1.12 & 0.213 & -0.333 \\ 1 & 0 & 0 \\ 0 & 1 & 0 \end{bmatrix} x(k) + \begin{bmatrix} 0.8 \\ 0 \\ 0 \end{bmatrix} \omega(k) \\
\qquad + \begin{bmatrix} -0.2193 & 0.0219 & 0.0844 \\ 0.2177 & -0.0032 & -0.0662 \\ 0.1298 & -0.0087 & -0.0381 \end{bmatrix} x(k-3) \\
y(k) = \begin{bmatrix} 1 & 0 & 0 \\ 0 & 1 & 0 \end{bmatrix} x(k) + \begin{bmatrix} 0 \\ 0.8 \end{bmatrix} \nu(k).
\end{cases}
$$

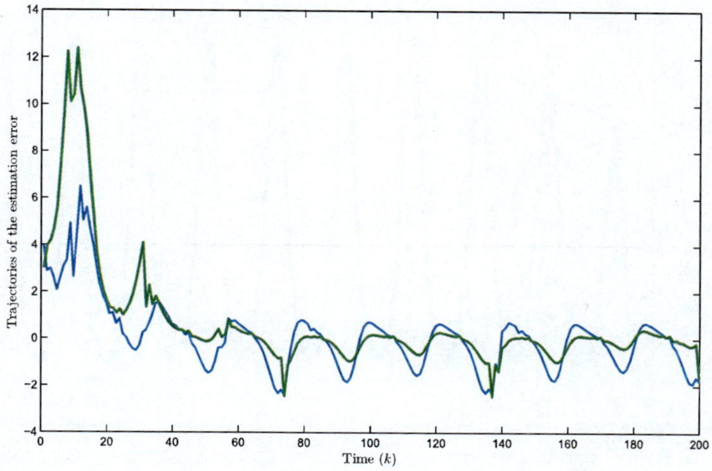

Fig. 5.3 The state trajectories of $e_1(k)$ and $e_2(k)$ under the TOD scheduling

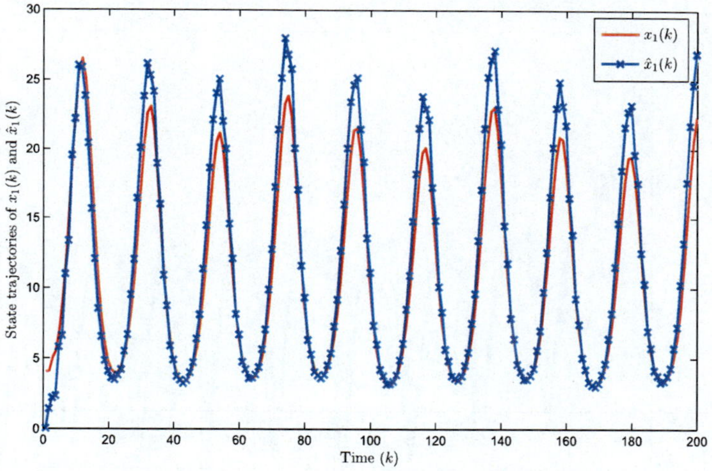

Fig. 5.4 The state trajectories of $x_1(k)$ and $\hat{x}_1(k)$ under the RR scheduling

The sensors of this system are grouped into two sensor nodes, and the weight matrices of the TOD protocol are set to be $Q_1 = 0.6$ and $Q_2 = 1.4$. The process and measurement noise sequences are assumed to be $\omega(k) = 0.4\cos(0.2k^{0.9})$ and $\nu(k) = 0.4\sin(0.2k^{0.9})$ ($k \in \mathbb{N}^+$), respectively. Hence, the matrices $S(k)$ and $R(k)$ could be selected as $S(k) = R(k) = 0.16I$. The initial state and constraint matrices are chosen as follows: $\psi(l) = [1\ 0.4\ 0.6\ 0.5\ 0.5]^T$ and $P(l) = \text{diag}\{1, 0.16, 0.36, 0.25, 0.25\}$ ($l = -1, 0$).

Based on the given algorithms and MATLAB software (YALMIP 3.0), the set of solutions to the optimization problem (5.28) subject to (5.16) in Corollary 5.1 and the optimization problem (5.29) subject to (5.27) in Corollary 5.2 are obtained,

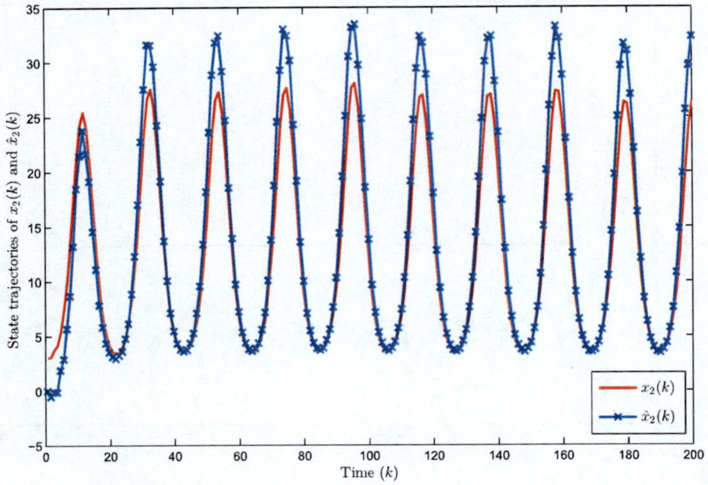

Fig. 5.5 The state trajectories of $x_2(k)$ and $\hat{x}_2(k)$ under the RR scheduling

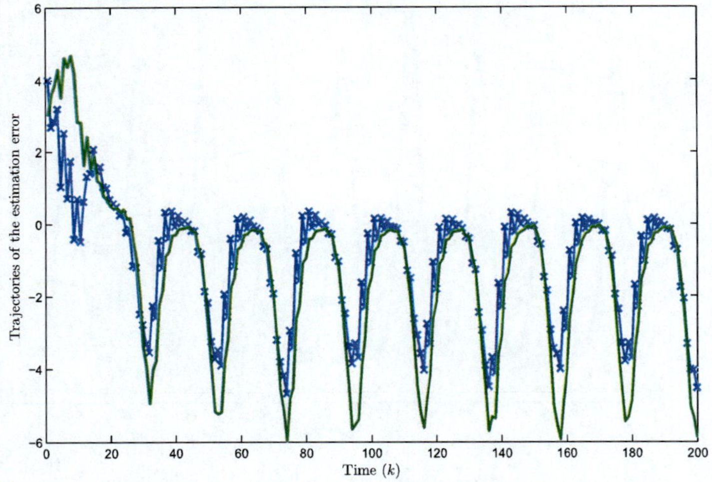

Fig. 5.6 The state trajectories of $e_1(k)$ and $e_2(k)$ under the RR scheduling

respectively. The simulation results are shown in Figs. 5.7, 5.8, 5.9 and 5.10. Figures 5.7, 5.8 and 5.9 depict the state responses and their estimations under the TOD scheduling and RR scheduling. Figure 5.10 plots the responses of $e(k)$ with respect to the TOD and RR protocols. All the simulation results have confirmed that the designed set-membership filters perform very well.

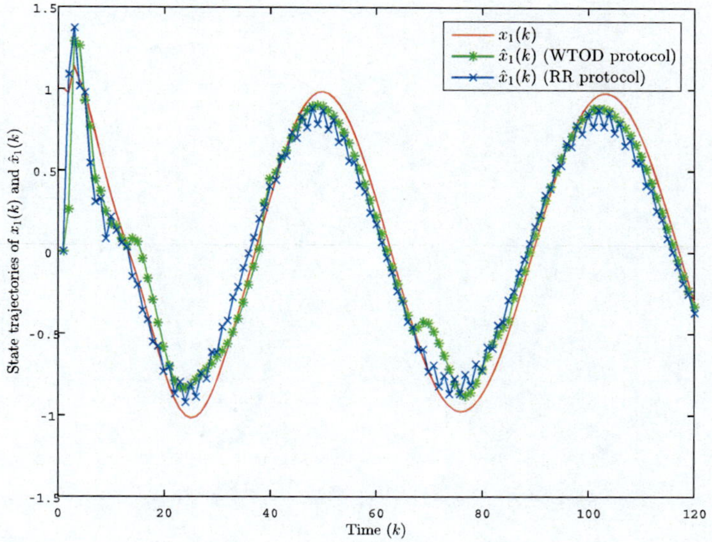

Fig. 5.7 The state trajectories of $x_1(k)$ and $\hat{x}_1(k)$ under the TOD and RR protocols

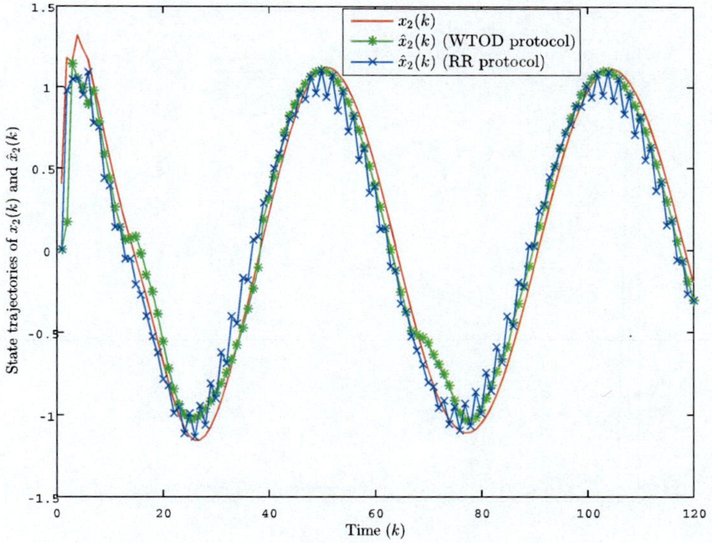

Fig. 5.8 The state trajectories of $x_2(k)$ and $\hat{x}_2(k)$ under the TOD and RR protocols

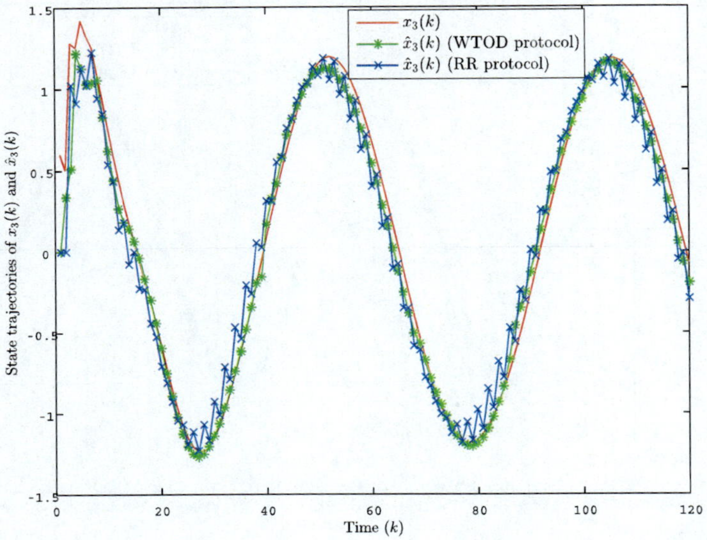

Fig. 5.9 The state trajectories of $x_3(k)$ and $\hat{x}_3(k)$ under the TOD and RR protocols

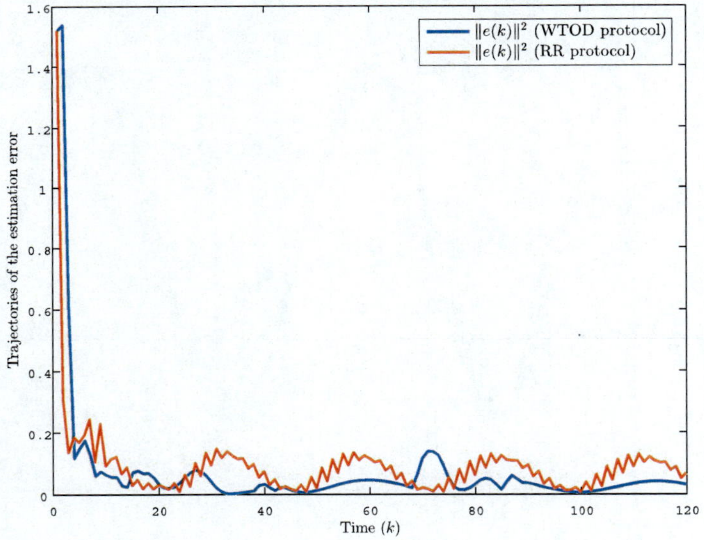

Fig. 5.10 Filtering errors for the networked-based test rig

5.4 Summary

In this chapter, the set-membership filtering problem has been investigated for a class of time-varying system with mixed-time-delays subject to the communication protocol scheduling. For the purpose of preventing the data from collisions, only one sensor node is permitted to get access to the shared network at each transmission instant. The transmission order of nodes is orchestrated by the communication protocol of the network. Two kinds of protocols (namely, the RR protocol and the TOD protocol) have been studied respectively. A time-varying filter has been designed to obtain the estimate of the plant subject to the non-Gaussian noises and protocol scheduling. Sufficient conditions have been derived for the designed filter to satisfy the prescribed $P(k)$-dependent constraint in terms of recursive matrix inequalities. Within the established theoretical framework, two optimization problems have been proposed by optimizing the constraint ellipsoid of the estimation error subject to the aforementioned protocols. Finally, two illustrative examples have been given to highlight the effectiveness of the proposed filter design strategy.

References

1. Yang, F., Li, Y.: Set-membership filtering for systems with sensor saturation. Automatica **45**(8), 1896–1902 (2009)
2. Guo, L., Wang, H.: Minimum entropy filtering for multivariate stochastic systems with non-Gaussian noises. IEEE Trans. Autom. Control **51**(4), 695–700 (2006)
3. Hu, J., Wang, Z., Gao, H., Stergioulas, L.K.: Robust sliding mode control for discrete stochastic systems with mixed time delays, randomly occurring uncertainties, and randomly occurring nonlinearities. IEEE Trans. Ind. Electron. **59**(7), 3008–3015 (2012)
4. Boyd, S., Ghaoui, L., Feron, E., Balakrishnan, V.: Linear Matrix Inequalities in System and Control Theory. SIAM Studies in Applied Mathematics, Philadelphia (1994)
5. El Ghaoui, L., Calafiore, G.: Robust filtering for discrete-time systems with bounded noise and parametric uncertainty. IEEE Trans. Autom. Control **46**(7), 1302–1313 (2001)
6. Zhao, Y., Liu, G., Rees, D.: Design of a packet-based control framework for networked control systems. IEEE Trans. Control Technol. **17**(4), 859–865 (2009)

Chapter 6
Recursive Filtering for Time-Varying Systems with Random Access Protocol

In this chapter, we intend to study the communication-protocol-based recursive filtering problem for time-varying systems. Recursive filtering scheme is a widely employed method which is developed in the sense of minimum filtering error variance. More specifically, such recursive filtering scheme works in a two-step process, namely the *prediction step* and the *correction step*. In the prediction step, the filter computes the one-step prediction of the system state based on the plant dynamics and the current state estimate. In the correction step, the state estimate would be updated by fusing the derived one-step prediction and the new measurement output with the well-designed weighted gain (i.e. the recursive filter gain). In the framework of recursive filtering, the recursive filter gain should be designed with the consideration of the resultant filtering performance. Obviously, the utilization of networks would give rise to certain complicated network-induced effects which could largely jeopardize the filtering performance of systems. As such, the recursive filtering problem of networked systems has attracted considerable research attention in the past decade.

This chapter is concerned with the recursive filtering problem for a class of networked linear time-varying systems where the communication between the sensors nodes and the remote filter is implemented via a shared network. The transmission order of sensor nodes is orchestrated by the RA Protocol scheduling, under which the selected nodes obtaining access to the network could be characterized by a sequence of i.i.d. variables. In this chapter, we would like to discuss the effects of RA Protocol scheduling on the resultant filtering error covariance and the ultimate boundedness of the filtering error covariance.

The rest of this chapter is organized as follows. In Sect. 6.1, the networked linear time-varying system with the RA Protocol scheduling is introduced and the corresponding recursive filtering problem is formulated. In Sect. 6.2, the desired filter gain matrix is recursively computed based on two Riccati-like difference equations.

L. Zou et al., *Communication-Protocol-Based Filtering and Control of Networked Systems*, Studies in Systems, Decision and Control 430,
https://doi.org/10.1007/978-3-030-97512-8_6

Then, the boundedness analysis issue of the filtering error covariance is studied for our developed recursive filtering scheme. Moreover, two numerical simulation examples are given in Sect. 6.3 to demonstrate the correctness and effectiveness of the main results. Finally, we conclude this chapter in Sect. 6.4.

6.1 Problem Formulation

6.1.1 The System Dynamics and Signal Transmission Behaviors

The plant under consideration is a linear time-varying system described by the following state-space model:

$$\begin{cases} x_{k+1} = A_k x_k + B_k \omega_k \\ y_k = C_k x_k + \nu_k \end{cases} \tag{6.1}$$

where $x_k \in \mathbb{R}^{n_x}$ and $y_k \in \mathbb{R}^{n_y}$ denote, respectively, the system state and the measurement output before transmitted through the communication network. $\omega_k \in \mathbb{R}^{n_\omega}$ and $\nu_k \in \mathbb{R}^{n_\nu}$ represent the process and measurement noises, respectively. The parameters A_k, B_k and C_k are real-valued time-varying matrices of appropriate dimensions.

The initial state x_0, the process noise ω_k and the measurement noise ν_k are mutually uncorrelated and have the following statistical properties:

$$\begin{aligned} &\mathbb{E}\{\omega_k\} = \mathbb{E}\{\nu_k\} = 0,\ \mathbb{E}\{x_0\} = \bar{x}_0,\ \mathbb{E}\{\omega_k \omega_k^T\} = Q_k, \\ &\mathbb{E}\{(x_0 - \bar{x}_0)(x_0 - \bar{x}_0)^T\} = P_{0|0},\ \mathbb{E}\{\nu_k \nu_k^T\} = R_k \end{aligned} \tag{6.2}$$

where $P_{0|0} > 0$, $Q_k > 0$ and $R_k > 0$ are known matrices with appropriate dimensions.

We are now ready to introduce the signal transmission over the communication network. Without loss of generality, it is assumed that sensors of the system are grouped into N sensor nodes according to their spatial distribution. As such, for technical analysis, the measurement output before transmitted can be rewritten as $y_k = \begin{bmatrix} y_{1,k}^T & y_{2,k}^T & \cdots & y_{N,k}^T \end{bmatrix}^T$ where $y_{i,k}$ ($i \in \{1, 2, \ldots, N\}$) is the measurement of the i-th sensor node before transmitted.

Now, let us introduce the signal transmission behaviors. In this chapter, we consider the case that the communication between the sensors and the remote filter is implemented via a network. In order to avoid possible data collision in the network-based communication, the widely-utilized RA Protocol is adopted to scheduling the signal transmissions. For the sake of examining the influence of communication protocols, let $\alpha(k) \in \{1, 2, \ldots, N\}$ be the selected sensor node obtaining access to the communication network at time instant k. As shown in Sect. 4.1,

$\alpha(k) \in \{1, 2, \ldots, N\}$ could be modeled as a sequence of random variables. The occurrence probability of $\alpha(k) = i$ ($\forall i \in \{1, 2, \ldots, N\}$) is given by

$$\text{Prob}\{\alpha(k) = i\} = p_i \tag{6.3}$$

where $p_i > 0$ ($i \in \{1, 2, \ldots, M\}$) is the occurrence probability that the node i is selected to transmit data via the communication network and $\sum_{i=1}^{M} p_i = 1$.

Assume that $\alpha(k)$ ($k \in \mathbb{N}^+$) is independent of all noise signals. Denote the measurement output after transmitted through the network by

$$\bar{y}_k \triangleq \left[\bar{y}_{1,k}^T \; \bar{y}_{2,k}^T \; \cdots \; \bar{y}_{N,k}^T\right]^T \in \mathbb{R}^{n_y}.$$

In this chapter, we adopt the ZI strategy to compensate the signals received by the filter. Accordingly, the updating rule of $\bar{y}_{i,k}$ ($k \in \mathbb{N}^+$, $i \in \{1, 2, \ldots, N\}$) subject to the RA Protocol scheduling is set to be

$$\bar{y}_{i,k} = \begin{cases} y_{i,k}, & \text{if } i = \alpha(k),\ k \geq 0 \\ 0, & \text{otherwise.} \end{cases} \tag{6.4}$$

According to the updating rule (6.4), the signal transmission model can be described by

$$\bar{y}_k = \begin{cases} \Psi_{\alpha(k)} y_k, & \text{if } k \geq 0 \\ 0, & \text{otherwise} \end{cases} \tag{6.5}$$

where $\Psi_{\alpha(k)} = \text{diag}\{\delta(\alpha(k) - 1)I, \delta(\alpha(k) - 2)I, \ldots, \delta(\alpha(k) - N)I\}$ $(1 \leq i \leq N)$ and $\delta(\cdot) \in \{0, 1\}$ is the Kronecker delta function.

6.1.2 The Structure of the Recursive Filter

In this chapter, we shall adopt the Kalman-type filtering approach to design a recursive filter for the linear time-varying system (6.1) subject to the RA Protocol scheduling described by (6.5). As we discussed before, the adopted Kalman-type filtering scheme works in a two-step process (i.e. the prediction step and the correction step). Under the framework of such filtering scheme, we consider a recursive filter with the following form:

$$\begin{cases} \hat{x}_{k+1|k} = A_k \hat{x}_{k|k} \\ \hat{x}_{k+1|k+1} = \hat{x}_{k+1|k} + K_{k+1}\left(\bar{y}_{k+1} - \Psi_{\alpha(k+1)} C_{k+1} \hat{x}_{k+1|k}\right) \end{cases} \tag{6.6}$$

where $\hat{x}_{k|k}$ is the estimate of x_k at time instant k with $\hat{x}_{0|0} = \bar{x}_0$, $\hat{x}_{k+1|k}$ is the one-step prediction at time instant k, and K_{k+1} is the filter gain to be determined.

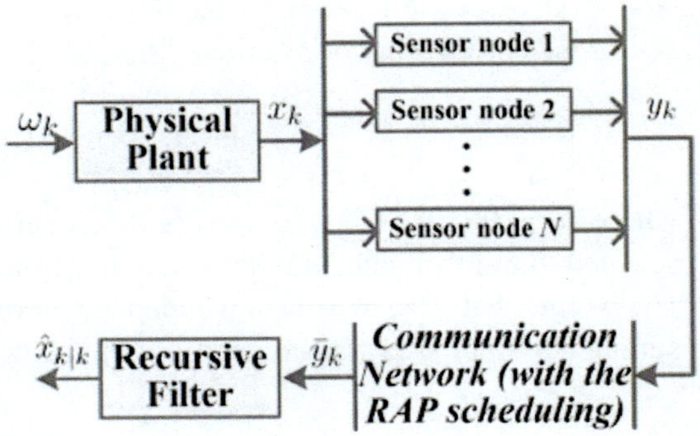

Fig. 6.1 Schematic structure for the plant and the filter over a network (with the RA Protocol scheduling)

Remark 6.1 Compared with the standard Kalman-type filtering algorithm, the correction step in (6.6) is implemented based on an "revised" innovation (i.e. $\bar{y}_{k+1} - \Psi_{\alpha(k+1)}C_{k+1}\hat{x}_{k+1|k}$). This is mainly due to the fact that $\bar{y}_{k+1} = \Psi_{\alpha(k+1)}y_{k+1}$, which implies that the valid information for state estimate updating in the correction step can be reflected by the scheduling matrix $\Psi_{\alpha(k+1)}$. In fact, the filter structure (6.6) can be used to handle the recursive filtering issue subject to other protocols (e.g. the RR protocol and TOD protocol).

The schematic diagram of the filtering system considered in this chapter is shown in Fig. 6.1. The objective of this chapter is to design a recursive filter of the structure (6.6) such that, for *all possible realizations* of the random sequence $\alpha(k)$, the filtering error covariance (i.e. $\mathbb{E}\{(x_{k+1} - \hat{x}_{k+1|k+1})(x_{k+1} - \hat{x}_{k+1|k+1})^T\}$) can be derived recursively and subsequently minimized.

Remark 6.2 Due to the RA Protocol scheduling behavior, the filtering performance is largely affected by the *stochastic parameter matrix* $\Psi_{\alpha(k)}$. The main difficulty of this chapter would be the handling of such a stochastic parameter matrix in design of the time-varying filter parameter K_{k+1} and the analysis on the boundedness issue of the filtering error covariance. It can be observed that the filter gain K_{k+1} should be calculated recursively based on the filtering error covariance $P_{k|k}$, and therefore, the boundedness of the filtering error covariance is very important to ensure the non-divergence of the filtering algorithm. Moreover, the boundedness of filtering error covariance often serves as an indispensable prerequisite to the boundedness guarantee of the filtering error in mean square. Note that, the problem studied in this chapter is distinct from the filtering issue with the sensor scheduling strategy. The protocol scheduling behavior is determined by the network agreements which are generated according to certain standards (e.g. IEEE 802 standards). Hence, the scheduling behavior could not be designed for the filtering task in this chapter. On the other hand, the sensor scheduling (as shown in [1, 2]) could be regarded as the "design variable". Hence, for the filtering problem with sensor scheduling, the filtering performance is related to the sensor selection strategy. In this chapter, we focus our attention on the recursive filtering problems with RA Protocol scheduling, in which the protocol scheduling behavior is modeled by the i.i.d. sequence of random variables.

6.2 Main Results

In this section, we aim to establish a unified framework to deal with the addressed recursive filtering problem under the RA Protocol scheduling. Before proceeding further, we recall the following lemma which will be used in the subsequent developments.

Lemma 6.1 (Jensens inequality) *[3] Let vector* $x_i \in \mathbb{R}^n$ *(*$i = 0, 1, \ldots, m$*) and scalar constants* $a_i > 0$ *(*$i = 0, 1, \ldots, m$*) be given. The following inequality*

$$f\left(\frac{1}{M}\sum_{i=1}^{m} a_i x_i\right) \leq \frac{1}{M}\sum_{i=1}^{m} a_i f(x_i) \tag{6.7}$$

holds for convex function $f : \mathbb{R}^n \to \mathbb{R}^n$ *where* $M = \sum_{i=1}^{m} a_i$.

Lemma 6.2 *For matrices* M, N, X *and* P *with appropriate dimensions, the following equations hold:*

$$\frac{\partial \text{tr}\{MXN\}}{\partial X} = M^T N^T, \quad \frac{\partial \text{tr}\{MX^T N\}}{\partial X} = NM,$$
$$\frac{\partial \text{tr}\{MXNX^T L\}}{\partial X} = M^T L^T X N^T + LMXN.$$

6.2.1 Design of the Filter Gain

Let us denote the one-step prediction error as $e_{k+1|k} = x_{k+1} - \hat{x}_{k+1|k}$ and the filtering error as $e_{k+1|k+1} = x_{k+1} - \hat{x}_{k+1|k+1}$. Subtracting (6.6) from (6.1), we have

$$e_{k+1|k} = A_k e_{k|k} + B_k \omega_k. \tag{6.8}$$

Similarly, the filtering error can be written as

$$e_{k+1|k+1} = \big(I - K_{k+1}\Psi_{\alpha(k+1)}C_{k+1}\big)e_{k+1|k} - K_{k+1}\Psi_{\alpha(k+1)}\nu_{k+1}. \tag{6.9}$$

In light of (6.8) and (6.9), the covariance for the one-step prediction error and filtering error are calculated in the following theorem.

Theorem 6.1 *Consider the filtering error system given by (6.9). The one-step prediction error covariance* $P_{k+1|k} \triangleq \mathbb{E}\{e_{k+1|k}e_{k+1|k}^T\}$ *and the filtering error covariance* $P_{k+1|k+1} \triangleq \mathbb{E}\{e_{k+1|k+1}e_{k+1|k+1}^T\}$ *(with the initial condition* $P_{0|0}$*) are given by the following difference equations*

$$P_{k+1|k} = A_k P_{k|k} A_k^T + B_k Q_k B_k^T, \tag{6.10}$$

$$P_{k+1|k+1} = \sum_{i=1}^{N} p_i \mathcal{K}_{i,k+1} P_{k+1|k} \mathcal{K}_{i,k+1}^T + \sum_{i=1}^{N} p_i K_{k+1} \Psi_i R_{k+1} \Psi_i K_{k+1}^T. \tag{6.11}$$

where $\mathcal{K}_{i,k+1} = I - K_{k+1} \Psi_i C_{k+1}$*. Moreover, the trace of the filtering error covariance* $P_{k+1|k+1}$ *is minimized by the following filter gain:*

$$K_{k+1} = P_{k+1|k} C_{k+1}^T \bar{\Psi} \left(\sum_{i=1}^{N} p_i \Psi_i \mathcal{R}_{k+1} \Psi_i \right)^{-1} \tag{6.12}$$

where $\mathcal{R}_{k+1} = C_{k+1} P_{k+1|k} C_{k+1}^T + R_{k+1}$ *and* $\bar{\Psi} = \mathrm{diag}\{p_1 I, p_2 I, \ldots, p_N I\}$.

Proof First, it is easy to conclude from (6.8) that the one-step prediction error covariance $P_{k+1|k}$ satisfies

$$P_{k+1|k} = A_k P_{k|k} A_k^T + B_k Q_k B_k^T. \tag{6.13}$$

Next, let us consider the filtering error covariance $P_{k+1|k+1}$. Noting (6.9), we have

$$\begin{aligned} P_{k+1|k+1} &= \mathbb{E}\{e_{k+1|k+1} e_{k+1|k+1}^T\} \\ &= \mathbb{E}\big\{\mathcal{K}_{\alpha(k+1),k+1} e_{k+1|k} e_{k+1|k}^T \mathcal{K}_{\alpha(k+1),k+1}^T + K_{k+1} \Psi_{\alpha(k+1)} v_{k+1} v_{k+1}^T \Psi_{\alpha(k+1)}^T K_{k+1}^T\big\}. \end{aligned} \tag{6.14}$$

On the other hand, it can be seen from the definition of $\Psi_{\alpha(k+1)}$ that

$$\Psi_{\alpha(k+1)} = \sum_{i=1}^{N} \delta(\alpha(k+1) - i) \Psi_i. \tag{6.15}$$

Since

$$\delta(\alpha(k+1) - i)\delta(\alpha(k+1) - j) = \begin{cases} 0, & i \neq j \\ \delta(\alpha(k+1) - i), & i = j \end{cases}$$

and $\mathbb{E}\{\delta(\alpha(k+1) - i)\} = \sum_{j=1}^{N} p_j \delta(j - i) = p_i$, we have

$$
\begin{aligned}
&\mathbb{E}\{\mathcal{K}_{\alpha(k+1),k+1}e_{k+1|k}e_{k+1|k}^T\mathcal{K}_{\alpha(k+1),k+1}^T\} \\
&=\mathbb{E}\Big\{\big(I-K_{k+1}\sum_{i=1}^{N}\delta(\alpha(k+1)-i)\Psi_i C_{k+1}\big)e_{k+1|k}e_{k+1|k}^T\big(I \\
&\quad -K_{k+1}\sum_{i=1}^{N}\delta(\alpha(k+1)-i)\Psi_i C_{k+1}\big)^T\Big\} \\
&=P_{k+1|k}-\mathbb{E}\Big\{\sum_{i=1}^{N}\delta(\alpha(k+1)-i)K_{k+1}\Psi_i C_{k+1}e_{k+1|k}e_{k+1|k}^T\Big\} \\
&\quad -\mathbb{E}\Big\{\sum_{i=1}^{N}\delta(\alpha(k+1)-i)e_{k+1|k}e_{k+1|k}^T C_{k+1}^T\Psi_i K_{k+1}^T\Big\} \\
&\quad +\mathbb{E}\Big\{\sum_{i=1}^{N}\delta(\alpha(k+1)-i)K_{k+1}\Psi_i C_{k+1}e_{k+1|k}e_{k+1|k}^T C_{k+1}^T\Psi_i K_{k+1}^T\Big\} \\
&=\sum_{i=1}^{N}p_i\big(I-K_{k+1}\Psi_i C_{k+1}\big)P_{k+1|k}\big(I-K_{k+1}\Psi_i C_{k+1}\big)^T. \qquad (6.16)
\end{aligned}
$$

Similarly, we have

$$
\mathbb{E}\{K_{k+1}\Psi_{\alpha(k+1)}v_{k+1}v_{k+1}^T\Psi_{\alpha(k+1)}^T K_{k+1}^T\}=\mathbb{E}\Big\{\sum_{i=1}^{N}p_i K_{k+1}\Psi_i R_{k+1}\Psi_i K_{k+1}^T\Big\}. \qquad (6.17)
$$

Substituting the inequalities (6.16)–(6.17) into (6.14) yields

$$
P_{k+1|k+1}=\sum_{i=1}^{N}p_i\mathcal{K}_{i,k+1}P_{k+1|k}\mathcal{K}_{i,k+1}^T+\sum_{i=1}^{N}p_i K_{k+1}\Psi_i R_{k+1}\Psi_i K_{k+1}^T.
$$

Next, let us show that the filter gain given by (6.12) is optimal in the sense that it minimizes the trace of the filtering error covariance $P_{k+1|k+1}$. Taking the partial derivative of $\mathrm{tr}\{P_{k+1|k+1}\}$ with respect to K_{k+1} and letting the derivative be zero, we have

$$
\frac{\partial \mathrm{tr}\{P_{k+1|k+1}\}}{\partial K_{k+1}}=-2\sum_{i=1}^{N}p_i\big(\mathcal{K}_{i,k+1}P_{k+1|k}C_{k+1}^T\Psi_i\big)+2K_{k+1}\sum_{i=1}^{N}p_i\big(\Psi_i R_{k+1}\Psi_i\big)=0. \qquad (6.18)
$$

Based on the above equation, the optimal filter gain K_{k+1} can be determined as

$$K_{k+1} = P_{k+1|k} C_{k+1}^T \bar{\Psi} \left(\sum_{i=1}^{N} p_i \Psi_i \mathcal{R}_{k+1} \Psi_i \right)^{-1}. \tag{6.19}$$

which is identical to (6.12). This completes the proof.

Remark 6.3 So far, we have completed the design issue of the recursive filter for time-varying systems with the RA Protocol scheduling. By using Lemma 6.1, it is easy to see that $\sum_{i=1}^{N} p_i \Psi_i R_{k+1} \Psi_i \geq (\sum_{i=1}^{N} p_i \Psi_i) R_{k+1} (\sum_{i=1}^{N} p_i \Psi_i) = \bar{\Psi} R_{k+1} \bar{\Psi} > 0$. As such, it can be concluded that the matrix $\sum_{i=1}^{N} p_i \Psi_i \big(C_{k+1} P_{k+1|k} C_{k+1}^T + R_{k+1}\big) \Psi_i$ is nonsingular. The computation of K_{k+1} is carried out by solving two discrete-time Riccati-like difference equations, which are suitable for online implementation.

Remark 6.4 In our proposed filter design, the information of $\alpha(k+1)$ is adopted in the structure of the recursive filter (as shown in (6.6)). The filter gain matrix K_{k+1} is calculated by minimizing the trace of the filtering error covariance $P_{k+1|k+1}$. According to the definition of $P_{k+1|k+1}$ (which is given in Theorem 6.1), the filter gain matrix is derived recursively in the sense of probability. Obviously, it is of both theoretical significance and practical importance for filter design to minimize $\mathrm{tr}\{P_{k+1|k+1}\}$. Another valuable strategy for the recursive filter design is to minimize the *conditional covariance matrix* $\mathbb{E}\{e_{k+1|k+1} e_{k+1|k+1}^T | \alpha(k+1)\}$, in which the filter gain K_{k+1} could be computed based on the exact value of $\alpha(k+1)$.

6.2.2 Boundedness Analysis of the Filtering Error Covariance

According to our designed filter gain K_{k+1} given by (6.12), the desired state estimate can be calculated recursively by using an online algorithm. Nevertheless, it should be noted that the implementation of such an online algorithm is determined by the non-divergence of the algorithm, which implies that the boundedness of the $P_{k+1|k+1}$ is of great importance for our developed recursive filtering algorithm. In this subsection, let us consider the boundedness of the filtering error covariance $P_{k+1|k+1}$. Throughout the rest of the chapter, the following assumption is made.

Assumption 6.1 Let $\bar{\Psi} = \mathrm{diag}\{p_1 I, p_2 I, \ldots, p_N I\}$. There exist real numbers $\underline{q}$, $\bar{q}$, $\bar{r}$ and $\underline{r}$, such that the following matrix inequalities are satisfied for every $k \geq 0$ and $i \in \{1, 2, \ldots . N\}$:

$$\underline{q} I \leq B_k Q_k B_k^T \leq \bar{q} I, \quad \Psi_i R_k \Psi_i \leq \bar{r} I, \quad \bar{\Psi} R_k \bar{\Psi} \geq \underline{r} I, \quad C_k^T C_k \leq \bar{c} I,$$
$$C_k C_k^T \leq \hat{c} I, \quad \underline{a} I \leq A_k A_k^T \leq \bar{a} I.$$

Theorem 6.2 *Under Assumption 6.1, there exists a positive constant $\underline{\varepsilon}$ such that the error covariance $P_{k|k}$ of the recursive filtering for systems (6.1) satisfies*

$$P_{k|k} \geq \underline{\varepsilon} I \tag{6.20}$$

for every $k > 0$ with the lower bound $\underline{\varepsilon}$ given as follows:

$$\underline{\varepsilon} = \left(\underline{q}^{-1} + \underline{r}^{-1}\bar{p}^2\bar{c}\right)^{-1}. \tag{6.21}$$

where $\bar{p} = \max_{i=1,2,\ldots,N}\{p_i\}$.

Proof Considering (6.11) and (6.12), we have

$$\begin{aligned}
P_{k+1|k+1} &= P_{k+1|k} - P_{k+1|k}C_{k+1}^T\bar{\Psi}\left(\sum_{i=1}^{N} p_i\Psi_i\mathcal{R}_{k+1}\Psi_i\right)^{-1}\bar{\Psi}C_{k+1}P_{k+1|k} \\
&= P_{k+1|k} - P_{k+1|k}C_{k+1}^T\bar{\Psi}\left(\Xi_{k+1} + \bar{\Psi}C_{k+1}P_{k+1|k}C_{k+1}^T\bar{\Psi}\right)^{-1}\bar{\Psi}C_{k+1}P_{k+1|k} \\
&= \left(P_{k+1|k}^{-1} + C_{k+1}^T\bar{\Psi}\Xi_{k+1}^{-1}\bar{\Psi}C_{k+1}\right)^{-1}
\end{aligned} \tag{6.22}$$

where $\Xi_{k+1} = \sum_{i=1}^{N} p_i\left(\tilde{\Psi}_i C_{k+1}P_{k+1|k}C_{k+1}^T\tilde{\Psi}_i + \Psi_i R_{k+1}\Psi_i\right)$ and $\tilde{\Psi}_i = \Psi_i - \bar{\Psi}$.

Noting that $\Xi_{k+1} > \sum_{i=1}^{N} p_i\Psi_i R_{k+1}\Psi_i \geq \bar{\Psi}R_{k+1}\bar{\Psi}$, it follows from (6.22) that

$$P_{k+1|k+1} \geq \left(P_{k+1|k}^{-1} + C_{k+1}^T\bar{\Psi}\left(\bar{\Psi}R_{k+1}\bar{\Psi}\right)^{-1}\bar{\Psi}C_{k+1}\right)^{-1}. \tag{6.23}$$

On the other hand, it is easy to see that

$$P_{k+1|k} \geq B_k Q_k B_k^T \geq \underline{q} I. \tag{6.24}$$

Subsequently, by considering (6.23) and (6.24), we have

$$\begin{aligned}
P_{k+1|k+1}^{-1} &\leq \underline{q}^{-1}I + C_{k+1}^T\bar{\Psi}\left(\bar{\Psi}R_{k+1}\bar{\Psi}\right)^{-1}\bar{\Psi}C_{k+1} \\
&\leq \underline{q}^{-1}I + \underline{r}^{-1}C_{k+1}^T\bar{\Psi}\bar{\Psi}C_{k+1} \leq \left(\underline{q}^{-1} + \underline{r}^{-1}\bar{p}^2\bar{c}\right)I
\end{aligned} \tag{6.25}$$

which implies that $P_{k|k} \geq \underline{\varepsilon} I$ for every $k > 0$. The proof is complete.

Theorem 6.2 provides a uniform lower bound of the filtering error covariance for the recursive filtering. In fact, Theorem 6.2 has indicated that the resultant filtering error covariance is always a positive definite matrix. This is mainly due to the existence of the process noise ω_k and the measurement noise ν_k. Next, we shall study the upper bound of the filtering error covariance.

Theorem 6.3 *Under the Assumption 6.1, there exists an upper bound μ_k such that the filtering error covariance $P_{k|k}$ satisfies*

$$P_{k|k} \leq \mu_k I, \quad k \geq 0 \tag{6.26}$$

where $\mu_k = \mu_0 \bar{a}^k + \bar{q} \sum_{i=0}^{k-1} \bar{a}^i$ *and* $\mu_0 = \lambda_{\max}\{P_{0|0}\}$.

Proof The proof of this theorem is performed by *mathematical induction*, which is divided into two steps, namely the initial step and the inductive step.

Initial step. For $k = 0$, it can be immediately known from the definition of μ_0 that

$$P_{0|0} \leq \lambda_{\max}\{P_{0|0}\} I = \mu_0 I. \tag{6.27}$$

Inductive step. Now that the assertion of this theorem is true for $t = 0$. Next, given that the assertion is true for $t = k$ (i.e. $P_{t|t} \leq \mu_t I$), we aim to show that the same assertion is true for $t = k + 1$ (i.e. $P_{t+1|t+1} \leq \mu_{t+1} I$). Obviously, it follows from (6.22) that

$$P_{k+1|k+1}^{-1} = P_{k+1|k}^{-1} + C_{k+1}^T \bar{\Psi} \Xi_{k+1}^{-1} \bar{\Psi} C_{k+1} \tag{6.28}$$

where $\Xi_{k+1} = \sum_{i=1}^{N} p_i \big(\tilde{\Psi}_i C_{k+1} P_{k+1|k} C_{k+1}^T \tilde{\Psi}_i + \Psi_i R_{k+1} \Psi_i\big)$ and $\tilde{\Psi}_i = \Psi_i - \bar{\Psi}$. Then, it is easy to see that

$$\begin{aligned} P_{k+1|k+1} &\leq P_{k+1|k} = A_k P_{k|k} A_k^T + B_k Q_k B_k^T \\ &\leq \mu_k A_k A_k^T + B_k Q_k B_k^T \\ &\leq \bar{a}\mu_k + \bar{q} I \\ &= \mu_0 \bar{a}^{k+1} + \bar{q} \sum_{i=0}^{k} \bar{a}^i \triangleq \mu_{k+1}. \end{aligned} \tag{6.29}$$

Hence, by the induction, it can be concluded that the assertion of this theorem is true for $k \geq 0$. The proof is complete.

Remark 6.5 By now, we have derived an upper bound function of the error covariance matrix for our developed recursive filtering algorithm. Obviously, the upper bound $\bar{a}$ has an important impact on our derived upper bound function μ_k. It can be found that the derived upper bound function μ_k is divergent if $\bar{a} \geq 1$. When $\bar{a} < 1$, the derived upper bound function would converge to a fixed value.

The following proposition gives a uniform upper bound of μ_k when $\bar{a} < 1$.

Proposition 6.1 *Under Assumption 6.1, if* $\bar{a} < 1$, *then the upper bound of the error covariance matrix* μ_k *is exponentially bounded with the uniform upper bound* $\mu_0 \bar{a} + \frac{\bar{q}}{1-\bar{a}}$ *for* $k > 0$.

Proof The proof is straightforward and is, therefore, omitted for conciseness.

In the case of $\bar{a} \geq 1$, the following theorem gives a sufficient condition to guarantee the boundedness of the error covariance matrix.

Theorem 6.4 *Let Assumption 6.1 hold. Suppose that there exist two positive scalars* $\sigma > 0$ *and* $m > 0$ *such that the following inequalities*

$$\begin{aligned}\sum_{i=k-m+1}^{k+1} &\left(\underline{\gamma}_i \phi^T(i,k+1) C_i^T \bar{\Psi}\big(\sigma\bar{a}\hat{c}(1-\underline{p})^2 I + \Theta_i\big)^{-1} \bar{\Psi} C_i \phi(i,k+1)\right) \\ &\geq \sigma^{-1} I, \qquad k \geq m-1 \qquad (6.30)\\ \mu_k &\leq \sigma, \qquad 0 \leq k < m-1 \qquad (6.31)\end{aligned}$$

hold for all $k \geq 0$, *where*

$$\underline{p} = \min_{1\leq i\leq N}\{p_i\},\ \tilde{\Psi}_j = \Psi_j - \bar{\Psi},\ \phi(k+1,k+1) = I,\ \phi(i,k+1) = A_i^{-1}A_{i+1}^{-1}\cdots A_k^{-1},$$

$$\mu_0 = \lambda_{\max}\{P_{0|0}\},\ \bar{\Psi} = \mathrm{diag}\{p_1 I, p_2 I, \ldots, p_N I\},\ \mu_k = \mu_0\bar{a}^k + \bar{q}\sum_{i=0}^{k-1}\bar{a}^i,$$

$$\Theta_i = \sum_{j=1}^{N} p_j(\tilde{\Psi}_j C_i B_{i-1} Q_{i-1} B_{i-1}^T C_i^T \tilde{\Psi}_j + \Psi_j R_i \Psi_j),\ \ \underline{\gamma}_i = (1+\bar{q}\underline{a}^{-1}\underline{\varepsilon}^{-1})^{i-k-1}.$$

Then, the error covariance satisfies $P_{k|k} \leq \sigma I$ *for* $k \geq 0$.

Proof The proof of this theorem is performed by mathematical induction as follows:

Initial step: Obviously, by employing Theorem 6.3 and (6.31), we have $P_{i|i} \leq \mu_i I \leq \sigma I$ for $0 \leq i < m-1$.

Inductive step: We know that the assertion of this theorem (i.e. $P_{i|i} \leq \sigma I$) is true for $i < m-1$. Next, given that the assertion is true for $i = k$, we aim to show that the same assertion is true for $i = k+1$. Since the assertion is true for $i = k$ (i.e. $P_{k|k} \leq \sigma I$), it follows from (6.10) that

$$P_{k+1|k} = A_k P_{k|k} A_k^T + B_k Q_k B_k^T \leq \sigma\bar{a}I + B_k Q_k B_k^T. \qquad (6.32)$$

On the other hand, according to the results proposed in Theorem 6.2, we have

$$\begin{aligned}P_{k+1|k} &= A_k P_{k|k} A_k^T + B_k Q_k B_k^T = A_k\big(P_{k|k} + A_k^{-1} B_k Q_k B_k^T A_k^{-T}\big)A_k^T \\ &\leq A_k\big(P_{k|k} + \bar{q}\underline{a}^{-1}\underline{\varepsilon}^{-1} P_{k|k}\big)A_k^T = (1+\bar{q}\underline{a}^{-1}\underline{\varepsilon}^{-1}) A_k P_{k|k} A_k^T. \qquad (6.33)\end{aligned}$$

Furthermore, it can be derived from (6.22) that

$$P_{k+1|k+1}^{-1} = P_{k+1|k}^{-1} + C_{k+1}^T\bar{\Psi}\left(\sum_{i=1}^{N} p_i(\Psi_i R_{k+1}\Psi_i + \tilde{\Psi}_i C_{k+1} P_{k+1|k} C_{k+1}^T \tilde{\Psi}_i)\right)^{-1} \bar{\Psi} C_{k+1}. \qquad (6.34)$$

Then, by substituting the inequalities (6.32) and (6.33) into (6.34), we have

$$
\begin{aligned}
P_{k+1|k+1}^{-1} \geq & P_{k+1|k}^{-1} + C_{k+1}^{T}\bar{\Psi}\big(\sigma\bar{a}\hat{c}(1-\underline{p})^{2}I + \Theta_{k+1}\big)^{-1}\bar{\Psi}C_{k+1} \\
\geq & (1+\bar{q}\underline{a}^{-1}\underline{\varepsilon}^{-1})^{-1}A_{k}^{-T}P_{k|k}^{-1}A_{k}^{-1} + C_{k+1}^{T}\bar{\Psi}\big(\sigma\bar{a}\hat{c}(1-\underline{p})^{2}I \\
& + \Theta_{k+1}\big)^{-1}\bar{\Psi}C_{k+1} \geq \cdots \\
\geq & \sum_{i=k-m+1}^{k+1}(1+\bar{q}\underline{a}^{-1}\underline{\varepsilon}^{-1})^{i-k-1}\phi^{T}(i,k+1)C_{i}^{T}\bar{\Psi}\big(\sigma\bar{a}\hat{c}(1-\underline{p})^{2}I \\
& + \Theta_{i}\big)^{-1}\bar{\Psi}C_{i}\phi(i,k+1) + (1+\bar{q}\underline{a}^{-1}\underline{\varepsilon}^{-1})^{-m-1}\phi^{T}(k-m,k+1) \\
& \times P_{k-m|k-m}^{-1}\phi(k-m,k+1) \\
\geq & \sum_{i=k-m+1}^{k+1}\underline{\gamma}_{i}\phi^{T}(i,k+1)C_{i}^{T}\bar{\Psi}\big(\sigma\bar{a}\hat{c}(1-\underline{p})^{2}I + \Theta_{i}\big)^{-1}\bar{\Psi}C_{i}\phi(i,k+1) \\
\geq & \sigma^{-1}I,
\end{aligned} \tag{6.35}
$$

which results in

$$
P_{k+1|k+1} \leq \sigma I. \tag{6.36}
$$

Hence, by induction, it can be concluded that the assertion of this theorem is true for $k \geq 0$. The proof is complete.

Remark 6.6 The derived inequality (6.30) in Theorem 6.4 is of the form similar to the *uniform observability condition* [4]. Considering the time-varying matrices $\underline{A}_k$ and $\underline{C}_k$, the pair $[\underline{C}_k, \underline{A}_k]$ is said to be *uniformly observable* if there exist a positive integer $0 < m < \infty$ and constants $\bar{\epsilon}, \underline{\epsilon}$ with $0 < \underline{\epsilon} \leq \bar{\epsilon} < \infty$ such that the condition $\bar{\epsilon}I \geq \sum_{i=k-m+1}^{k+1}\big(\overrightarrow{\psi}^{T}(i,k+1)\underline{C}_i^T\underline{C}_i\overrightarrow{\psi}(i,k+1)\big) \geq \underline{\epsilon}I$ holds for all $k \geq m-1$, where $\overrightarrow{\psi}(k+1,k+1) = I$ and $\overrightarrow{\psi}(i,k+1) = \underline{A}_i^{-1}\underline{A}_{i+1}^{-1}\cdots\underline{A}_k^{-1}$. Obviously, when the sensor nodes of the plant could simultaneously get access to the network and transmit signals, the inequality (6.30) reduces to the well-known uniform observability condition based on Assumption 6.1. So far, we have investigated the recursive filtering problems for linear time-varying systems subject to the RA Protocol scheduling. It can be observed from the filter design that all the important factors contributing to the system complexity are reflected in the main results. These factors include: (1) the time-varying system parameters; (2) the noise information (characterized by Q_k and R_k); (3) the initial condition of the system (i.e. $P_{0|0}$); and (4) the information about the RA Protocol scheduling (determined by (6.3)).

Remark 6.7 It is worth pointing out that, the recursive filtering technology developed in this chapter is not a global optimal filtering scheme. The main purpose of this chapter is to develop the recursive filtering scheme with a verifiable boundedness condition of the filtering error covariance matrix rather than design the optimal state estimation technology. If we adopt the well-known Kalman filtering to deal with the filtering problem of the linear time-varying system (6.1) subject to the RA Protocol scheduling, the observability Gramian matrix would be related to the scheduling matrices in a sliding time window with fixed finite length. Hence, it is quite difficult to verify the corresponding uniform observability condition. Nevertheless, the observability-like condition proposed in Theorem 6.4 is verifiable. In other words, for the problem considered in this chapter, our developed filtering scheme possesses the advantage that the non-divergent of the filtering algorithm is verifiable compared with the Kalman filtering.

Remark 6.8 The main results of this chapter can be regarded as the analysis and synthesis for the linear time-varying system with a random parameter $\alpha(k)$. Such an issue has so far attracted considerable attention due to its clear engineering insight. Typical examples include the research results shown in [5, 6], which concern the design problems of linear-quadratic-Gaussian (LQG) controllers for linear systems with random parameters. Compared with the results in [5, 6], the effects on the system dynamics induced by the random parameter $\alpha(k)$ in this chapter is more complex, which, in turn, increases the difficulty on the filter design. Moreover, we have analyzed the effects on the filtering performance induced by $\alpha(k)$ (i.e. the boundedness analysis issue proposed in Sect. 6.2.2).

6.3 Illustrative Examples

In this section, two numerical examples are presented to demonstrate the effectiveness of the proposed recursive filtering algorithm for system (6.1) subject to the RA Protocol scheduling.

Example 6.1 Consider a linear time-varying system (6.1) with the following system parameters:

$$A_k = \begin{bmatrix} 0.2\sin(0.04\pi k) & 0.5 & 0.5 \\ 0.6 & 0.7 & 0.7 \\ 0.2 & -0.1 & 0.5 \end{bmatrix}, \quad B_k = \begin{bmatrix} 2\cos(0.04\pi k) & 0.1 & 0.2 \\ 0.2 & 0.1 & -0.1 \\ 0.2 & -0.1 & 0.2 \end{bmatrix},$$

$$Q_k = I, \quad C_k = \begin{bmatrix} 0.5 & 0.7 & -0.6 \\ -0.3 & -0.2 & 0.4 \end{bmatrix}, \quad R_k = I, \quad P_{0|0} = 25I.$$

The sensors of this system are grouped into two sensor nodes and the occurrence probability about the RA Protocol scheduling is taken to be:

$$\begin{cases} \text{Prob}\{\alpha(k) = 1\} = 0.3 \\ \text{Prob}\{\alpha(k) = 1\} = 0.7. \end{cases}$$

Then, by constructing the corresponding filter and applying Theorem 6.1, the values of $\{K_k\}_{k>0}$ are derived in Table 6.1:

Table 6.1 The values of $\{K_k\}_{k>0}$

k	1	2	3	…
K_k	$\begin{bmatrix} 0.6985 & -1.4845 \\ 1.0501 & -1.9042 \\ 0.2063 & -0.1469 \end{bmatrix}$	$\begin{bmatrix} 0.7134 & -1.5770 \\ 1.057 & -2.1362 \\ 0.2376 & -0.4515 \end{bmatrix}$	$\begin{bmatrix} 0.6822 & -1.2494 \\ 0.9263 & -1.4926 \\ 0.1307 & -0.2022 \end{bmatrix}$	…

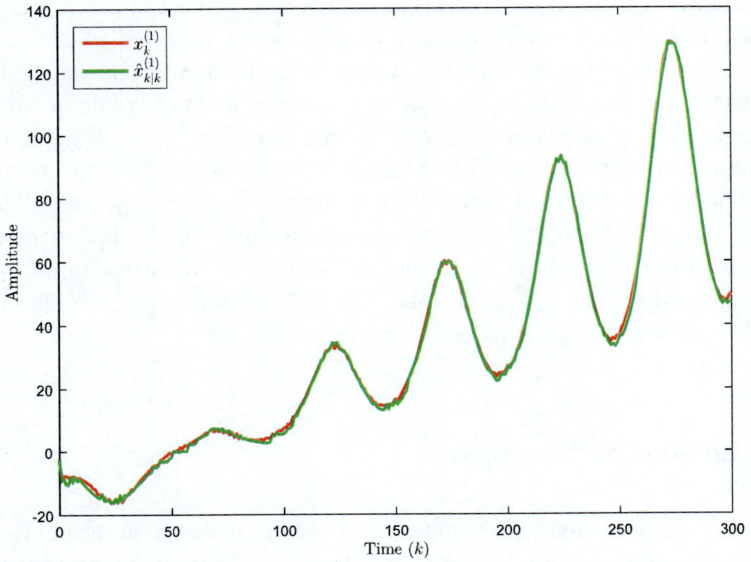

Fig. 6.2 The state trajectories of $x_k^{(1)}$ and $\hat{x}_{k|k}^{(1)}$ under the RA Protocol scheduling

Set the simulation run length to be 300. Based on the derived filter gain matrices, numerical simulation results are given in Figs. 6.2, 6.3, 6.4, 6.5 and 6.6. Figures 6.2, 6.3 and 6.4 plot the state trajectories and their estimates under the RA Protocol scheduling. Figure 6.5 depicts the trajectory of the filtering error $e_{k|k}$ and Fig. 6.6 shows the selected sensor node obtaining access to the communication network under the RA Protocol scheduling. All the simulation results confirm that the filtering performance is well achieved.

Example 6.2 In this example, we shall verify correctness of the uniform lower bound and uniform upper bound derived in this chapter through an *unstable second-order* system

$$\begin{cases} x_{k+1} = \begin{bmatrix} 1.01 & 0.02 \\ -0.02 & 1.02 \end{bmatrix} x_k + \begin{bmatrix} 0.48 \\ 0.46 \end{bmatrix} \omega_k \\ y_k = \begin{bmatrix} 1.66 & 0 \\ 0 & 1.68 \end{bmatrix} x_k + \nu_k \end{cases}$$

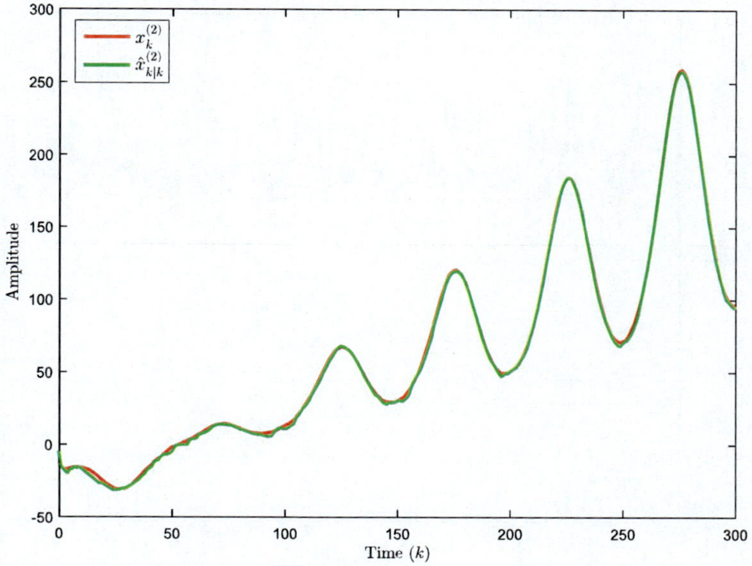

Fig. 6.3 The state trajectories of $x_k^{(2)}$ and $\hat{x}_{k|k}^{(2)}$ under the RA Protocol scheduling

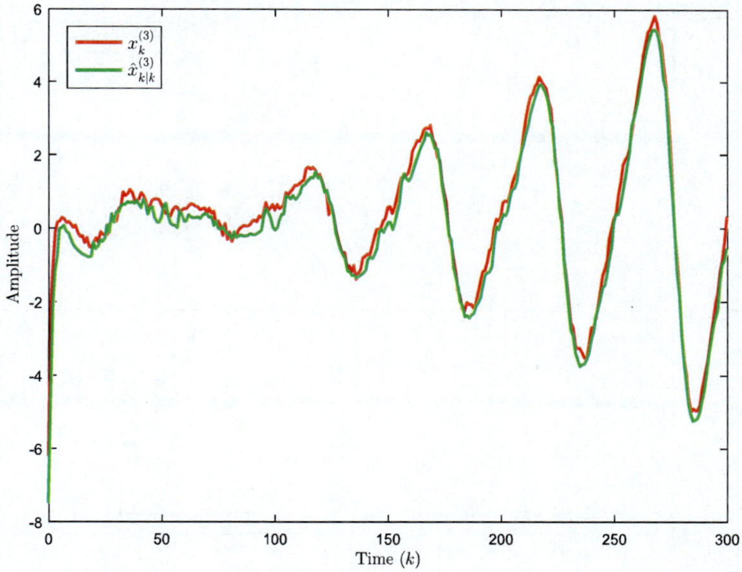

Fig. 6.4 The state trajectories of $x_k^{(3)}$ and $\hat{x}_{k|k}^{(3)}$ under the RA Protocol scheduling

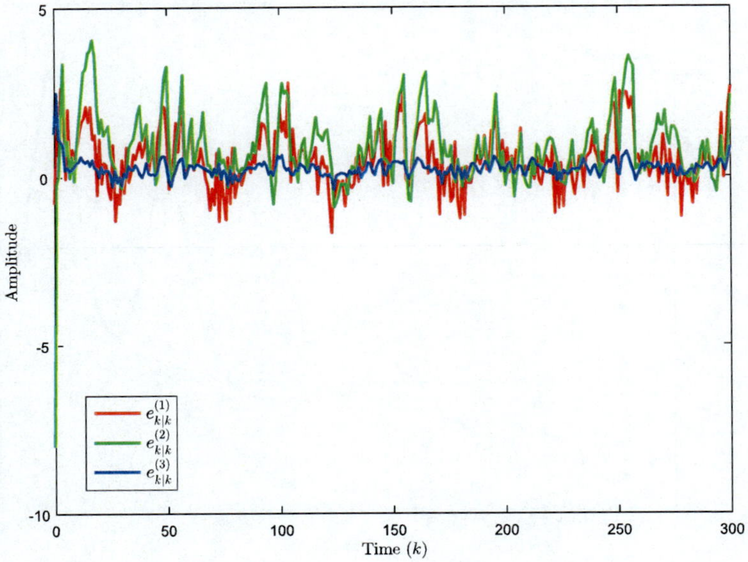

Fig. 6.5 The filtering errors under the RA Protocol scheduling

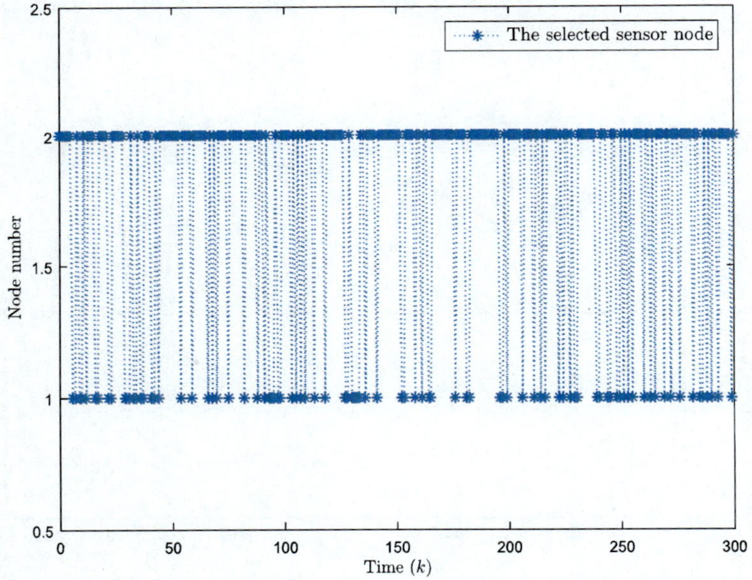

Fig. 6.6 The selected node obtaining access to the communication network under the RA Protocol scheduling

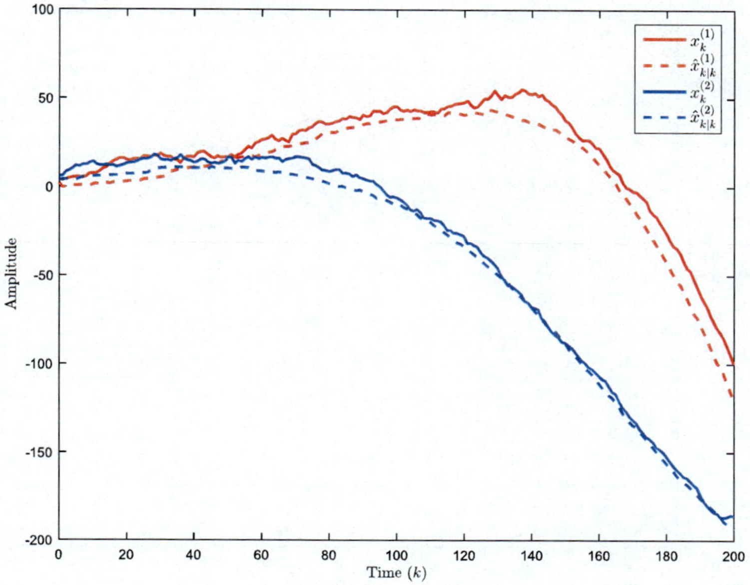

Fig. 6.7 The state trajectories of the plant and the filter under the RA Protocol scheduling

with the covariance matrix $P_{0|0} = 12I$, $Q_k = 6I$ and $R_k = 6I$. The sensors are grouped into two nodes and the corresponding occurrence probabilities are selected as $p_1 = 0.5$ and $p_2 = 0.5$. It is easy to check that the this system is an *unstable* system. By employing Theorem 6.2, we obtain the uniform lower bound of the filtering error covariance (i.e. $P_{k|k} \geq 0.7949I$). On the other hand, considering the inequality (6.30) with $\sigma = 20$ and $m = 2$, we have

$$\sum_{i=k-m+1}^{k+1} \underline{\gamma}_i \phi^T(i, k+1) C_i^T \bar{\Psi} \left(\sigma \bar{a} \hat{c}(1 - \underline{p})^2 I + \Theta_i\right)^{-1} \bar{\Psi} C_i \\ \times \phi(i, k+1) = \text{diag}\{0.0499, 0.0509\},$$

which implies that $\sum_{i=k-m+1}^{k+1} \underline{\gamma}_i \phi^T(i, k+1) C_i^T \bar{\Psi} \left(\sigma \bar{a} \hat{c}(1 - \underline{p})^2 I + \Theta_i\right)^{-1} \bar{\Psi} C_i$ $\phi(i, k+1) \geq \sigma^{-1} I$. Furthermore, it is easy to check that $\mu_0 = 13.8720$ and $\mu_1 = 15.8204$, which means that the inequality (6.31) is satisfied. Then, it can be derived that the $P_{k|k} \leq 20I$ in virtue of Theorem 6.4. The results are shown in Figs. 6.7 and 6.8. The simulation results have confirmed our theoretical analysis on the uniform lower bound and the uniform upper bound of the filtering covariance $P_{k|k}$.

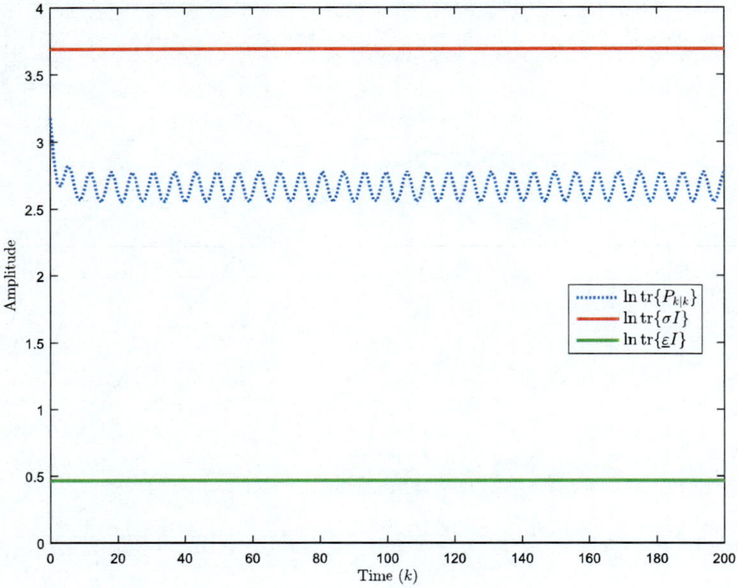

Fig. 6.8 The trace of $P_{k|k}$ and the their corresponding bounds

6.4 Summary

In this chapter, the recursive filtering problem has been addressed for a class of linear time-varying systems subject to the scheduling of the so-called RA Protocol. The scheduling behavior of the RA Protocol has been modeled by the i.i.d. sequence of random variables with known occurrence probabilities. The corresponding recursive filter has been presented to generate the state estimates. The filter gain has been calculated recursively to guarantee a minimized trace of the filtering error covariance. Furthermore, the boundedness problem of the filtering error covariance has been studied. The uniform lower and upper bounds have been derived. Finally, two examples have been provided to illustrate the correctness and effectiveness of the filtering algorithm developed in this chapter.

As we have discussed in Sects. 3.1 and 4.1, there are generally two modeling methods to describe the RA Protocol scheduling behavior. In this chapter, we consider the case that the RA Protocol scheduling behavior is modeled by a sequence of i.i.d. random variables. In fact, the results of this chapter can be easily extended to the case that the RA Protocol scheduling behavior is characterized by the discrete-time Markov chain. In practical applications, the selection of the modeling methods for the RA Protocol is always determined by the collected statistical information about the underlying network status.

References

1. Shi, L., Cheng, P., Chen, J.: Sensor data scheduling for optimal state estimation with communication energy constraint. Automatica **47**(8), 1693–1698 (2011)
2. Wu, J., Jia, Q.-S., Johansson, K.H., Shi, L.: Event-based sensor data scheduling: trade-off between communication rate and estimation quality. IEEE Trans. Autom. Control **58**(4), 1041–1046 (2013)
3. Needham, T.: A visual explanation of Jensen's inequality. Am. Math. Mon. **100**(8), 768–771 (1993)
4. Deyst, J., Price, C.: Conditions for asymptotic stability of the discrete minimum-variance linear estimator. IEEE Trans. Autom. Control **13**(6), 702–705 (1968)
5. Ku, R.T., Athans, M.: Further results on the uncertainty threshold principle. IEEE Trans. Autom. Control **22**(5), 866–868 (1977)
6. Athans, M., Ku, R., Gershwin, S.: The uncertainty threshold principle: some fundamental limitations of optimal decision making under dynamic uncertainty. IEEE Trans. Autom. Control **22**(3), 491–495 (1977)

Chapter 7
Filtering of Communication-Based Train Control Systems with CSMA Protocol

With experiencing fast economic growth, population expansion and urbanization worldwide, especially in some major developing countries, the demand for a safer, more efficient and comfortable mass transit system is very urgent. Railway system is a good choice for either urban commuting or intercity transport, as they meet the increasing needs for low emissions and high capacity. In traditional railway systems, the track-circuit technology is commonly employed to realize the data exchange between the individual operating train and the train control center [1] and such kind of data exchange has proven to be reliable in early years. However, such track-circuit based control (TBTC) system gives rise to high detection tolerance, thereby leading to long operation headway for each train to make sure there is no possibility for potential collisions. In other words, the track-circuit based technique would probably result in low operation efficiency in the train control center. Moreover, the short operation headway of the TBTC system needs a short length of the track circuit, which would give rise to high cost of the TBTC system.

On the other hand, in response to the rapidly increasing demand for a highly efficient urban public transportation system, the communication-based train control (CBTC) system has been developed by utilizing modern radio frequency (RF) technology. Compared with the TBTC system, the CBTC system could achieve much higher detection resolution and moving block by applying wireless data communication (instead of the track-circuit-based technology). In the working process of the CBTC systems, trains should "report" their locations, velocities, identities and other operation information to the train control centers with sufficient frequency, and then the train control centers notify the corresponding trains of the relevant movement authorities (MAs) according to the working conditions of metro systems. Trains are allowed to keep moving forward with the required velocities after receiving their MAs.

L. Zou et al., *Communication-Protocol-Based Filtering and Control of Networked Systems*, Studies in Systems, Decision and Control 430,
https://doi.org/10.1007/978-3-030-97512-8_7

For a CBTC system, the exact information about the locations and velocities of trains are of essential importance to the train control centers for the purpose of safe operations. Nevertheless, it is practically difficult to acquire such information because of the "inaccurate measurements" induced by measurement noises and the "unreliable communication" caused by underlying data transmission schemes. In particular, the location of a train for most of CBTC systems are measured by the combination of balises and speed odometers which could cause location errors especially when the train is running on wet or leafy rail tracks. Moreover, the signal transmissions between the trains and the train control center are implemented based on network communication technology which would give rise to data dropouts and communication delays. Such "inaccurate measurements" and "unreliable communication" will lead to inaccurate train status information and unplanned traction/braking commands, which will affect the safety and Quality of Service (QoS) of CBTC systems.

In order to obtain accurate information about trains, in this chapter, we intend to develop a filtering scheme to generate the estimates on the location and velocity of a train in the CBTC system. Noting that the wireless communication process is of great importance for the CBTC systems, in this chapter, we consider the case that the signal transmission over the wireless communication is scheduling by the well-known carrier-sense multiple access (CSMA) protocol (which is a widely employed protocol in wireless communication network). The purpose of the problem addressed is to design a suitable filter such that the filtering error is exponentially ultimately bounded (with a certain asymptotic upper bound) in mean square.

The rest of this chapter is organized as follows. In Sect. 7.1, the filtering problem is introduced for a CBTC system with the p-persistent CSMA protocol. In Sect. 7.2, we deal with the ultimate boundedness problem of the filtering error in mean square by adopting the stochastic analysis approach. Furthermore, two optimization problems are proposed for the design issue of the desired filter parameters. In Sect. 7.3, the effectiveness of the main results is demonstrated by a numerical simulation example. Finally, conclusions are drawn in Sect. 7.4.

7.1 Problem Formulation and Preliminaries

In this section, we first give an overview of the communication-based train control (CBTC) system and introduce some preliminaries related to the dynamic model of the train. Then, we will describe the problem setup.

7.1.1 Overview of the CBTC System and the Importance of Accurate Location and Velocity

Let's give a brief description of the so-called CBTC system. As shown in Fig. 7.1, a typical CBTC system consists of three distinctive parts, namely the trains, the wayside access points (WAPs) and the Zone Controller (ZC). In such a system, the

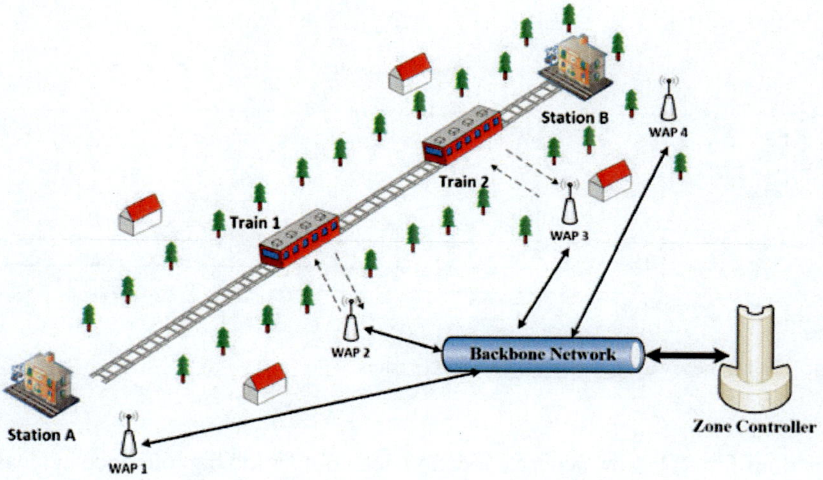

Fig. 7.1 A typical communication-based train control system

communication between the train and the WAPs (i.e. the *train–ground communication*) is realized through the wireless local area networks (WLANs). The WAPs collect the information about the train status (e.g. the identity, location, direction and velocity) via the train–ground communication, and then transmit the information to the ZC through the backbone network (which is always realized through a wired network). Based on the received information, the ZC makes the timetables for each train according to the location information of the trains in its area [2]. Unfortunately, due to the "unreliable" communication through the WLANs, the signals collected by WAPs are unavoidably subject to certain negative impacts brought from the wireless communication (e.g. protocol-induced constraints), which largely affect the "precision" of information about the location and velocity.

As the essential basis of the timetable design task for the ZC, the information of the location and velocity is of great importance to guarantee the security and dependability of CBTC systems. To protect the train and onboard passengers from happening serious hazards or even catastrophes, so-called automated train protection (ATP) systems are employed to ensure the safety of CBTC systems. ATP systems are formed by two parts. The first part is the onboard equipment including the vehicle onboard controller (VOBC), speed sensor, acceleration sensor, balise transmission module, etc. The other is the wayside part, which belongs to the ZC. By fusing the data from the axle counter and track mounted balises, VOBC generates real-time data about the train's location and speed, and then transmits them to the ZC via the data communication system (DCS). In return, based on these train operation data, including location and speed, the ZC can calculate and send back the limits of movement authorities (LMAs) to trains, which must be respected by every train. Following the received LMAs, VOBC can calculate the safety profile and regulate the train to obey it. If the actual speed is against to the safety profile, braking will be applied. So the precise location and speed data is the fundament in achieving

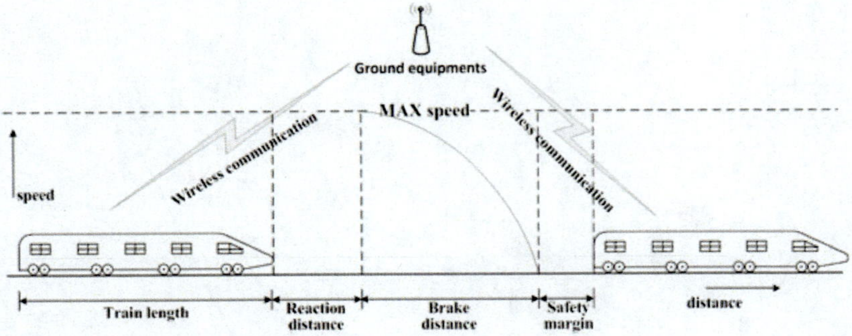

Fig. 7.2 The headway distance of the CBTC system

functions for ATP. Low accurate location detection or too big tolerance in measuring speed could lead to errors in generating LMAs or malfunctions in regulating safety profile, which are serious threats for the safety of CBTC systems.

Let's now show the impacts of inaccurate location and velocity on the headway of a CBTC system. The typical headway time represents the minimum time interval between two neighboring trains which the signaling will permit, so that the train ahead does not affect the following train. The headway distance denotes the minimum distance between two neighboring trains that the signal will permit. Headway time/distance is the most commonly utilized index to describe the railway capacity within the signaling profession. Reducing the headway time is beneficial to improve the line capacity. The headway distance of a CBTC system is calculated by considering the information about the train location, train speed and braking performance. The headway time is computed based on the derived headway distance and the train speed. If there is no communication delay in the system, the headway distance could be shown in Fig. 7.2.

As shown in Fig. 7.2, if we could obtain the exact information about the train location and train speed, the headway distance can be calculated as follows [3]:

$$H_D = L_{train} + L_{reaction} + L_{brake} + L_{safe}$$

where L_{train} is the train length, $L_{reaction}$ is the train running distance during the reaction time of onboard system, L_{brake} represents the brake distance at current train speed and L_{safe} denotes the safety margin. However, due to the effects of "inaccurate measurements" and "unreliable communication", the precise location of the train is almost impossible to achieve. For the purpose of safe operation, the train control system always calculates the headway by replacing the train length with the train safety envelope which is defined as the train length at a high confidence level on the premise of safety. The train safety envelope l_S can be computed as follows:

$$l_S = L_{train} + L_{error}$$

where L_{error} is the distance error that is defined as the difference between the actual train location and the measurement. Assuming that the maximum speed is $v_{\max}$, we can calculate the headway time H_T as follows:

$$\begin{aligned} H_T &= \frac{l_S + L_{reaction} + L_{brake} + L_{safe}}{v_{\max}} \\ &= \frac{L_{train} + L_{error} + L_{reaction} + L_{brake} + L_{safe}}{v_{\max}}. \end{aligned}$$

Obviously, we can reduce headway time H_T by decreasing the values of L_{error}, which implies that the accurate information about the train location and train velocity is of great importance for CBTC systems. In this chapter, we would like to develop an effective method for the ZC to obtain accurate information about the location and velocity of the train by using the filtering scheme.

7.1.2 *Mathematical Model of a Moving Train*

In this chapter, we would like to design the desired filter based on the model-based framework. Before starting designing the filter, let us first establish the mathematical model of a moving train with n cars ($n \geq 2$) linked by couplers. The diagram of the train is shown in Fig. 7.3. The dynamic characteristic of the train is described by the following Newton's motion equations:

$$\begin{cases} \dot{x}_i(t) = v_i(t), & i = 1, 2, \ldots, n \\ m_1\dot{v}_1(t) = u_1(t) - f_1(t) - f_{r,1}(t) \\ m_i\dot{v}_i(t) = u_i(t) + f_{i-1}(t) - f_i(t) - f_{r,i}(t), & i = 2, 3, \ldots, n-1 \\ m_n\dot{v}_n(t) = u_n(t) + f_{n-1}(t) - f_{r,n}(t) \end{cases} \tag{7.1}$$

where $x_i(t), v_i(t), m_i$ and $u_i(t)$ ($i = 1, 2, \ldots, n$) represent, respectively, the position, the velocity, the mass and the traction force of the ith car in the train. $f_i(t)$ ($i = 1, 2, \ldots, n$) denotes the in-train force from the coupler that connects the ith car and the $(i+1)$th car. $f_{r,i}(t)$ ($i = 1, 2, \ldots, n$) represents the resistive force that acts on the ith car satisfying $|f_{r,i}(t)| \leq \delta_i$ with a known constant positive scalar δ_i, where $|f_{r,i}(t)|$ represent the absolute value of $f_{r,i}(t)$.

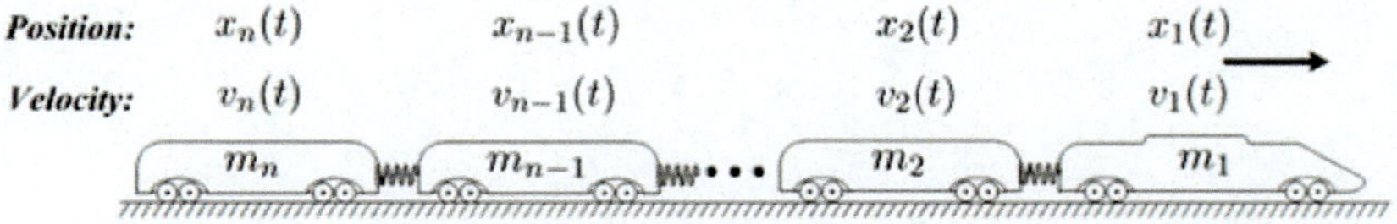

Fig. 7.3 The diagram of the train

Remark 7.1 There are two design schemes for the composition of the traction force: push-pull driving (PPD) design and distributed driving (DD) design. In the PPD design, only the first and the last cars of the train have their traction forces (i.e. $u_2(t) = u_3(t) = \cdots = u_{n-1}(t) = 0$). For the DD design, every car has its own traction force.

Remark 7.2 The resistive force $f_{r,i}(t)$ $(i = 1, 2, \ldots, n)$ exists consistently during the train operation which is caused by certain specified conditions. Generally speaking, the resistive force can be expressed as the sum of the ramp resistance, cure resistance, wind resistance and rolling mechanical resistance. According to [4], the resistive force $f_{r,i}(t)$ $(i = 1, 2, \ldots, n)$ could be described as follows:

$$f_{r,i}(t) = \begin{cases} (c_0 + c_1 v_1(t))m_1 + c_2 \sum_{i=1}^{n} m_i v_1^2(t) + \omega_1(t), & i = 1 \\ (c_0 + c_1 v_i(t))m_i + \omega_i(t), & i = 2, 3, \ldots, n \end{cases}$$

where the coefficients c_0, c_1 and c_2 are obtained by experimental test. $\omega_i(t)$ denotes the ramp resistance of the ith car satisfying $|\omega_i(t)| \leq \omega_{\max}$. Supposing that the speed limit is $v_{\max}$ (i.e. $v_i(t) \leq v_{\max}$). Hence, it is easy to see that $|f_{r,i}(t)| \leq \delta_i$, where $\delta_1 = (c_0 + c_1 v_{\max})m_1 + c_2 \sum_{i=1}^{n} m_i v_{\max}^2 + \omega_{\max}$ and $\delta_i = (c_0 + c_1 v_{\max})m_i + \omega_{\max}$ $(i = 2, 3, \ldots, n)$. As such, in this chapter, it is assumed that these resistances are norm-bounded scalars (i.e. $|f_{r,i}(t)| \leq \delta_i$) in order to reflect the reality.

The behavior of couples can be characterized by a spring model where the restoring force is given by

$$f_i(t) = k\left(x_i(t) - x_{i+1}(t) - \vartheta - l\right) \tag{7.2}$$

where $k > 0$ is a known constant positive scalar representing the stiffness coefficient, ϑ denotes the slack length of the couple, and l represents the length of a car.

Denoting

$$\begin{aligned} x(t) &= \left[x_1(t)\ x_2(t)\ \ldots\ x_n(t)\right]^T, \quad v(t) = \left[v_1(t)\ v_2(t)\ \ldots\ v_n(t)\right]^T, \\ \vec{x}(t) &= \left[x^T(t)\ v^T(t)\right]^T, \quad \vec{f}_r(t) = \left[f_{r,1}(t)\ f_{r,2}(t)\ \ldots\ f_{r,n}(t)\right]^T, \\ h &= \left[k(\vartheta + l)\ 0\ 0 \cdots 0\ -k(\vartheta + l)\right]^T, \quad \vec{h} = \left[0\ h^T\right]^T, \\ \mathscr{M} &= \mathrm{diag}\{I_n, \mathrm{diag}\{m_1^{-1}, m_2^{-1}, \ldots, m_n^{-1}\}\}, \end{aligned}$$

the dynamics of the train can be rewritten as follows:

$$\dot{\vec{x}}(t) = \vec{A}\,\vec{x}(t) + \vec{B}\,\vec{u}(t) + \vec{E}\,\vec{f}_r(t) + \mathscr{M}\,\vec{h} \tag{7.3}$$

where

$$\overrightarrow{A} = \mathscr{M}\begin{bmatrix} 0 & I \\ A & 0 \end{bmatrix}, \ \overrightarrow{B} = \mathscr{M}\begin{bmatrix} 0 \\ B \end{bmatrix}, \ \overrightarrow{E} = \mathscr{M}\begin{bmatrix} 0 \\ -I \end{bmatrix},$$

$$B = \begin{cases} I, & \text{for DD design,} \\ \begin{bmatrix} 1\,0\,0 \cdots 0\,0 \\ 0\,0\,0 \cdots 0\,1 \end{bmatrix}^T, & \text{for PPD design.} \end{cases}$$

$$A = \begin{bmatrix} -k & k & 0 & 0 & \cdots & \cdots & \cdots & 0 \\ k & -2k & k & 0 & 0 & \cdots & \cdots & 0 \\ 0 & k & -2k & k & 0 & \cdots & \cdots & 0 \\ \vdots & \ddots & \ddots & \ddots & \ddots & \ddots & \cdots & \vdots \\ 0 & \cdots & \cdots & \cdots & k & -2k & k & 0 \\ 0 & \cdots & \cdots & \cdots & \cdots & k & -2k & k \\ 0 & \cdots & \cdots & \cdots & \cdots & \cdots & k & -k \end{bmatrix},$$

$$\overrightarrow{u}(t) = \begin{cases} \left[u_1^T(t)\ u_2^T(t)\ \ldots\ u_n^T(t)\right]^T, & \text{for DD design,} \\ \left[u_1^T(t)\ u_n^T(t)\right]^T, & \text{for PPD design.} \end{cases}$$

Note that the position and velocity can be easily measured by sensors in the train with a certain sampling period. Assume that the sampling period is T. The measurement output and the signal to be estimated of system (7.3) are modeled by

$$\begin{cases} \overrightarrow{y}(kT) = \tilde{C}\overrightarrow{x}(kT) + \omega(kT) \\ \overrightarrow{z}(kT) = \tilde{M}\overrightarrow{x}(kT) \end{cases} \tag{7.4}$$

where $\overrightarrow{y}(kT)$, $\overrightarrow{z}(kT)$ and $\omega(kT)$ ($k \in \mathbb{N}^+$) denote, respectively, the measurement output, the signal to be estimated and the measurement noise of the kth sampling instant. $\tilde{C} = \text{diag}\{\tilde{c}_1 I, \tilde{c}_2 I\}$ is the weight matrix of the measurement data. The constants $\tilde{c}_1$ and $\tilde{c}_2$ represent the weights of the position and velocity, respectively. $\tilde{M}$ is a known matrix with appropriate dimensions. Without loss of generality, it is assumed that $\omega(kT)$ is the zero-mean Gaussian white noise with covariance $\Omega \triangleq \bar{\Omega}^T\bar{\Omega}$.

For technical convenience, we now transform the system (7.3) into a discrete-time model according the sampling period T. Suppose that $\tilde{u}(k) = \overrightarrow{u}(kT)$ and $\tilde{f}_r(k) = \overrightarrow{f}_r(kT)$ for $kT \leq t < (k+1)T$. It follows from (7.3) and (7.4) that

$$\begin{cases} \tilde{x}(k+1) = \tilde{A}\tilde{x}(k) + \tilde{B}\tilde{u}(k) + \tilde{E}\tilde{f}_r(k) + \tilde{H}\overrightarrow{h} \\ \tilde{y}(k) = \tilde{C}\tilde{x}(k) + \tilde{\omega}(k) \\ \tilde{z}(k) = \tilde{M}\tilde{x}(k) \end{cases} \tag{7.5}$$

where

$$\tilde{x}(k) = \vec{x}(kT),\ \tilde{u}(k) = \vec{u}(kT),\ \tilde{f}_r(k) = \vec{f}_r(kT),\ \tilde{y}(k) = \vec{y}(kT),\ \tilde{z}(k) = \vec{z}(kT),$$

$$\tilde{\omega}(k) = \omega(kT),\ \tilde{A} = e^{\vec{A}T},\ \tilde{B} = \int_0^T e^{\vec{A}t}dt\vec{B},\ \tilde{H} = \int_0^T e^{\vec{A}t}dt\mathscr{M},\ \tilde{E} = \int_0^T e^{\vec{A}t}dt\vec{E}.$$

7.1.3 *Signal Transmissions over the Train–Ground Communication*

As shown in Sect. 7.1.1, the signal transmissions over the train–ground communication are implemented through the "unreliable" wireless communication. In this section, we would like to consider the effects induced by the train–ground communication. Without loss of generality, let us consider the case that the communication between each WAP and trains in its coverage area is scheduled by certain protocol according to the IEEE 802.11p standard. As show in [5], the IEEE 802.11p is a draft amendment to the IEEE 802.11 standards and has been applied to fast changing vehicular networks including the train–ground communication in CBTC systems. Furthermore, IEEE 802.11p could be modeled as the p-persistent CSMA protocol [6]. In such a protocol, *only one train* is permitted to communicate with the corresponding WAP at each transmission instant. Each train would transmit the signal with the probability p $(0 < p \leq 1)$ if the communication channel is sensed idle. Otherwise, the train waits until next available transmission instant. For the sake of simplicity, we assume that the measurement output is transmitted at each sampling instant. Then, the transmission model of the system (7.5) subject to the p-persistent CSMA protocol is given by

$$\bar{y}(k) = \alpha(k)\tilde{y}(k) \tag{7.6}$$

where $\bar{y}(k)$ represents the measurement signal received by the ZC via WLANs and $\alpha(k) \in \{0, 1\}$ is a Bernoulli distributed stochastic variable indicating whether the communication channel is idle at present. We assumed that $\alpha(k)$ is unrelated to the measurement noise $\tilde{\omega}(k)$. The probability distribution of $\alpha(k)$ can be calculated as follows: [7]:

$$\begin{cases} \text{Prob}\{\alpha(k) = 1\} = p(1-p)^{N(k)-1}, \\ \text{Prob}\{\alpha(k) = 0\} = 1 - p(1-p)^{N(k)-1}, \end{cases} \tag{7.7}$$

where $N(k)$ is the number of trains in the coverage area of certain WAP at time k. Obviously, $N(k)$ is a bounded integer variable due to the fact that trains must keep a distance far enough between each other in order to guarantee safety.

Assuming that $\underline{N} \leq N(k) \leq \overline{N}$ where $\overline{N}$ and $\underline{N}$ are known constant positive integers, it follows from (7.7) that

$$\begin{cases} \bar{\alpha} \triangleq \mathbb{E}\{\alpha(k)\} = \bar{p} + \Delta_1(k) \\ \mathbb{E}\left\{(\alpha(k) - \bar{\alpha})^2\right\} = \hat{p} + \Delta_2(k) \end{cases} \tag{7.8}$$

where

$$\begin{aligned} \bar{p} =& \frac{1}{2}\Big(p(1-p)^{\underline{N}-1} + p(1-p)^{\overline{N}-1}\Big), \ \ \Delta_1(k) = p(1-p)^{N(k)-1} - \bar{p}, \\ \hat{p} =& \frac{1}{2}\Big(p(1-p)^{\underline{N}-1}\big(1-p(1-p)^{\underline{N}-1}\big) + p(1-p)^{\overline{N}-1}\big(1-p(1-p)^{\overline{N}-1}\big)\Big), \\ \Delta_2(k) =& p(1-p)^{N(k)-1}\left(1-p(1-p)^{N(k)-1}\right) - \hat{p}. \end{aligned}$$

It is easy to verify that

$$\begin{cases} |\Delta_1(k)| \leq \bar{\Delta}_p \\ |\Delta_2(k)| \leq \hat{\Delta}_p \end{cases} \tag{7.9}$$

where

$$\begin{aligned} \bar{\Delta}_p =& \frac{p(1-p)^{\underline{N}-1} - p(1-p)^{\overline{N}-1}}{2} \\ \hat{\Delta}_p =& \frac{1}{2}\Big(p(1-p)^{\underline{N}-1}\big(1-p(1-p)^{\underline{N}-1}\big) - p(1-p)^{\overline{N}-1}\big(1-p(1-p)^{\overline{N}-1}\big)\Big). \end{aligned}$$

Remark 7.3 It is worth mentioning that the identity information of each train is transmitted to the ZC simultaneously in the transmission. As such, the value of the stochastic variable $\alpha(k)$ is available to the ZC, but the exact number of trains $N(k)$ in the coverage area of a WAP could not be obtained.

7.1.4 Structure of the Filter

Consider the following filter for the discrete-time stochastic system (7.5) with the CSMA protocol scheduling (7.6):

$$\begin{cases} \hat{x}(k+1) = \tilde{A}\hat{x}(k) + \tilde{B}\tilde{u}(k) + \tilde{H}\overrightarrow{h} + K\left(\bar{y}(k) - \alpha(k)\tilde{C}\hat{x}(k)\right) \\ \hat{z}(k) = \tilde{M}\hat{x}(k) \end{cases} \tag{7.10}$$

where K is the filter parameter to be designed.

Let the filtering error be $e(k) \triangleq x(k) - \hat{x}(k)$ and the output filtering error be $\tilde{z}(k) \triangleq z(k) - \hat{z}(k)$. Then, the filtering error dynamics for the discrete-time system (7.5) is obtained as follows:

$$\begin{cases} e(k+1) = \big(\tilde{A} - \alpha(k)K\tilde{C}\big)e(k) - \alpha(k)K\tilde{\omega}(k) + \tilde{E}\tilde{f}_r(k) \\ \qquad\quad = \big(\tilde{A} - (\bar{p} + \Delta_1(k) + \tilde{\alpha}(k))K\tilde{C}\big)e(k) + \tilde{E}\tilde{f}_r(k) \\ \qquad\qquad - \big(\bar{p} + \Delta_1(k) + \tilde{\alpha}(k)\big)K\tilde{\omega}(k) \\ \tilde{z}(k) = \tilde{M}e(k) \end{cases} \tag{7.11}$$

where $\tilde{\alpha}(k) = \alpha(k) - \bar{\alpha}(k)$.

We are now in a position to state the main goal of this chapter. We are interested in investigating the ultimate boundedness of the filtering error dynamics and designing the filter gain matrix according to two performance optimization problems. More specifically, our objectives in this chapter are to:

(1) derive sufficient conditions for the dynamical system (7.11) under which the ultimate boundedness is guaranteed for the filtering error $e(k)$ in mean square;
(2) design the filter gain matrix K in order to obtain the minimization of the ultimate bound and the fastest decay rate of the output filtering error.

7.2 Main Results

In this section, the ultimate boundedness is analyzed for the filtering error $e(k)$ in mean square.

7.2.1 Ultimate Boundedness Analysis of the Filtering Error

The following theorem provides a sufficient condition under which the dynamics of the filtering error is exponentially ultimately bounded in mean square.

Theorem 7.1 *Let the filter parameter K be given. Assume that there exist a positive definite matrices $P \in \mathbb{R}^{2n\times 2n}$, three positive scalars λ_1, λ_2 and γ satisfying*

$$\bar{\Gamma} \triangleq \begin{bmatrix} \bar{\Gamma}_{11} & \bar{\Gamma}_{12} & \bar{\Gamma}_{13} \\ * & \bar{\Gamma}_{22} & \bar{\Gamma}_{23} \\ * & * & \bar{\Gamma}_{33} \end{bmatrix} < 0 \tag{7.12}$$

where

$$\begin{aligned}
&\bar{\Gamma}_{11} = (\tilde{A} - \bar{p}K\tilde{C})^T P(\tilde{A} - \bar{p}K\tilde{C}) - (1-\gamma)P + \bar{\Delta}_p^2\lambda_1 I + (\hat{p} + \hat{\Delta}_p))\tilde{C}^T K^T PK\tilde{C},\\
&\bar{\Gamma}_{12} = -(\tilde{A} - \bar{p}K\tilde{C})^T PK\tilde{C}, \quad \bar{\Gamma}_{13} = (\tilde{A} - \bar{p}K\tilde{C})^T P\tilde{E}, \quad \bar{\Gamma}_{22} = \tilde{C}^T K^T PK\tilde{C} - \lambda_1 I,\\
&\bar{\Gamma}_{23} = -\tilde{C}^T K^T P\tilde{E}, \quad \bar{\Gamma}_{33} = \tilde{E}^T P\tilde{E} - \lambda_2 I.
\end{aligned}$$

Then, the dynamics of the filtering error $e(k)$ is ultimately bounded in mean square subject to the measurement noise $\tilde{\omega}(k)$ and the resistive force $\tilde{f}_r(k)$.

Proof To analyze the ultimate boundedness of the filtering error $e(k)$, we choose the following Lyapunov-like function:

$$V(k) = e^T(k)Pe(k). \tag{7.13}$$

The corresponding difference of $V(k)$ along the trajectory of system (7.11) can be calculated as follows:

$$\begin{aligned}
&\Delta V(k) \triangleq e^T(k+1)Pe(k+1) - e^T(k)Pe(k)\\
&= \Big((\tilde{A} - \bar{p}K\tilde{C})e(k) - \Delta_1(k)K\tilde{C}e(k) - \tilde{\alpha}(k)K\big(\tilde{C}e(k) + \tilde{\omega}(k)\big) - \big(\bar{p} + \Delta_1(k)\big)K\tilde{\omega}(k)\\
&\quad + \tilde{E}\tilde{f}_r(k)\Big)^T P\Big((\tilde{A} - \bar{p}K\tilde{C})e(k) - \Delta_1(k)K\tilde{C}e(k) - \tilde{\alpha}(k)K\big(\tilde{C}e(k) + \tilde{\omega}(k)\big) + \tilde{E}\tilde{f}_r(k)\\
&\quad - \big(\bar{p} + \Delta_1(k)\big)K\tilde{\omega}(k)\Big) - e^T(k)Pe(k).
\end{aligned} \tag{7.14}$$

Since $\mathbb{E}\{\tilde{\omega}(k)\} = 0$ and $\mathbb{E}\{\tilde{\omega}(k)\tilde{\omega}^T(k)\} = \Omega$, it follows from (7.14) that

$$\begin{aligned}
\mathbb{E}\{\Delta V(k)\} =& \mathbb{E}\Big\{e^T(k)\Big((\tilde{A} - \bar{p}K\tilde{C})^T P(\tilde{A} - \bar{p}K\tilde{C}) - P\Big)e(k) + \Delta_1^2(k)e^T(k)\tilde{C}^T K^T\\
&\times PK\tilde{C}e(k) - 2\Delta_1(k)e^T(k)(\tilde{A} - \bar{p}K\tilde{C})^T PK\tilde{C}e(k) + (\hat{p} + \Delta_2(k))\\
&\times (e^T(k)\tilde{C}^T K^T PK\tilde{C}e(k) + \tilde{\omega}^T(k)K^T PK\tilde{\omega}(k)) + \tilde{f}_r^T(k)\tilde{E}^T P\tilde{E}\tilde{f}_r(k)\\
&+ (\bar{p} + \Delta_1(k))^2\tilde{\omega}^T(k)K^T PK\tilde{\omega}(k) + 2e^T(k)(\tilde{A} - \bar{p}K\tilde{C})^T P\tilde{E}\tilde{f}_r(k)\\
&- 2\Delta_1(k)e^T(k)\tilde{C}^T K^T P\tilde{E}\tilde{f}_r(k)\Big\}\\
=& \mathbb{E}\{\vartheta^T(k)\Gamma\vartheta(k)\} + ((\bar{p} + \Delta_1(k))^2 + \hat{p} + \Delta_2(k))\mathrm{tr}\{\bar{\Omega}^T K^T PK\bar{\Omega}\}
\end{aligned} \tag{7.15}$$

where

$$\begin{aligned}
&\Gamma_{11} = (\tilde{A} - \bar{p}K\tilde{C})^T P(\tilde{A} - \bar{p}K\tilde{C}) - P + (\hat{p} + \Delta_2(k))\tilde{C}^T K^T PK\tilde{C},\\
&\Gamma_{12} = -(\tilde{A} - \bar{p}K\tilde{C})^T PK\tilde{C}, \ \Gamma_{13} = (\tilde{A} - \bar{p}K\tilde{C})^T P\tilde{E}, \ \Gamma_{22} = \tilde{C}^T K^T PK\tilde{C},
\end{aligned}$$

$$\Gamma_{23} = -\tilde{C}^T K^T P \tilde{E},\ \Gamma_{33} = \tilde{E}^T P \tilde{E},\ \vartheta(k) = \begin{bmatrix} e(k) \\ \Delta_1(k)e(k) \\ \tilde{f}_r(k) \end{bmatrix},\ \Gamma = \begin{bmatrix} \Gamma_{11} & \Gamma_{12} & \Gamma_{13} \\ * & \Gamma_{22} & \Gamma_{23} \\ * & * & \Gamma_{33} \end{bmatrix}.$$

Noting that $|\Delta_1(k)| \leq \bar{\Delta}_p$ and $|\Delta_2(k)| \leq \hat{\Delta}_p$, it can be derived from (7.15) that

$$\mathbb{E}\{\Delta V(k)\} \leq \mathbb{E}\{\vartheta^T(k)\Gamma\vartheta(k)\} + \varepsilon \text{tr}\{\bar{\Omega}^T K^T P K \bar{\Omega}\} \tag{7.16}$$

where $\varepsilon = p(1-p)^{\underline{N}-1}$. Moreover, one can infer from (7.16) that

$$\begin{aligned}\mathbb{E}\{\Delta V(k)\} \leq & \mathbb{E}\{\vartheta^T(k)\Gamma\vartheta(k)\} + \varepsilon \text{tr}\{\bar{\Omega}^T K^T P K \bar{\Omega}\} - \lambda_2 \|\tilde{f}_r(k)\|^2 + \lambda_2 \|\tilde{f}_r(k)\|^2 \\ & + \lambda_1 \Delta_1^2(k)\|e(k)\|^2 - \lambda_1 \Delta_1^2(k)\|e(k)\|^2 + \varepsilon \text{tr}\{\bar{\Omega}^T K^T P K \bar{\Omega}\} \\ \leq & -\gamma \mathbb{E}\{V(k)\} + \bar{\varepsilon}\end{aligned} \tag{7.17}$$

where $\bar{\varepsilon} = \varepsilon \text{tr}\{\bar{\Omega}^T K^T P K \bar{\Omega}\} + \lambda_2 \sum_{i=1}^{n} \delta_i^2$. Hence, for any scalar $\rho > 0$, it follows that

$$\begin{aligned}&\mathbb{E}\{\rho^{k+1} V(k+1)\} - \mathbb{E}\{\rho^k V(k)\} \\ &= \rho^{k+1}\big(\mathbb{E}\{V(k+1)\} - \mathbb{E}\{V(k)\}\big) + \rho^k(\rho - 1)\mathbb{E}\{V(k)\} \\ &\leq \rho^k (\rho - 1 - \rho\gamma)\, \mathbb{E}\{V(k)\} + \rho^{k+1}\bar{\varepsilon}.\end{aligned} \tag{7.18}$$

It is obvious that $0 < \gamma < 1$. Letting $\rho = \rho_0 = \frac{1}{1-\gamma}$ and summing up both sides of (7.18) from 0 to $\kappa - 1$ with respect to κ, we obtain

$$\mathbb{E}\{\rho_0^\kappa V(t)\} - \mathbb{E}\{V(0)\} \leq \frac{\rho_0(1-\rho_0^\kappa)}{1-\rho_0}\bar{\varepsilon} \tag{7.19}$$

which implies that

$$\begin{aligned}\mathbb{E}\{V(\kappa)\} &\leq \rho_0^{-\kappa}\left(\mathbb{E}\{V(0)\} + \frac{\rho_0}{1-\rho_0}\bar{\varepsilon}\right) - \frac{\rho_0}{1-\rho_0}\bar{\varepsilon} \\ &= (1-\gamma)^\kappa \left(\mathbb{E}\{V(0)\} - \frac{\bar{\varepsilon}}{\gamma}\right) + \frac{\bar{\varepsilon}}{\gamma}.\end{aligned} \tag{7.20}$$

Finally, it follows readily from Definition 2.1 that the error dynamical system (7.11) is exponentially ultimately bounded in mean square where the asymptotic upper bound of the filtering error can be computed as $\frac{\bar{\varepsilon}}{\gamma\lambda_{\min}\{P\}}$. The proof is complete.

Remark 7.4 In Theorem 7.1, a sufficient condition has been derived under which the filtering error $e(k)$ is exponentially ultimately bounded in mean square subject to the measurement noise $\tilde{\omega}(k)$ and the resistive force $\tilde{f}_r(k)$. It is worth mentioning that a guaranteed ultimate bound of the filtering error dynamics can be associated with a kind of filtering performance about the "attenuation" against the effect of perturbations. On the other hand, it can be seen from (7.12) that the decay rate of the filtering error (i.e. $1-\gamma$) should be large enough to guarantee the ultimate boundedness of $e(k)$. From a practical point of view, the decay rate and the ultimate bound in mean square can be regarded as two important performance indices for the filtering performance.

7.2.2 Optimization Problems

In this subsection, we focus our attention on the design problem of the filter gain matrix k by solving two optimization problems.

Optimization Problem 7.1 Minimization of the ultimate bound in mean square of the filtering error dynamics for most accurate filtering performance.

Corollary 7.1 *For system (7.11), let a scalar $1 > \theta > 0$ be given. Suppose that there exist two positive definite matrices $P \in \mathbb{R}^{2n\times 2n}$, $Q \in \mathbb{R}^{2n\times 2n}$, two positive scalars λ_1, λ_2, and a matrix $\mathscr{K} \in \mathbb{R}^{2n\times 2n}$ satisfying*

$$\Xi = \begin{bmatrix} \Xi_{11} & 0 & 0 & \Xi_{14} & \Xi_{15} \\ * & \Xi_{22} & 0 & \Xi_{24} & 0 \\ * & * & \Xi_{33} & \Xi_{34} & 0 \\ * & * & * & \Xi_{44} & 0 \\ * & * & * & * & \Xi_{55} \end{bmatrix} < 0 \tag{7.21}$$

$$\begin{bmatrix} -Q & \bar{\Omega}^T \mathscr{K}^T \\ * & -P \end{bmatrix} < 0 \tag{7.22}$$

$$P \geq \tilde{M}^T \tilde{M} \tag{7.23}$$

where

$$\Xi_{11} = \bar{\Delta}_p^2 \lambda_1 I - \theta P, \quad \Xi_{14} = \tilde{A}^T P - \bar{p}\tilde{C}^T \mathscr{K}^T, \quad \Xi_{15} = \tilde{C}^T \mathscr{K}^T, \quad \Xi_{22} = -\lambda_1 I,$$
$$\Xi_{24} = -\tilde{C}^T \mathscr{K}^T, \quad \Xi_{33} = -\lambda_2 I, \quad \Xi_{34} = \tilde{E}^T P, \quad \Xi_{44} = -P,$$
$$\Xi_{55} = -\frac{P}{\hat{p} + \hat{\Delta}_p}, \quad \varepsilon = p(1-p)^{\underline{N}-1}.$$

Then, the dynamics of the filtering error $e(k)$ is ultimately bounded in mean square subject to the measurement noise $\tilde{\omega}(k)$ and the resistive force $\tilde{f}_r(k)$. Furthermore, the decay rate of the filtering error is less than θ, and the minimum of the asymptotic upper bound of $\mathbb{E}\{\|e(k)\|^2\}$ can be derived by solving the following minimization problem

$$\min\left\{\varepsilon \mathrm{tr}\{Q\} + \lambda_2 \sum_{i=1}^{n} \delta_i^2\right\} \tag{7.24}$$

subject to the matrix inequality constraints (7.21)–(7.23). The desired filter gain can be obtained by

$$K = P^{-1}\mathscr{K}. \tag{7.25}$$

Proof Based on the Schur Complement Lemma and the desired filter gain in (7.25), it can be concluded that the inequality (7.12) holds with the condition $\gamma = 1 - \theta$ if and only if the inequality (7.21) is satisfied. Hence, it follows from Theorem 7.1 that the dynamics of the filtering error $e(k)$ is ultimately bounded in mean square.

On the other hand, along similar lines as the proof of Theorem 7.1, it can be derived that

$$\mathbb{E}\{V(k)\} \leq \theta^k \left(\mathbb{E}\{V(0)\} - \frac{\bar{\varepsilon}}{1-\theta}\right) + \frac{\bar{\varepsilon}}{1-\theta} \tag{7.26}$$

where

$$\bar{\varepsilon} = \varepsilon \mathrm{tr}\{\bar{\Omega}^T \mathscr{K}^T P^{-1} \mathscr{K} \bar{\Omega}\} + \lambda_2 \sum_{i=1}^{n} \delta_i^2.$$

Moreover, one can infer from (7.22) that

$$\bar{\Omega}^T \mathscr{K}^T P^{-1} \mathscr{K} \bar{\Omega} \leq Q. \tag{7.27}$$

Therefore, it follows from (7.26) and (7.27) that

$$\mathbb{E}\{V(k)\} \leq \theta^k \mathbb{E}\{V(0)\} - \frac{\theta^k \bar{\varepsilon}}{1-\theta} + \frac{\varepsilon \mathrm{tr}\{Q\} + \lambda_2 \sum_{i=1}^{n} \delta_i^2}{1-\theta}. \tag{7.28}$$

Since $\|\tilde{z}(k)\|^2 \leq V(k)$, we have

$$\mathbb{E}\{\|\tilde{z}(k)\|^2\} \leq \mathbb{E}\{V(k)\} \leq \theta^k \left(\mathbb{E}\{V(0)\} - \frac{\bar{\varepsilon}}{1-\theta}\right) + \frac{\varepsilon \mathrm{tr}\{Q\} + \lambda_2 \sum_{i=1}^{n} \delta_i^2}{1-\theta}. \tag{7.29}$$

As such, we can conclude that the asymptotic bound of $\mathbb{E}\{\|e(k)\|^2\}$ is

$$\frac{\varepsilon \mathrm{tr}\{Q\} + \lambda_2 \sum_{i=1}^{n} \delta_i^2}{1-\theta}$$

and the minimum of this asymptotic bound can be derived by minimizing

$$\varepsilon \mathrm{tr}\{Q\} + \lambda_2 \sum_{i=1}^{n} \delta_i^2$$

which is equivalent to (7.24). The proof is complete.

Optimization Problem 7.2 Optimization of the decay rate of the filtering error dynamics for fastest convergence performance.

Corollary 7.2 *For system (7.11), let a scalar $\varpi > 0$ be given. Suppose that a non-singular matrix $R \in \mathbb{R}^{2n\times 2n}$, four positive scalars $\lambda_1 > 0$, $\lambda_2 > 0$, $\bar{\gamma} > 1$, $\rho > 0$, and a matrix $\mathscr{K} \in \mathbb{R}^{2n\times 2n}$ satisfying*

$$\Theta = \begin{bmatrix} \Theta_{11} & 0 & 0 & \Theta_{14} & \Theta_{15} \\ * & \Theta_{22} & 0 & \Theta_{24} & 0 \\ * & * & \Theta_{33} & \Theta_{34} & 0 \\ * & * & * & \Theta_{44} & 0 \\ * & * & * & * & \Theta_{55} \end{bmatrix} < 0 \tag{7.30}$$

$$\begin{bmatrix} -\varpi(\bar{\gamma}-1) + \lambda_2 \sum_{i=1}^{n} \delta_i^2 & \Upsilon \\ * & -\varepsilon^{-1} I \end{bmatrix} < 0 \tag{7.31}$$

$$\begin{bmatrix} -R - R^T + \rho I & \bar{\gamma}\tilde{M}^T \\ * & -\rho I \end{bmatrix} \le 0 \tag{7.32}$$

where

$$\Theta_{11} = -R - R^T + \bar{\gamma} I + \bar{\Delta}_p^2 \lambda_1 I,\ \Theta_{14} = \tilde{A}^T R^T - \bar{p}\tilde{C}^T \mathscr{K}^T,\ \Theta_{15} = \tilde{C}^T \mathscr{K}^T,$$
$$\Theta_{22} = -\lambda_1 I,\ \Theta_{24} = -\tilde{C}^T \mathscr{K}^T,\ \Theta_{33} = -\lambda_2 I,\ \Theta_{34} = \tilde{E}^T R^T,\ \Theta_{44} = -I,$$
$$\Theta_{55} = -\frac{I}{\hat{p} + \hat{\Delta}_p},\ \Lambda_i = \left[\hbar(i-1)\ \hbar(i-2)\ \ldots\ \hbar(i-2n)\right]^T,$$
$$\Upsilon = \left[\Lambda_1^T \bar{\Omega}^T \mathscr{K}^T\ \Lambda_2^T \bar{\Omega}^T \mathscr{K}^T\ \ldots\ \Lambda_{2n}^T \bar{\Omega}^T \mathscr{K}^T\right],\ \varepsilon = p(1-p)^{\underline{N}-1},$$

and $\hbar(\cdot) \in \{0, 1\}$ is the Kronecker delta function. Then, the dynamics of the output filtering error $\tilde{z}(k)$ is exponentially ultimately bounded in mean square with the decay rate $\bar{\gamma}^{-1}$ subject to the measurement noise $\tilde{\omega}(k)$ and the resistive force $\tilde{f}_r(k)$. Furthermore, the upper bound of $\mathbb{E}\{\|\tilde{z}(k)\|^2\}$ is less than ϖ and the optimum of the decay rate can be derived by solving the following maximization problem

$$\max\{\bar{\gamma}\} \tag{7.33}$$

subject to the matrix inequality constraints (7.30)–(7.32). The desired filter gain can be obtained by

$$K = R^{-1}\mathscr{K}. \tag{7.34}$$

Proof Choose the following Lyapunov-like function:

$$V(k) = e^T(k)R^T Re(k). \tag{7.35}$$

Along the similar lines as the proof of Theorem 7.1, it can be derived from (7.34) that

$$\mathbb{E}\{\Delta V(k)\} = \mathbb{E}\{\vartheta^T(k)\bar{\Theta}\vartheta(k)\} - \left(1 - \frac{1}{\bar{\gamma}}\right)\mathbb{E}\{V(k)\} + \hat{\varepsilon} \tag{7.36}$$

where

$$\begin{aligned}
\bar{\Theta} &= \text{diag}\left\{-\bar{\gamma}^{-1}R^T R + \bar{\Delta}_p^2\lambda_1 I, -\lambda_1 I, -\lambda_2 I\right\} \\
&\quad + \begin{bmatrix}\Theta_{14}\\ \Theta_{24}\\ \Theta_{34}\end{bmatrix}\begin{bmatrix}\Theta_{14}\\ \Theta_{24}\\ \Theta_{34}\end{bmatrix}^T + \frac{1}{\hat{p} + \hat{\Delta}_p}\begin{bmatrix}\Theta_{15}\\ 0\\ 0\end{bmatrix}\begin{bmatrix}\Theta_{15}\\ 0\\ 0\end{bmatrix}^T \\
\hat{\varepsilon} &= \varepsilon\text{tr}\left\{\bar{\Omega}^T\mathscr{K}^T\mathscr{K}\bar{\Omega}\right\} + \lambda_2\sum_{i=1}^{n}\delta_i^2 \\
&= \varepsilon\sum_{i=1}^{2n}\Lambda_i^T\bar{\Omega}^T\mathscr{K}^T\mathscr{K}\bar{\Omega}\Lambda_i + \lambda_2\sum_{i=1}^{n}\delta_i^2 \\
&= \varepsilon\Upsilon\Upsilon^T + \lambda_2\sum_{i=1}^{n}\delta_i^2.
\end{aligned}$$

Since

$$-\bar{\gamma}^{-1}R^T R \le -R^T - R + \bar{\gamma}I,$$

it follows from the Schur Complement Lemma, (7.30) and (7.36) that $\bar{\Theta} < 0$, which implies that

$$\mathbb{E}\{\Delta V(k)\} \le -\left(1 - \frac{1}{\bar{\gamma}}\right)\mathbb{E}\{V(k)\} + \hat{\varepsilon}. \tag{7.37}$$

Then, it can be concluded from Theorem 7.1 that the dynamics of the filtering error $e(k)$ is ultimately bounded in mean square.

On the other hand, one can infer from (7.32) that

$$R^T R \ge \rho\left(R + R^T - \rho I\right) \ge \bar{\gamma}^2\tilde{M}^T\tilde{M}. \tag{7.38}$$

Hence, we have

$$\mathbb{E}\{\|\tilde{z}(k)\|^2\} \leq \mathbb{E}\{V(k)\} \leq \bar{\gamma}^{-k-2}\left(E\{V(0)\} - \frac{\bar{\gamma}}{\bar{\gamma}-1}\right) + \frac{\hat{\varepsilon}}{\bar{\gamma}-1}. \tag{7.39}$$

According to the inequality (7.31) that

$$-\varpi(\bar{\gamma}-1) + \lambda_2 \sum_{i=1}^{n} \delta_i^2 + \varepsilon \Upsilon \Upsilon^T < 0$$

which implies that the upper bound of $\mathbb{E}\{\|\tilde{z}(k)\|^2\}$ is less than ϖ. Furthermore, the decay rate of the filtering error in mean square is $\bar{\gamma}^{-1}$. Therefore, the minimization of the decay rate can be derived by maximizing $\bar{\gamma}$ which is equivalent to the condition (7.33). The proof is complete.

Remark 7.5 In Corollaries 7.1 and 7.2, the design problem of the filter for the CBTC system is considered and two kinds of optimization problems are studied (i.e. optimization problems for the ultimate bound and the decay rate of the filtering error in mean square). The main result established in Theorem 7.1, Corollaries 7.1 and 7.2 contains all the information about the system parameters, the p-persistent CSMA protocol, the measurement noise and the bound of the resistive force.

7.3 An Illustrative Example

To verify the effectiveness of the proposed filtering schemes, an illustrative numerical example is conducted based on a high-speed train with 6 cars. The parameters are given in Table 7.1, which are chosen from the experimental results of the Japanese Shinkansen high-speed train [4]. The parameters of the CSMA protocol are selected to be $p = 0.5$, $\underline{N} = 1$ and $\overline{N} = 3$.

The matrices $\tilde{C}$ and $\tilde{M}$ are selected to be $\tilde{C} = \text{diag}\{10^{-3}I, 3.6I\}$ and $\tilde{M} = \text{diag}\{0.01I, 0.1I\}$. Let the sampling period and the covariance of the noise $\omega(k)$ be 0.1 s and $\Omega = \text{diag}\left\{10^{-6}I, 3.6^2 I\right\}$, respectively. The corresponding control commands are presented in Table 7.2.

Now, let us consider Optimization Problem 7.1 with the given decay rate $\theta = 0.98$. By applying Corollary 7.1, the minimization problem (7.24) subject to the inequality constraints (7.21)–(7.23) can be solved and the filter gain matrix is given in the following box:

Using the MATLAB software, the simulation results about the filtering issue are shown in Figs. 7.4, 7.5, 7.6 and 7.7, where Figs. 7.4 and 7.6 plot the position tracking trajectories and velocity tracking trajectories, respectively. Figures 7.5 and 7.7 show the position tracking errors and velocity tracking errors of the train, respectively.

Conventionally, the information about the position and velocity is always measured by the combination of balises and speed odometers. The ZC would receive such

Table 7.1 Parameters of the high-speed train

Symbol	Value	Unit
m_1, m_6	80×10^3	kg
m_2, m_3, m_4, m_5	40×10^3	kg
k	80×10^3	$\mathrm{N} \cdot \mathrm{m}^{-1}$
ϑ	1	m
l	20	m
T	0.1	s

Table 7.2 Control commands of the high-speed train

Commands	Samples (0.1 s)	Traction force(N)
Command 1	$0 \rightarrow 3000$	$\tilde{u}_1(k) = \tilde{u}_6(k) = 2.8 \times 10^4$ $\tilde{u}_2(k) = \tilde{u}_3(k) = \tilde{u}_4(k)$ $= \tilde{u}_5(k) = 1.4 \times 10^4$
Command 2	$3000 \rightarrow 6250$	$\tilde{u}_1(k) = \tilde{u}_6(k) = 2.48 \times 10^4$ $\tilde{u}_2(k) = \tilde{u}_3(k) = \tilde{u}_4(k)$ $= \tilde{u}_5(k) = 1.24 \times 10^4$
Command 3	$6250 \rightarrow 9920$	$\tilde{u}_1(k) = \tilde{u}_6(k) = -0.28 \times 10^4$ $\tilde{u}_2(k) = \tilde{u}_3(k) = \tilde{u}_4(k)$ $= \tilde{u}_5(k) = -0.14 \times 10^4$

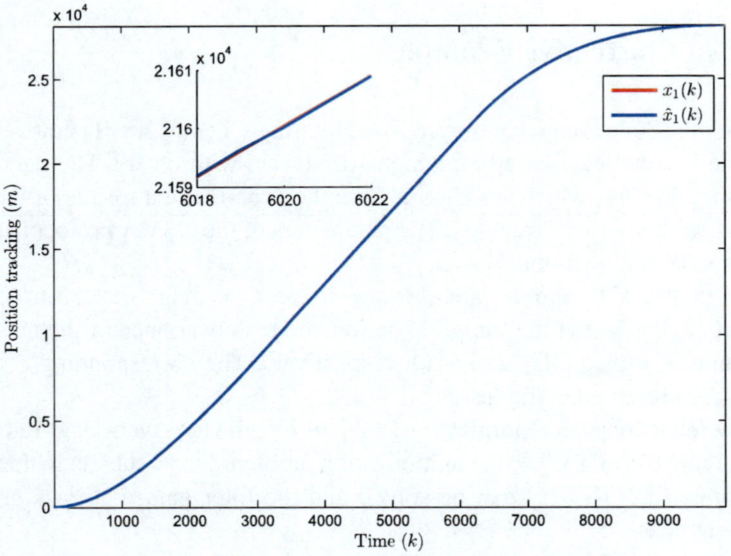

Fig. 7.4 The position tracking trajectories ($\tilde{x}_1(k)$ and $\hat{x}_1(k)$)

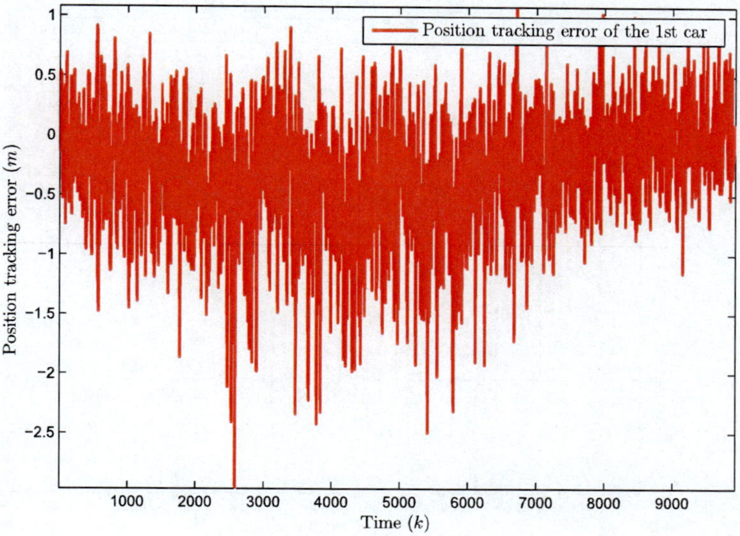

Fig. 7.5 The position tracking error $(e_1(k))$

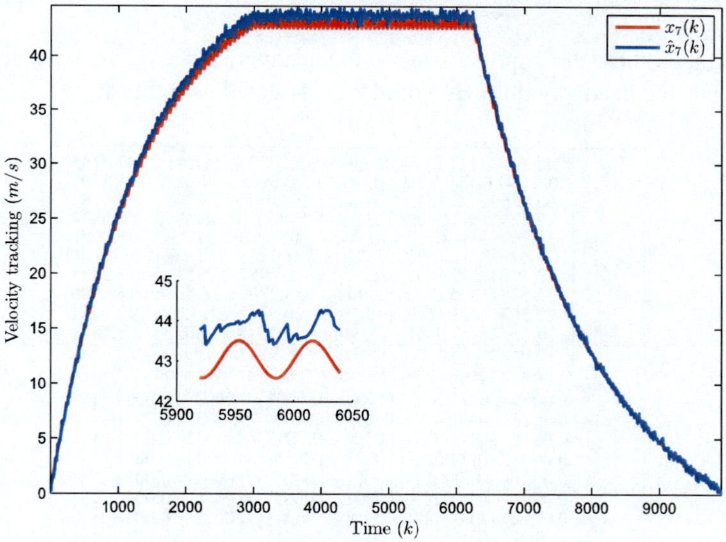

Fig. 7.6 The velocity tracking trajectories $\tilde{x}_7(k)$ and $\hat{x}_7(k)$)

information via network-based communication. The tracking errors associated with the direct measurements and our proposed filtering scheme are shown in Table 7.3. Obviously, the filtering scheme presented in this chapter could improve the accuracy of the information about the position and velocity.

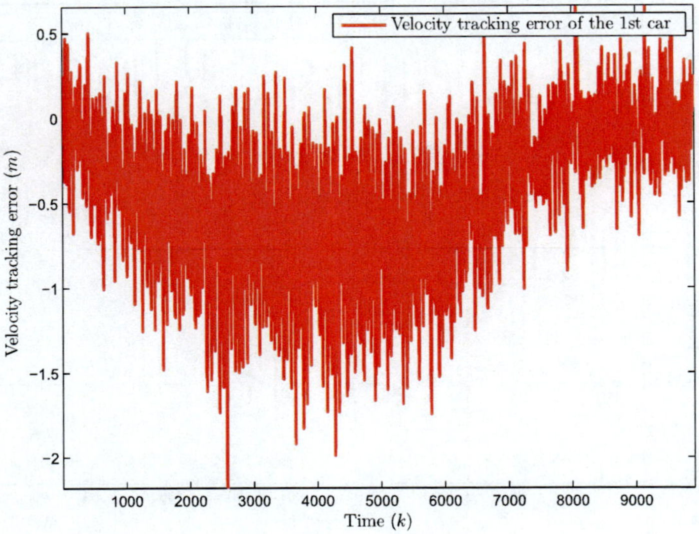

Fig. 7.7 The velocity tracking error ($e_7(k)$)

Next, let's show the improvement on the headway time. As we have discussed in Sect. 7.1.1, the headway time H_T could be calculated as follows:

$$K = \begin{bmatrix} 185.6246 & 111.0990 & 81.3980 & 55.0893 & 51.2714 & 37.4185 \\ 106.7411 & 157.1306 & 105.4820 & 70.3985 & 54.0893 & 31.2701 \\ 60.6869 & 98.4843 & 159.3407 & 103.1487 & 68.2454 & 37.5287 \\ 37.5243 & 68.2405 & 103.1544 & 159.3371 & 98.4892 & 60.6898 \\ 31.2678 & 54.0902 & 70.4012 & 105.4870 & 157.1327 & 106.7408 \\ 37.4227 & 51.2692 & 55.1015 & 81.3948 & 111.1046 & 185.6200 \\ 13.1934 & 25.6005 & 20.5542 & 19.4957 & 15.5023 & 12.5453 \\ 32.6444 & -5.0637 & 25.6867 & 15.3862 & 18.5095 & 16.0706 \\ 25.3634 & 23.1113 & -3.5958 & 20.5772 & 21.0055 & 19.0684 \\ 19.0572 & 21.0461 & 20.5142 & -3.5314 & 23.0700 & 25.3801 \\ 16.0650 & 18.5500 & 15.3193 & 25.7547 & -5.1082 & 32.6582 \\ 12.5558 & 15.4636 & 19.5560 & 20.4897 & 25.6408 & 13.1819 \\ 0.0056 & 0.0015 & 0.0005 & -0.0001 & 0.0002 & 0.0007 \\ 0.0012 & 0.0064 & 0.0005 & 0.0000 & -0.0001 & 0.0001 \\ 0.0011 & -0.0003 & 0.0059 & -0.0003 & -0.0002 & 0.0007 \\ 0.0007 & -0.0002 & -0.0003 & 0.0059 & -0.0003 & 0.0011 \\ 0.0001 & -0.0001 & 0.0000 & 0.0005 & 0.0064 & 0.0012 \\ 0.0007 & 0.0002 & -0.0001 & 0.0005 & 0.0015 & 0.0056 \\ 0.0262 & -0.0068 & -0.0018 & -0.0003 & 0.0023 & 0.0041 \\ -0.0099 & 0.0554 & -0.0085 & -0.0041 & -0.0019 & -0.0036 \\ -0.0085 & -0.0073 & 0.0569 & -0.0059 & -0.0036 & -0.0065 \\ -0.0065 & -0.0036 & -0.0059 & 0.0569 & -0.0073 & -0.0085 \\ -0.0036 & -0.0019 & -0.0041 & -0.0085 & 0.0554 & -0.0099 \\ 0.0041 & 0.0023 & -0.0003 & -0.0018 & -0.0068 & 0.0262 \end{bmatrix}.$$

$$H_T = \frac{L_{train} + L_{error} + L_{reaction} + L_{brake} + L_{safe}}{v_{\max}}.$$

The safety margin is selected from [3] as follows: $L_{safe} = 50$. The brake distance is given by $L_{brake} = \frac{v_{\max}^2}{2b}$ where b is the deceleration rate. $L_{reaction}$ is the train

Table 7.3 Accuracy comparison results

	Tracking errors ($\sum_{k=0}^{9920} \|\tilde{z}(k)\|^2$)
Direct measurements	1705.6
Filtering scheme	263.1801

Table 7.4 Headway time comparison results

Maximum speed $v_{\max}$ (km/h)	64.857	81.071	97.285	113.499
H_T (Direct measurements)	40.309	37.004	35.003	34.124
H_T (Filtering scheme)	33.309	31.362	30.049	29.113

running distance during the reaction time of onboard system. A reasonable values of $L_{reaction}$ is given by $L_{reaction} = v_{\max} T_{reaction}$ where $T_{reaction}$ denotes the reaction time of onboard system. Assume the reaction time of onboard system is 1 s (i.e. $T_{reaction} = 1$ s).

The headway time with the direct measurements and our proposed filtering scheme is shown in Table 7.4. It can be observed from Table 7.4 that our proposed filtering scheme could significantly reduce headway time H_T.

7.4 Summary

This chapter has addressed the filtering problem for a train with n cars linked by couples. The dynamics of the train has been modeled by a continuous-time system and then reformulated as a discrete-time system. Considering the data transmission of the train–ground communication, the transmission behavior of the train subject to the p-persistent CSMA protocol has been described by a Bernoulli distributed sequence whose probability distribution is dependent on the number of competing nodes (trains). The corresponding filter has then been developed to generate the estimate of states. A sufficient condition has been derived to guarantee the ultimate boundedness of filtering error in mean square. In order to calculate the desired filter gains under different requirements, two optimization problems have been solved to obtain the minimum of the asymptotic bound and the decay rate, respectively. A simulation example has been used to illustrate the effectiveness of the proposed design method.

References

1. Pascoe, R.D., Eichorn, T.N.: What is communication-based train control? IEEE Veh. Technol. Mag. **4**(4), 16–21 (2009)
2. Zhu, L., Yu, F.R., Ning, B., Tang, T.: Cross-layer handoff design in MIMO-enabled WLANs for communication-based train control (CBTC) systems. IEEE J. Sel. Areas Commun. **30**(4), 719–728 (2012)
3. Li, W., Tang, T., Liang, D.: Impact of wireless communication delay on train headway of communication based train control system. In: Proceedings of the 19th International Conference on Intelligent Transportation Systems, Nov. 1–4, Brazil, pp. 360–365 (2016)
4. Yang, C., Sun, Y.: Mixed $\mathscr{H}_2/\mathscr{H}_\infty$ cruise controller design for high speed train. Int. J. Control **74**(9), 905–920 (2001)
5. Zhu, L., Yu, F.R., Ning, B.: A seamless handoff scheme for train-ground communication systems in CBTC. In: Proceedings of the 2010 IEEE 72nd Vehicular Technology Conference Fall, Ottawa, pp. 1–5 (2010)
6. Wang, Y., Ahmed, A., Krishnamachari, B., Psounis, K.: IEEE 802.11p performance evaluation and protocol enhancement. In: Proceedings of the 2008 IEEE International Conference on Vehicular Electronics and Safety Columbus, OH, USA, pp. 317–322 (2008)
7. Tabbara, M., Nešić, D.: Input-output stability of networked control systems with stochastic protocols and channels. IEEE Trans. Autom. Control **53**(5), 1160–1175 (2008)

Chapter 8
Observer-Based $\mathscr{H}_\infty$ Control of Time-Varying Systems with Random Access Protocol

In this chapter, we focus our attention on the control problem of networked systems subject to the effects induced by the communication protocol. As we have discussed in Chap. 3, almost all practical systems are subject to time-varying parameters variations (e.g. the temperature fluctuation, operating point shifting, graduate aging of system components, etc.) in reality, which motivates researchers to investigate the transient dynamics of time-varying systems over a finite horizon. Finite-horizon $\mathscr{H}_\infty$ is one of the most studied control issues which has received considerable research attention in the past two decades due mainly to the insight of their engineering applications.

In this chapter, let us consider the $\mathscr{H}_\infty$ control problem for a class of linear time-varying networked systems subject to the RA protocol scheduling. Two communication networks (i.e. the sensor-to-controller network and the controller-to-actuator network) are employed to implement the signal transmissions between the plant and the controller. In order to prevent the data from collisions during transmission, only one sensor and one actuator are permitted to get access to the communication networks at each transmission instant. The aim of the problem addressed is to design an observer-based controller such that the $\mathscr{H}_\infty$ performance of the closed-loop system is guaranteed over a given finite horizon.

The rest of this chapter is organized as follows. In Sect. 8.1, a networked control system (NCS) with time-varying parameters and two communication networks are introduced and the problem under consideration is formulated. In Sect. 8.2, the design problem of the observer-based controller is solved in terms of the solution to two coupled backward recursive difference equations (RDEs). Furthermore, two numerical simulation examples are given in Sect. 8.3 to illustrate the effectiveness of the controller design scheme. Finally, we conclude this chapter in Sect. 8.4.

L. Zou et al., *Communication-Protocol-Based Filtering and Control of Networked Systems*, Studies in Systems, Decision and Control 430,
https://doi.org/10.1007/978-3-030-97512-8_8

8.1 Problem Formulation

8.1.1 Plant and Network-Based Communication

Consider a NCS with two communication networks shown in Fig. 8.1. The signal transmission is implemented between the plant and remote controller via two communication networks: the S/C (sensor-to-controller) network and the C/A (controller-to-actuator) network. The plant is a discrete time-varying system defined on the finite horizon ($k \in [0, N-1]$) of the form

$$\begin{cases} x(k+1) = A(k)x(k) + B(k)\bar{u}(k) + D(k)\nu(k) \\ y(k) = C(k)x(k) + E(k)\nu(k) \\ z(k) = M(k)x(k) \end{cases} \tag{8.1}$$

where $x(k) \in \mathbb{R}^{n_x}$, $\bar{u}(k) \in \mathbb{R}^{n_u}$, $y(k) \in \mathbb{R}^{n_y}$ and $z(k) \in \mathbb{R}^{n_z}$ denote, respectively, the state vector, the control input after transmitted through the C/A network, the measurement output before transmitted through the S/C network and the signal to be controlled. $\nu(k) \in l_2([0, N-1]; \mathbb{R}^{n_\nu})$ is the disturbance input. $A(k)$, $B(k)$, $C(k)$, $D(k)$, $E(k)$ and $M(k)$ are known time-varying matrices with appropriate dimensions.

Let us now discuss the effect induced by communication networks (i.e. S/C network and C/A network). In this chapter, it is assumed that there is no packet dropout occurring during the data transmissions through communication networks. For technical analysis, we first write

$$y(k) = \left[y_1^T(k)\ y_2^T(k)\ \ldots\ y_{n_y}^T(k)\right]^T,\ \bar{u}(k) = \left[\bar{u}_1^T(k)\ \bar{u}_2^T(k)\ \ldots\ \bar{u}_{n_u}^T(k)\right]^T.$$

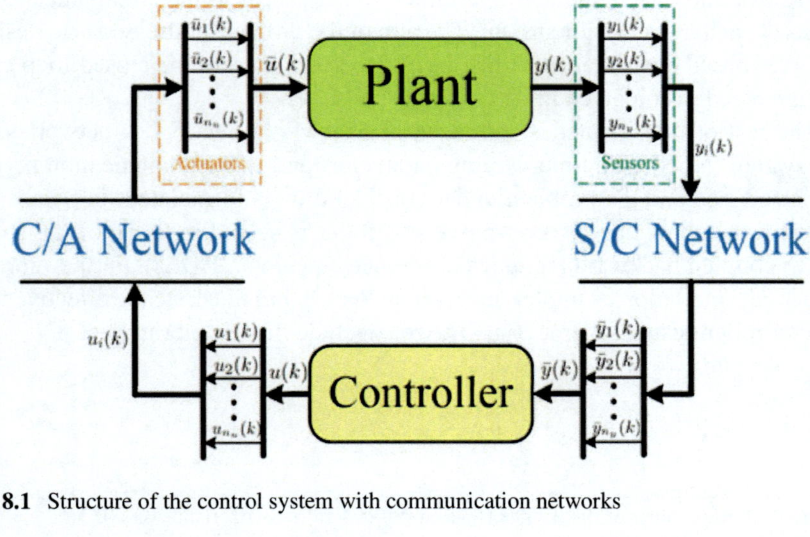

Fig. 8.1 Structure of the control system with communication networks

where $y_i(k)$ is the measurement of the ith sensor, $\bar{u}_i(k)$ is the input signal of the jth actuator.

In this chapter, the RA Protocol is utilized to determine which sensor (or actuator) obtains access to the S/C network (C/A network). According to the RA Protocol scheduling, only one sensor is allowed to get access to the S/C network and only one actuator obtains access to the C/A network. For the sake of examining the influence of the RA Protocol constraints, let $\alpha(k) \in \{1, 2, \ldots, n_y\}$ ($\beta(k) \in \{1, 2, \ldots, n_u\}$) denote the selected sensor (actuator) obtaining access to the S/C network (C/A network) at the time instant k, where $\beta(k)$ and $\alpha(k)$ are independent of each other. In this chapter, we consider the case that the RA Protocol scheduling behavior is modeled by a Markov chain. As we discussed in Sect. 3.1.3, under the effects of RA Protocol scheduling, $\beta(k)$ and $\alpha(k)$ can be governed by Markov chains with the transition probability matrices $\mathscr{P}_1(k)$ and $\mathscr{P}_2(k)$, respectively. The transition probability matrices $\mathscr{P}_1(k) = \left[p_1^{ij}(k)\right]_{n_y \times n_y}$ and $\mathscr{P}_2(k) = \left[p_2^{ij}(k)\right]_{n_u \times n_u}$ are defined as follows:

$$\begin{cases} p_1^{ij}(k) \triangleq \text{Prob}\left(\beta(k+1) = j | \beta(k) = i\right) \\ p_2^{ij}(k) \triangleq \text{Prob}\left(\alpha(k+1) = j | \alpha(k) = i\right) \end{cases} \tag{8.2}$$

Let $\bar{y}(k) \triangleq \left[\bar{y}_1^T(k)\ \bar{y}_2^T(k)\ \ldots\ \bar{y}_{n_y}^T(k)\right]^T \in \mathbb{R}^{n_y}$ be the measurement output after transmitted through S/C network and $u(k) \triangleq \left[u_1^T(k)\ u_2^T(k)\ \ldots\ u_{n_u}^T(k)\right] \in \mathbb{R}^{n_u}$ be the control input before transmitted through C/A network. Considering the ZOH strategy, the updating rule for $\bar{y}_i(k)$ $(i = 1, 2, \ldots, n_y)$ and $\bar{u}_i(k)$ $(i = 1, 2, \ldots, n_u)$ subject to the RA Protocol is set to be

$$\bar{y}_i(k) = \begin{cases} y_i(k) + E_{1,i}\omega(k), & \text{if } i = \alpha(k) \\ \bar{y}_i(k-1), & \text{otherwise} \end{cases} \tag{8.3}$$

$$\bar{u}_i(k) = \begin{cases} u_i(k) + E_{2,i}\omega(k), & \text{if } i = \beta(k) \\ \bar{u}_i(k-1), & \text{otherwise} \end{cases} \tag{8.4}$$

where $\omega(k) \in l_2\left([0, N-1]; \mathbb{R}^{n_\omega}\right)$ is the network-induced disturbance, and $E_{1,i}(k) \in \mathbb{R}^{1 \times n_\omega}$ $(i = 1, 2, \ldots, n_y)$ and $E_{2,j}(k) \in \mathbb{R}^{1 \times n_\omega}$ $(j = 1, 2, \ldots, n_u)$ are known, real, time-varying matrices.

$$\begin{aligned} E_1(k) &\triangleq \left[E_{1,1}^T(k)\ E_{1,2}^T(k)\ \ldots\ E_{1,n_y}^T(k)\right]^T \\ E_2(k) &\triangleq \left[E_{2,1}^T(k)\ E_{2,2}^T(k)\ \ldots\ E_{2,n_u}^T(k)\right]^T \end{aligned} \tag{8.5}$$

we have

$$\begin{cases} \bar{y}(k) = \Phi_{\alpha(k)}^y\big(y(k) + E_1(k)\omega(k)\big) + (I - \Phi_{\alpha(k)}^y)\bar{y}(k-1) \\ \bar{u}(k) = \Phi_{\beta(k)}^u\big(u(k) + E_2(k)\omega(k)\big) + (I - \Phi_{\beta(k)}^u)\bar{u}(k-1) \end{cases} \tag{8.6}$$

where

$$\begin{cases} \Phi^y_{\alpha(k)} = \mathrm{diag}\{\bar{\delta}^1_{\alpha(k)}, \bar{\delta}^2_{\alpha(k)}, \ldots, \bar{\delta}^{n_y}_{\alpha(k)}\} \\ \Phi^u_{\beta(k)} = \mathrm{diag}\{\bar{\delta}^1_{\beta(k)}, \bar{\delta}^2_{\beta(k)}, \ldots, \bar{\delta}^{n_u}_{\beta(k)}\} \end{cases}$$

and $\bar{\delta}^b_a \triangleq \delta(a-b)$ in which $\delta(\cdot)$ is the Kronecker delta function.

Denoting $\tilde{x}(k) \triangleq \left[x^T(k)\ \bar{y}^T(k-1)\ \bar{u}^T(k-1)\right]^T$, the time-varying system (8.1) with the RA Protocol scheduling can be reformulated as follows:

$$\begin{cases} \tilde{x}(k+1) = \bar{A}_{\alpha(k),\beta(k)}(k)\tilde{x}(k) + \bar{B}_{\beta(k)}(k)u(k) + \bar{D}_{\alpha(k),\beta(k)}(k)\bar{\nu}(k) \\ \bar{y}(k) = \bar{C}_{\alpha(k)}(k)\tilde{x}(k) + \bar{E}_{\alpha(k)}(k)\bar{\nu}(k) \\ z(k) = \bar{M}(k)\tilde{x}(k) \end{cases} \tag{8.7}$$

where

$$\bar{A}_{\alpha(k),\beta(k)}(k) = \begin{bmatrix} A(k) & 0 & B(k)\left(I - \Phi^u_{\beta(k)}\right) \\ \Phi^y_{\alpha(k)}C(k) & I - \Phi^y_{\alpha(k)} & 0 \\ 0 & 0 & I - \Phi^u_{\beta(k)} \end{bmatrix},$$

$$\bar{B}_{\beta(k)}(k) = \left[\left(\Phi^u_{\beta(k)}\right)^T B^T(k)\ 0\ \left(\Phi^u_{\beta(k)}\right)^T\right]^T,\ \bar{\nu}(k) = \left[\nu^T(k)\ \omega^T(k)\right]^T,$$

$$\bar{D}_{\alpha(k),\beta(k)}(k) = \begin{bmatrix} D(k) & B(k)\Phi^u_{\beta(k)}E_2(k) \\ \Phi^y_{\alpha(k)}E(k) & \Phi^y_{\alpha(k)}E_1(k) \\ 0 & \Phi^u_{\beta(k)}E_2(k) \end{bmatrix},\ \bar{M}(k) = \left[M(k)\ 0\ 0\right],$$

$$\bar{E}_{\beta(k)}(k) = \left[\Phi^y_{\alpha(k)}E(k)\ \Phi^y_{\alpha(k)}E_1(k)\right],\ \bar{C}_{\alpha(k)}(k) = \left[\Phi^y_{\alpha(k)}C(k)\ I - \Phi^y_{\alpha(k)}\ 0\right].$$

For analysis convenience, we now reformulate the system (8.7) by mapping the two stochastic processes $\beta(k)$ and $\alpha(k)$ to one Markov chain. The following proposition can be easily accessible from Lemma 1 of [1].

Proposition 8.1 *The Markov chains $\beta(k)$ and $\alpha(k)$ of the system (8.7) can be mapped to the sequence $r(k) \in \mathscr{R} \triangleq \{1, 2, \ldots, n_y n_u\}$ by the following mapping $\Theta(\cdot,\cdot)$:*

$$r(k) = \Theta(\beta(k), \alpha(k)) \triangleq \beta(k) + (\alpha(k) - 1)n_u. \tag{8.8}$$

Moreover, if $r(k)$ is given, the values of $\beta(k)$ and $\alpha(k)$ can be derived by $\phi_1(r(k))$ and $\phi_2(r(k))$:

$$\begin{cases} \alpha(k) = \phi_1(r(k)) \triangleq \left\lfloor \dfrac{r(k)-1}{n_u} \right\rfloor + 1 \\ \beta(k) = \phi_2(r(k)) \triangleq \mathrm{mod}(r(k)-1, n_u) + 1. \end{cases} \tag{8.9}$$

Obviously, there is a *one-to-one correspondence* between the variable $r(k)$ and the pair $(\beta(k), \alpha(k))$. According to the mapping described by (8.8), the transition probability matrix $\bar{\mathscr{P}} = \left[\bar{p}_{ij}(k)\right]_{n_u n_y \times n_u n_y}$ of the Markov chain $r(k)$ is obtained as follows:

$$\begin{aligned}
&\bar{p}_{ij}(k) = \text{Prob}(r(k+1) = j | i = r(k)) \\
&= \text{Prob}\big(\alpha(k+1) = \phi_1(j)\big|\alpha(k) = \phi_1(i)\big)\text{Prob}\big(\beta(k+1) = \phi_2(j)\big|\beta(k) = \phi_2(i)\big) \\
&= p_1^{\phi_2(i)\phi_2(j)}(k)\, p_2^{\phi_1(i)\phi_1(j)}(k)
\end{aligned} \tag{8.10}$$

where $p_1^{ij}(k)$ and $p_2^{ij}(k)$ have been defined in (8.2). Based on Proposition 8.1, the augmented system (8.7) can be rewritten as follows:

$$\begin{cases}
\tilde{x}(k+1) = \overrightarrow{A}_{r(k)}(k)\tilde{x}(k) + \overrightarrow{B}_{r(k)}(k)u(k) + \overrightarrow{D}_{r(k)}(k)\bar{\nu}(k) \\
\bar{y}(k) = \overrightarrow{C}_{r(k)}(k)\tilde{x}(k) + \overrightarrow{E}_{r(k)}(k)\bar{\nu}(k) \\
z(k) = \bar{M}(k)\tilde{x}(k)
\end{cases} \tag{8.11}$$

where

$$\begin{aligned}
&\overrightarrow{A}_{r(k)}(k) = \bar{A}_{\phi_1(r(k)),\phi_2(r(k))}(k), \quad \overrightarrow{B}_{r(k)}(k) = \bar{B}_{\phi_2(r(k))}(k), \quad \overrightarrow{C}_{r(k)}(k) = \bar{C}_{\phi_1(r(k))}(k), \\
&\overrightarrow{D}_{r(k)}(k) = \bar{D}_{\phi_1(r(k)),\phi_2(r(k))}(k), \quad \overrightarrow{E}_{r(k)}(k) = \bar{E}_{\phi_1(r(k))}(k).
\end{aligned}$$

8.1.2 Observer-Based Controller

The observer-based control scheme for the system (8.11) is described by

$$\begin{cases}
\hat{x}(k+1) = \overrightarrow{A}_{r(k)}(k)\hat{x}(k) + \overrightarrow{B}_{r(k)}(k)K_{r(k)}(k)\hat{x}(k) \\
\qquad\qquad + L_{r(k)}(k)\left(\bar{y}(k) - \overrightarrow{C}_{r(k)}(k)\hat{x}(k)\right) \\
u(k) = K_{r(k)}(k)\hat{x}(k)
\end{cases} \tag{8.12}$$

where $\hat{x}(k) \in \mathbb{R}^n$ is the state estimate of the system (8.11) with $n = n_x + n_y + n_u$, and the time-varying matrices $K_{r(k)}(k)$ and $L_{r(k)}(k)$ are controller parameters to be designed. Let the observer error be $e(k) = \tilde{x}(k) - \hat{x}(k)$. Then, the error dynamics can be obtained from (8.11) to (8.12) as follows:

$$e(k+1) = \big(\overrightarrow{A}_{r(k)}(k) - L_{r(k)}(k)\overrightarrow{C}_{r(k)}(k)\big)e(k)\big(\overrightarrow{D}_{r(k)}(k) - L_{r(k)}(k)\overrightarrow{E}_{r(k)}(k)\big)\bar{\nu}(k). \tag{8.13}$$

By defining variable $\eta(k) = \left[\tilde{x}^T(k) \; e^T(k) \right]^T$, we obtain the closed-loop system as follows:

$$\begin{cases} \eta(k+1) = \mathscr{A}_{r(k)}(k)\eta(k) + \mathscr{D}_{r(k)}(k)\bar{\nu}(k) \\ z(k) = \mathscr{M}(k)\eta(k) \end{cases} \tag{8.14}$$

where

$$\mathscr{A}_{r(k)}(k) = \begin{bmatrix} \mathcal{A}_{11}(k) & \mathcal{A}_{12}(k) \\ 0 & \mathcal{A}_{22}(k) \end{bmatrix}, \quad \mathscr{D}_{r(k)}(k) = \begin{bmatrix} \mathcal{D}_{11}(k) \\ \mathcal{D}_{21}(k) \end{bmatrix}, \quad \mathscr{M}(k) = \left[\bar{M}(k) \; 0 \right],$$
$$\mathcal{A}_{11}(k) = \overrightarrow{A}_{r(k)}(k) + \overrightarrow{B}_{r(k)}(k)K_{r(k)}(k), \; \mathcal{A}_{12}(k) = -\overrightarrow{B}_{r(k)}(k)K_{r(k)}(k),$$
$$\mathcal{A}_{22}(k) = \overrightarrow{A}_{r(k)}(k) - L_{r(k)}(k)\overrightarrow{C}_{r(k)}(k), \; \mathcal{D}_{11}(k) = \overrightarrow{D}_{r(k)}(k),$$
$$\mathcal{D}_{21}(k) = \overrightarrow{D}_{r(k)}(k) - L_{r(k)}(k)\overrightarrow{E}_{r(k)}(k).$$

We are now in the position to state the problem addressed in this chapter as follows. We aim to design appropriate controller parameters $K_{r(k)}(k)$ and $L_{r(k)}(k)$ such that, for the given positive scalar γ, the closed-loop system (8.14) satisfies the following $\mathscr{H}_\infty$ performance requirement:

$$J \triangleq \mathbb{E}\left\{ \sum_{k=0}^{N-1} \left(\|z(k)\|^2 - \gamma^2 \|\bar{\nu}(k)\|^2 \right) \right\} - \gamma^2 \eta^T(0) W \eta(0) \leq 0, \; \forall (\bar{\nu}(k), \eta(0)) \neq 0 \tag{8.15}$$

where W is a given positive definite matrix.

Remark 8.1 In (8.15), an $\mathscr{H}_\infty$ performance index is proposed in the sense of mathematical expectation. Roughly speaking, such an index can better describe the weighted average $\mathscr{H}_\infty$ performance of the closed-loop system over the possible modes with individual weights (transition probabilities), and is, therefore, particularly suitable when the overall control performance becomes a concern. Also, such an index could facilitate the subsequent mathematical analysis and enhance the feasibility of the controller design problem. On the other hand, one could also ask for a certain $\mathscr{H}_\infty$ performance requirement to be met for all possible modes, and this might be useful for considering the worst-case performance at the cost of increasing the computational complexity.

8.2 Main Results

Lemma 8.1 ([2]) *Let $\mathscr{U}$, $\mathscr{V}$ and $\mathscr{W}$ be known nonzero matrices with appropriate dimensions. The solution X to $\min_X \|\mathscr{U} X \mathscr{W} - \mathscr{V}\|_F$ is $\mathscr{U}^\dagger \mathscr{V} \mathscr{W}^\dagger$.*

Lemma 8.2 *Consider the system (8.1) with the RA Protocol constraints governed by the Markov chain $r(k)$ whose transition probability matrix is $\bar{\mathscr{P}}$ defined in (8.10). With the updating rule (8.6), let the controller parameters $\{K_i(k)\}_{0\le k\le N-1}$, $\{L_i(k)\}_{0\le k\le N-1}$, the disturbance attenuation level $\gamma > 0$ and the positive definite matrix W be given. For any disturbances $\{\omega(k)\}_{0\le k\le N-1}$ and $\{\nu(k)\}_{0\le k\le N-1}$, the augmented system (8.14) satisfies the $\mathscr{H}_\infty$ performance requirement if there exist a family of non-negative definite matrices $P_i(k)$ ($i \in \mathscr{R}$, $0 \le k \le N-1$) (with final condition $P_i(N) = 0$) satisfying the following backward recursive difference equations (RDEs)*

$$
\begin{aligned}
P_i(k) = {} & \mathscr{A}_i^T(k)\bar{P}_i(k+1)\mathscr{A}_i(k) + \mathscr{M}^T(k)\mathscr{M}(k) \\
& + \mathscr{A}_i^T(k)\bar{P}_i(k+1)\mathscr{D}_i(k)\Delta_i^{-1}(k)\mathscr{D}_i^T(k)\bar{P}_i(k+1)\mathscr{A}_i(k)
\end{aligned} \tag{8.16}
$$

subject to

$$
\begin{cases}
\Delta_i(k) = \gamma^2 I - \mathscr{D}_i^T(k)\bar{P}_i(k+1)\mathscr{D}_i(k) > 0 \\
P_i(0) \le \gamma^2 W
\end{cases} \tag{8.17}
$$

where

$$
\bar{P}_i(k+1) = \sum_{j\in\mathscr{R}} \bar{p}_{ij}(k)P_j(k+1) \tag{8.18}
$$

and the nonempty finite set $\mathscr{R}$ is defined in Proposition 8.1.

Proof By defining

$$
Y_{r(k)}(k) \triangleq \mathbb{E}\big\{\eta^T(k+1)P_{r(k+1)}(k+1)\eta(k+1) - \eta^T(k)P_i(k)\eta(k)|i = r(k)\big\} \tag{8.19}
$$

and noticing (8.14), we have

$$
\begin{aligned}
Y_{r(k)}(k) = {} & \mathbb{E}\Big\{\big(\mathscr{A}_i(k)\eta(k) + \mathscr{D}_i(k)\bar{\nu}(k)\big)^T \bar{P}_i(k+1)\big(\mathscr{A}_i(k)\eta(k) + \mathscr{D}_i(k)\bar{\nu}(k)\big) \\
& - \eta^T(k)P_i(k)\eta(k)|i = r(k)\Big\} \\
= {} & \mathbb{E}\Big\{\eta^T(k)\Big(\mathscr{A}_i^T(k)\bar{P}_i(k+1)\mathscr{A}_i(k) - P_i(k)\Big)\eta(k) + 2\eta^T(k)\mathscr{A}_i^T(k)\bar{P}_i(k+1) \\
& \times \mathscr{D}_i(k)\bar{\nu}(k) + \bar{\nu}^T(k)\mathscr{D}_i^T(k)\bar{P}_i(k+1)\mathscr{D}_i(k)\bar{\nu}(k)|i = r(k)\Big\}.
\end{aligned} \tag{8.20}
$$

Adding the following zero term

$$
\|z(k)\|^2 - \gamma^2\|\bar{\nu}(k)\|^2 - \big(\|z(k)\|^2 - \gamma^2\|\bar{\nu}(k)\|^2\big) \tag{8.21}
$$

to the right-hand side of (8.20) and then taking the mathematical expectation, we have

$$\begin{aligned}Y_{r(k)}(k) =& \mathbb{E}\Big\{\gamma^2\|\bar{\nu}(k)\|^2 + \eta^T(k)\big(\mathscr{M}^T(k)\mathscr{M}(k) + \mathscr{A}_i^T(k)\bar{P}_i(k+1)\mathscr{A}_i(k)\\&- P_i(k)\big)\eta(k) - \|z(k)\|^2 + 2\eta^T(k)\mathscr{A}_i^T(k)\bar{P}_i(k+1)\mathscr{D}_i(k)\bar{\nu}(k) - \bar{\nu}^T(k)\big(\gamma^2 I\\&- \mathscr{D}_i^T(k)\bar{P}_i(k+1)\mathscr{D}_i(k)\big)\bar{\nu}(k)|i = r(k)\Big\}.\end{aligned} \tag{8.22}$$

Applying the completing squares method results in

$$\begin{aligned}Y_{r(k)}(k) =& \mathbb{E}\Big\{\eta^T(k)\big(\mathscr{A}_i^T(k)\bar{P}_i(k+1)\mathscr{A}_i(k) - P_i(k) + \mathscr{M}^T(k)\mathscr{M}(k)\big)\eta(k)\\&+ (\bar{\nu}^*(k))^T\Delta_i(k)\bar{\nu}^*(k) - \big(\bar{\nu}(k) - \bar{\nu}^*(k)\big)^T\Delta_i(k)\big(\bar{\nu}(k) - \bar{\nu}^*(k)\big)|i = r(k)\Big\}\\&- \mathbb{E}\Big\{\|z(k)\|^2 - \gamma^2\|\bar{\nu}(k)\|^2\Big\}\end{aligned} \tag{8.23}$$

where

$$\bar{\nu}^*(k) = \Delta_{r(k)}^{-1}(k)\mathscr{D}_{r(k)}^T(k)\bar{P}_{r(k)}(k+1)\mathscr{A}_{r(k)}(k)\eta(k).$$

Taking the sum on both sides of (8.23) from 0 to $N-1$, we obtain

$$\begin{aligned}&\mathbb{E}\{\eta^T(N)P_{r(N)}(N)\eta(N) - \eta^T(0)P_{r(0)}(0)\eta(0)\}\\&= \mathbb{E}\Big\{-\sum_{k=0}^{N-1}\big(\bar{\nu}(k) - \bar{\nu}^*(k)\big)^T\Delta_{r(k)}(k)\big(\bar{\nu}(k) - \bar{\nu}^*(k)\big)\\&\quad - \sum_{k=0}^{N-1}\big(\|z(k)\|^2 - \gamma^2\|\bar{\nu}(k)\|^2\big)\Big\}.\end{aligned} \tag{8.24}$$

Since $\Delta_i(k) > 0$, $P_i(0) \le \gamma^2 W$ and $P_i(N) = 0$, we have

$$\begin{aligned}J =& \mathbb{E}\Big\{-\sum_{k=0}^{N-1}\big(\bar{\nu}(k) - \bar{\nu}^*(k)\big)^T\Delta_{r(k)}(k)\big(\bar{\nu}(k) - \bar{\nu}^*(k)\big) + \eta^T(0)\big(P_i(0)\\&- \gamma^2 W\big)\eta(0)\Big\} \le 0\end{aligned} \tag{8.25}$$

which means the pre-specified $\mathscr{H}_\infty$ performance is satisfied. The proof of is complete.

So far, we have conducted the $\mathscr{H}_\infty$ performance analysis in terms of the solvability of a backward RDE in Lemma 8.2. In the next stage, let us propose an approach for computing the appropriate controller parameters $K_{r(k)}(k)$ and $L_{r(k)}(k)$ in each step under the worst situation, i.e. $\bar{\nu}(k) = \bar{\nu}^*(k) = \Delta_{r(k)}^{-1}(k)\mathscr{D}_{r(k)}^T(k)\bar{P}_{r(k)}(k+$

$1)\mathscr{A}_{r(k)}(k)\eta(k)$. For this purpose, we rewrite the augmented system (8.14) as follows:

$$\eta(k+1) = \left(\bar{\mathscr{A}}_{r(k)}(k) + \mathscr{D}_{r(k)}(k)\mathscr{T}_{r(k)}(k)\right)\eta(k) + \bar{\mathscr{B}}_{r(k)}(k)u(k) + \bar{\mathscr{I}}L_{r(k)}(k)\tilde{y}(k) \tag{8.26}$$

where $\bar{\mathscr{A}}_{r(k)}(k) = \text{diag}\{\overrightarrow{A}_{r(k)}(k),\ \overrightarrow{A}_{r(k)}(k)\}$, $\mathscr{T}_{r(k)}(k) = \Delta^{-1}_{r(k)}(k)\mathscr{D}^T_{r(k)}(k)\bar{P}_{r(k)}(k+1)\mathscr{A}_{r(k)}(k)$, $\bar{\mathscr{B}}_{r(k)}(k) = \left[\overrightarrow{B}^T_{r(k)}(k)\ 0\right]^T$, $\bar{\mathscr{I}} = [0\ -I]^T$, $\tilde{y}(k) = \overrightarrow{C}_{r(k)}(k)e(k)$ and $u(k)$ is defined in (8.12). Then, we define the following cost functional:

$$\bar{J}_{\bar{\nu}^*} \triangleq \mathbb{E}\left\{\sum_{k=0}^{N-1}\left(\|z(k)\|^2 + \varepsilon_1\|u(k)\|^2 + \varepsilon_2\|L_{r(k)}(k)\tilde{y}(k)\|^2\right)\right\} \tag{8.27}$$

where ε_1 and ε_2 are known constants introduced for more flexibility in the controller parameter design.

Theorem 8.1 *Consider the system (8.1) with the RA Protocol associating with (8.10) and the updating rule (8.6). Let the disturbance attenuation level $\gamma > 0$ and the positive definite matrix W be given. The augmented system (8.14) satisfies the $\mathscr{H}_\infty$ performance requirement if there exist solutions $\{(P_i(k), Q_i(k), K_i(k), L_i(k))\}_{0\le k\le N-1}$ ($i \in \mathscr{R}$) satisfying the backward RDE (8.16) as well as the following backward RDE:*

$$\begin{cases} Q_i(k) = \left(\bar{\mathscr{A}}_i(k) + \mathscr{D}_i(k)\mathscr{T}_i(k)\right)^T\bar{Q}_i(k+1)\left(\bar{\mathscr{A}}_i(k) + \mathscr{D}_i(k)\mathscr{T}_i(k)\right) + \mathscr{M}^T(k)\mathscr{M}(k) \\ \qquad + \bar{\Upsilon}(k) + \tilde{\mathscr{I}}^T K_i^T(k)\bar{\mathscr{B}}_i^T(k)\bar{Q}_i(k+1)\bar{\mathscr{I}}L_i(k)\tilde{C}_i(k) \\ \qquad + \tilde{C}_i^T(k)L_i^T(k)\bar{\mathscr{I}}^T\bar{Q}_i(k+1)\bar{\mathscr{B}}_i(k)K_i(k)\tilde{\mathscr{I}} \\ Q_i(N) = 0 \end{cases} \tag{8.28}$$

subject to

$$\begin{cases} P_i(0) \le \gamma^2 W,\ P_i(k) \ge 0,\ P_i(N) = Q_i(N) = 0 \\ \Delta_i(k) = \gamma^2 I - \mathscr{D}_i^T(k)\bar{P}_i(k+1)\mathscr{D}_i(k) > 0 \\ \bar{\Delta}_{1,i}(k) = \bar{\mathscr{B}}_i^T(k)\bar{Q}_i(k+1)\bar{\mathscr{B}}_i(k) + \varepsilon_1 I > 0 \\ \bar{\Delta}_{2,i}(k) = \bar{\mathscr{I}}^T\bar{Q}_i(k+1)\bar{\mathscr{I}} + \varepsilon_2 I > 0 \end{cases} \tag{8.29}$$

with the controller parameters given as follows:

$$K_i(k) = -\mathscr{P}_i(k)\tilde{\mathscr{I}}^\dagger,\quad L_i(k) = -\mathscr{Q}_i(k)\tilde{\mathscr{C}}_i^\dagger(k) \tag{8.30}$$

where

$$\bar{\Upsilon}(k) = -\Upsilon_1(k) - \Upsilon_2(k) + \Upsilon_3(k) + \Upsilon_4(k), \quad \bar{P}_i(k+1) = \sum_{j\in\mathscr{R}} \bar{p}_{ij}(k)P_j(k+1),$$

$$\tilde{C}_i(k) = \begin{bmatrix} 0 & \vec{C}_i(k) \end{bmatrix}, \quad \bar{Q}_i(k+1) = \sum_{j\in\mathscr{R}} \bar{p}_{ij}(k)Q_j(k+1), \quad \tilde{\mathscr{I}} = \begin{bmatrix} I & -I \end{bmatrix},$$

$$\Upsilon_1(k) = \bar{\mathscr{A}}_i^T(k)\bar{Q}_i(k+1)\bar{\mathscr{B}}_i(k)\bar{\Delta}_{1,i}^{-1}(k)\bar{\mathscr{B}}_i^T(k)\bar{Q}_i(k+1)\bar{\mathscr{A}}_i(k),$$

$$\Upsilon_2(k) = \bar{\mathscr{A}}_i^T(k)\bar{Q}_i(k+1)\bar{\mathscr{I}}\bar{\Delta}_{2,i}^{-1}(k)\bar{\mathscr{I}}^T\bar{Q}_i(k+1)\bar{\mathscr{A}}_i(k),$$

$$\begin{aligned}\Upsilon_3(k) = {} & \mathscr{T}_i^T(k)\mathscr{D}_i^T(k)\bar{Q}_i(k+1)\bar{\mathscr{B}}_i(k)K_i(k)\tilde{\mathscr{I}} + \tilde{\mathscr{I}}^T K_i^T(k)\bar{\mathscr{B}}_i^T(k)\\ & \times \bar{Q}_i(k+1)\mathscr{D}_i(k)\mathscr{T}_i(k),\end{aligned}$$

$$\begin{aligned}\Upsilon_4(k) = {} & \mathscr{T}_i^T(k)\mathscr{D}_i^T(k)\bar{Q}_i(k+1)\bar{\mathscr{I}}L_i(k)\tilde{C}_i(k) + \tilde{C}_i^T(k)L_i^T(k)\bar{\mathscr{I}}^T\\ & \times \bar{Q}_i(k+1)\mathscr{D}_i(k)\mathscr{T}_i(k),\end{aligned}$$

$$\mathscr{P}_i(k) = \bar{\Delta}_{1,i}^{-1}(k)\bar{\mathscr{B}}_i^T(k)\bar{Q}_i(k+1)\bar{\mathscr{A}}_i(k), \quad \mathscr{Q}_i(k) = \bar{\Delta}_{2,i}^{-1}(k)\bar{\mathscr{I}}^T\bar{Q}_i(k+1)\bar{\mathscr{A}}_i(k),$$

and the nonempty finite set $\mathscr{R}$ is defined in Proposition 8.1.

Proof First, it follows from Lemma 8.2 that, if there exists solutions $P_i(k)$ to (8.28) such that $\Delta_i(k) > 0$ and $P_i(0) < \gamma^2 W$, then system (8.14) achieves the pre-specified $\mathscr{H}_\infty$ performance. In this case, the worst-case disturbance can be expressed as $\bar{v}^*(k) = \mathscr{T}_{r(k)}\eta(k)$. In what follows, by employing the worst-case disturbance, we aim to provide a design scheme of the controller parameters $K_i(k)$ and $L_i(k)$. For this purpose, we define

$$\bar{Y}_{r(k)}(k) \triangleq \mathbb{E}\{\eta^T(k+1)Q_{r(k+1)}\eta(k+1) - \eta^T(k)Q_i(k)\eta(k) | i = r(k)\}. \tag{8.31}$$

Then, we have

$$\begin{aligned}\bar{Y}_{r(k)}(k) = {} & \mathbb{E}\Big\{\eta^T(k)\big(\bar{\mathscr{A}}_i(k) + \mathscr{D}_i(k)\mathscr{T}_i(k)\big)^T\bar{Q}_i(k+1)\big(\bar{\mathscr{A}}_i(k) + \mathscr{D}_i(k)\mathscr{T}_i(k)\big)\eta(k)\\ & - \eta^T(k)Q_i(k)\eta(k) + 2\eta^T(k)\big(\bar{\mathscr{A}}_i(k) + \mathscr{D}_i(k)\mathscr{T}_i(k)\big)^T\bar{Q}_i(k+1)\bar{\mathscr{B}}_i(k)u(k)\\ & + u^T(k)\bar{\mathscr{B}}_i^T(k)\bar{Q}_i(k+1)\bar{\mathscr{B}}_i(k)u(k) + 2\eta^T(k)\big(\bar{\mathscr{A}}_i(k) + \mathscr{D}_i(k)\mathscr{T}_i(k)\big)^T\\ & \times \bar{Q}_i(k+1)\bar{\mathscr{I}}L_i(k)\tilde{y}(k) + \tilde{y}^T(k)L_i^T(k)\bar{\mathscr{I}}^T\bar{Q}_i(k+1)\bar{\mathscr{I}}L_i(k)\tilde{y}(k) + 2u^T(k)\\ & \times \bar{\mathscr{B}}_i^T(k)\bar{Q}_i(k+1)\bar{\mathscr{I}}L_i(k)\tilde{y}(k) | i = r(k)\Big\}.\end{aligned} \tag{8.32}$$

Since

$$u(k) = K_{r(k)}(k)\tilde{\mathscr{I}}\eta(k), \tilde{y}(k) = \tilde{C}_{r(k)}(k)\eta(k), Q_i(N) = 0,$$

it follows from the cost function (8.27) that

$$\bar{J}_{\bar{\nu}^*} = \sum_{k=0}^{N-1} \Big(\bar{Y}_{r(k)}(k) + \mathbb{E}\Big\{ \|z(k)\|^2 + \varepsilon_1 \|u(k)\|^2 + \varepsilon_2 \|L_{r(k)}(k)\tilde{y}(k)\|^2 + \eta^T(0) Q_{r(0)}(0)\eta(0) \Big\}\Big). \tag{8.33}$$

Completing the square with respect to $u(k)$ and $\tilde{y}(k)$, it can be derived from (8.29) that

$$\begin{aligned} \bar{J}_{\bar{\nu}^*} = & \mathbb{E}\{\eta^T(0) Q_{r(0)}(0)\eta(0)\} + \sum_{k=0}^{N-1} \mathbb{E}\Big\{ \big(u(k) + u_i^*(k)\big)^T \bar{\Delta}_{1,i}(k) \big(u(k) + u_i^*(k)\big) \\ & + \big(L_i(k)\tilde{y}(k) + \tilde{y}_i^*(k)\big)^T \bar{\Delta}_{2,i}(k) \big(L_i(k)\tilde{y}(k) + \tilde{y}_i^*(k)\big) \big| i = r(k) \Big\} \end{aligned} \tag{8.34}$$

where

$$\begin{aligned} u_i^*(k) &= \bar{\Delta}_{1,i}^{-1}(k) \bar{\mathscr{B}}_i^T(k) \bar{Q}_i(k+1) \bar{\mathscr{A}}_i(k)\eta(k) = \mathscr{P}_i(k)\eta(k), \\ \tilde{y}_i^*(k) &= \bar{\Delta}_{2,i}^{-1}(k) \mathscr{I}^T \bar{Q}_i(k+1) \bar{\mathscr{A}}_i(k)\eta(k) = \mathscr{Q}_i(k)\eta(k). \end{aligned}$$

Consider the following inequality:

$$\begin{aligned} \bar{J}_{\bar{\nu}^*} = & \mathbb{E}\{\eta^T(0) Q_{r(0)}(0)\eta(0)\} + \sum_{k=0}^{N-1} \mathbb{E}\Big\{ \big(u(k) + u_i^*(k)\big)^T \bar{\Delta}_{1,i}(k) \big(u(k) + u_i^*(k)\big) \\ & + \big(L_i(k)\tilde{y}(k) + \tilde{y}_i^*(k)\big) \bar{\Delta}_{2,i}(k) \big(L_i(k)\tilde{y}(k) + \tilde{y}_i^*(k)\big) \big| i = r(k) \Big\} \\ \leq & \sum_{k=0}^{N-1} \mathbb{E}\Big\{ \|K_i(k)\tilde{\mathscr{I}} + \mathscr{P}_i(k)\|_F^2 \|\bar{\Delta}_{1,i}(k)\|_F \|\eta(k)\|^2 + \|L_i(k)\tilde{\mathscr{C}}_i(k) + \mathscr{Q}_i(k)\|_F^2 \\ & \times \|\bar{\Delta}_{2,i}(k)\|_F \|\eta(k)\|^2 \Big\} + \mathbb{E}\{\eta^T(0) Q_{r(0)}(0)\eta(0)\}. \end{aligned} \tag{8.35}$$

For the purpose of suppressing the cost function (8.27), the controller parameters $K_i(k)$ and $L_i(k)$ can be selected in each iteration backward as follows:

$$\begin{cases} K_i(k) = \arg \min\limits_{K_i^*(k)} \Big\| K_i^*(k)\tilde{\mathscr{I}} + \mathscr{P}_i(k) \Big\|_F \\ L_i(k) = \arg \min\limits_{L_i^*(k)} \Big\| L_i^*(k)\tilde{\mathscr{C}}_i(k) + \mathscr{Q}_i(k) \Big\|_F . \end{cases} \tag{8.36}$$

Then, it follows from Lemma 8.1 that (8.30) is the solution of the optimization problem (8.36). The proof is complete.

Next, we proceed to examine the impact from the RA Protocol constraints on the design of the control parameter with hope to simplify the computational algorithm. To be specific, we now study the "special" structure of the matrix $K_i(k)$ $(i \in \mathscr{R})$ according to the controller parameters given by (8.30). Firstly, let us focus our attention on the structure of $\bar{\mathscr{B}}_i(k)$. It is easy to obtain that $\bar{\mathscr{B}}_i(k) = \left[\bar{B}^T_{\phi_2(i)}(k)\ 0\right]^T = \mathcal{B}(k)\Phi^u_{\phi_2(i)}$ where $\mathcal{B}(k) = \left[B^T(k)\ 0\ I\ 0\right]^T$. Therefore, we have

$$\bar{\Delta}_{1,i}(k) = \bar{\mathscr{B}}_i^T(k)\bar{Q}_i(k+1)\bar{\mathscr{B}}_i(k) + \varepsilon_1 I = \Phi^u_{\phi_2(i)}\mathcal{B}^T(k)\bar{Q}_i(k+1)\mathcal{B}(k)\Phi^u_{\phi_2(i)} + \varepsilon_1 I. \tag{8.37}$$

Denoting $\bar{\Omega}(k) \triangleq \mathcal{B}^T(k)\bar{Q}_i(k+1)\mathcal{B}(k) \triangleq \left[\Omega_{ij}(k)\right]_{n_u \times n_u}$, it follows from (8.37) that

$$\bar{\Delta}_{1,i}(k) = \text{diag}\Big\{\underbrace{\varepsilon_1, \varepsilon_1, \ldots, \varepsilon_1}_{\phi_2(i)-1}, \bar{\varepsilon}_1, \underbrace{\varepsilon_1, \varepsilon_1, \ldots, \varepsilon_1}_{n_u-\phi_2(i)}\Big\} \tag{8.38}$$

where $\bar{\varepsilon}_1 = \varepsilon_1 + \Omega_{\phi_2(i)\phi_2(i)}(k)$. Then, $K_{r(k)}(k)$ can be rewritten as follows:

$$\begin{aligned} K_{r(k)}(k) &= -\mathscr{P}_{r(k)}(k)\tilde{\mathscr{I}}^\dagger = -\bar{\Delta}^{-1}_{1,r(k)}(k)\bar{\mathscr{B}}^T_{r(k)}(k)\bar{Q}_{r(k)}(k+1)\bar{\mathscr{A}}_{r(k)}(k)\tilde{\mathscr{I}}^\dagger \\ &= -\bar{\Delta}^{-1}_{1,r(k)}(k)\Phi^u_{\phi_2(r(k))}\mathcal{B}^T(k)\bar{Q}_{r(k)}(k+1)\bar{\mathscr{A}}_{r(k)}(k)\tilde{\mathscr{I}}^\dagger \\ &= -\Phi^u_{\beta(k)}\bar{\Delta}^{-1}_{1,r(k)}(k)\mathcal{B}^T(k)\bar{Q}_{r(k)}(k+1)\bar{\mathscr{A}}_{r(k)}(k)\tilde{\mathscr{I}}^\dagger \end{aligned} \tag{8.39}$$

where $\beta(k)$ is derived by (8.9). Subsequently, denoting

$$\begin{aligned} \mathscr{K}(k) &\triangleq \bar{\Delta}^{-1}_{1,r(k)}(k)\mathcal{B}^T(k)\bar{Q}_{r(k)}(k+1)\bar{\mathscr{A}}_{r(k)}(k)\tilde{\mathscr{I}}^\dagger \\ &\triangleq \left[\mathscr{K}_1^T(k)\ \mathscr{K}_2^T(k)\ \ldots\ \mathscr{K}_{n_u}^T(k)\right]^T, \end{aligned}$$

one has the special structure of $K_{r(k)}(k)$ as follows:

$$K_{r(k)}(k) = \Big[\underbrace{0\,0\cdots 0}_{\beta(k)-1}\ \ \mathscr{K}^T_{\beta(k)}(k)\ \ \underbrace{0\,0\cdots 0}_{n_u-\beta(k)}\Big]^T \tag{8.40}$$

where the zero entries are the reflection of the adoption of the scheduled RA Protocol.

Remark 8.2 It is worth mentioning that the special structure of $K_{r(k)}(k)$ in (8.40) is generated mainly due to the matrix $\phi^u_{\beta(k)}$ indicating the selected actuator obtaining access to the C/A network at time instant k. Under the scheduled RA Protocol, only $u_{\beta(k)}(k)$ is transmitted to the actuator and others (i.e. $u_i(k)$ $(i \neq \beta(k))$) have no effect on the update of $\bar{u}(k)$ at time instant k. In other words, only the calculation of $u_{\beta(k)}(k) = \mathscr{K}_{\beta(k)}(k)\hat{x}(k)$ is "necessary" for the controller design purpose of the time-varying system (8.1) at time instant k. Therefore, the structure of the matrix $K_{r(k)}(k)$ is reasonable for suppressing the cost function (8.27) and the zero entries in (8.40) would simplify the design algorithm to a great extent.

By means of Theorem 8.1, we can summarize the Finite-Horizon Observer-Based $\mathscr{H}_\infty$ Controller Design (FHOBCD) algorithm as follows:

Algorithm FHOBCD:

Step 1. Set $k = N - 1$, $i = r(N-1) \in \mathscr{R}$. Then $P_i(N) = 0$ and $Q_i(N) = 0$ are available.

Step 2. Calculate the matrices $\bar{\Delta}_{1,i}(k)$ and $\bar{\Delta}_{2,i}(k)$ by, (8.29) respectively. If $\bar{\Delta}_{1,i}(k)$ and $\bar{\Delta}_{2,i}(k)$ are all positive definite, then the controller parameters $K_i(k)$ and $L_i(k)$ can be solved by (8.30), and go to the next step, else jump to *Step 6*.

Step 3. Compute the matrix $\Delta_i(k)$. If $\Delta_i(k)$ is positive definite, then step to the next procedure, else jump to *Step 6*.

Step 4. Solve the backward RDEs of (8.16) and (8.28) to get $P_i(k)$ and $Q_i(k)$, respectively.

Step 5. If $k \neq 0$, set $k = k - 1$, $i = r(k) \in \mathscr{R}$ and go back to *Step 2*, else turn to the next step.

Step 6. If the condition $\{\bar{\Delta}_{r,i}(k) > 0\ (r = 1, 2),\ \Delta_i(k) > 0,\ P_i(0) \leq \gamma^2 W\}$ is not satisfied, this algorithm is infeasible. Stop.

Remark 8.3 In this chapter, the finite-horizon observer-based $\mathscr{H}_\infty$ controller is designed for a time-varying systems with the RA Protocol by solving coupled backward RDEs. Note that Lemma 8.2 and Theorem 8.1 are proved mainly by the "completing the square" technique which results in very little conservatism. It can be observed from Algorithm FHOBCD that, in the controller design procedure, all the important factors contributing to the system complexity have been reflected which include: (1) the time-varying system parameters; (2) the transition probabilities of the RA Protocol; and (3) the prescribed disturbance attenuation level.

8.3 Illustrative Examples

In this section, we present two examples to illustrate the effectiveness of the proposed $\mathscr{H}_\infty$ controller design scheme.

Example 8.1 Consider system (8.1) with the following parameters:

$$A(k) = \begin{bmatrix} 0.44 + \frac{\sin(1.2k)}{10} & -0.40 & 0.65 \\ -0.50 & 0.64 + \frac{\cos(1.2k)}{10} & 0.50 \\ 0.60 & -0.30 & -0.4 - 0.3e^{-\frac{k}{5}} \end{bmatrix}, \quad B(k) = \begin{bmatrix} 0.64 & 0.65 \\ 0.58 & 0.52 \\ 0.3 & 0.25 \end{bmatrix},$$

$$C(k) = \begin{bmatrix} 0.65 & -0.30 & -0.40 \\ 0.55 & 0.40 & -0.45 \end{bmatrix}, \quad D(k) = \begin{bmatrix} 0.04 \\ -0.06 \\ 0.05 \end{bmatrix}, \quad E(k) = \begin{bmatrix} 0.01 \\ 0.02 \end{bmatrix},$$

$$M(k) = \begin{bmatrix} 0.2 & -0.1 & 0.2 \end{bmatrix}.$$

The matrices $\mathscr{P}_1(k)$ and $\mathscr{P}_2(k)$ are taken to be

$$\mathscr{P}_1(k) = \begin{bmatrix} 0.6 + 0.1(-1)^k & 0.4 - 0.1(-1)^k \\ 0.55 & 0.45 \end{bmatrix},$$
$$\mathscr{P}_2(k) = \begin{bmatrix} 0.6 + 0.1(-1)^k & 0.4 - 0.1(-1)^k \\ 0.55 & 0.45 \end{bmatrix}.$$

The matrices $E_1(k)$ and $E_2(k)$ are defined as follows:

$$E_1(k) = \begin{bmatrix} 0.3 & -0.2 \end{bmatrix}^T, \; E_2(k) = \begin{bmatrix} 0.2 & -0.3 \end{bmatrix}^T.$$

In this example, the $\mathscr{H}_\infty$ performance level γ, positive definite matrix W and time-horizon N are selected as 0.99, $2.2I$ and 80, respectively. The scalars ε_1 and ε_2 are selected as $\varepsilon_1 = 0.01$ and $\varepsilon_2 = 0.01$, respectively. The exogenous disturbance inputs are selected as

$$\nu(k) = 0.8\sin(1.2k), \quad \omega(k) = 0.6\cos(k).$$

Using the given algorithm and MATLAB software, the set of solutions to backward RDEs in Theorem 8.1 are obtained and the controller parameters can be achieved. The simulation results are shown in Figs. 8.2, 8.3 and 8.4, where Figs. 8.2 and 8.3 plot the possible realizations of the Markovian jumping modes $\alpha(k)$ and $\beta(k)$, respectively, and Fig. 8.4 depicts the output trajectories of the open-loop and closed-loop system. It can be seen that the designed controller performs very well.

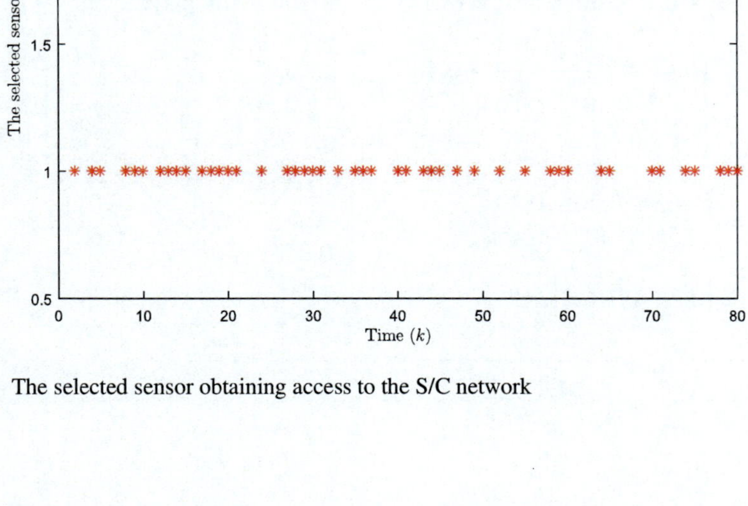

Fig. 8.2 The selected sensor obtaining access to the S/C network

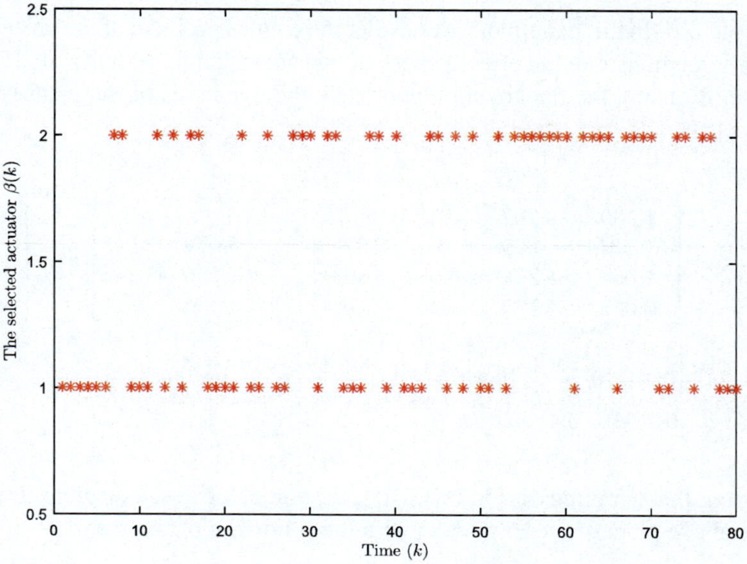

Fig. 8.3 The selected actuator obtaining access to the C/A network

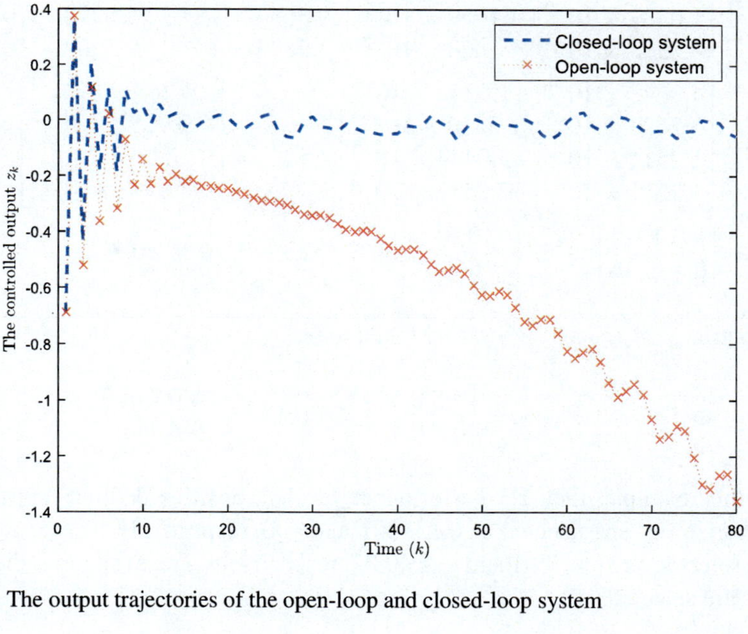

Fig. 8.4 The output trajectories of the open-loop and closed-loop system

Example 8.2 In this example, we consider the control problem of the unstable batch reactor to demonstrate the effectiveness of our presented finite-horizon $\mathscr{H}_\infty$ control scheme. By using the linearization approach, the dynamics of the plant is given as follows [3]:

$$\begin{cases} \dfrac{dx}{dt} = \begin{bmatrix} 1.380 & -0.2077 & 6.715 & -5.676 \\ -0.5814 & -4.29 & 0 & 0.675 \\ 1.067 & 4.273 & -6.654 & 5.893 \\ 0.048 & 4.273 & 1.343 & -2.104 \end{bmatrix} x + \begin{bmatrix} 0 & 0 \\ 5.679 & 0 \\ 1.136 & -3.146 \\ 1.136 & 0 \end{bmatrix} u + \begin{bmatrix} 0.4 \\ -0.3 \\ 0.3 \\ -0.5 \end{bmatrix} \nu \\ y = \begin{bmatrix} 1 & 0 & 1 & -1 \\ 0 & 1 & 0 & 0 \end{bmatrix} x + \begin{bmatrix} 0.6 \\ 0.4 \end{bmatrix} \nu \\ z = \begin{bmatrix} 0.3 & -0.3 & 0.4 & -0.3 \end{bmatrix} x. \end{cases}$$

Letting the sampling period be 0.01 s, we can achieve the discrete-time system for the above plant of the form (8.1) with the following parameters:

$$A = \begin{bmatrix} 1.0142 & -1.8283 \times 10^{-3} & 6.5064 \times 10^{-2} & -5.4647 \times 10^{-2} \\ -5.73 \times 10^{-3} & 9.5815 \times 10^{-1} & -1.4498 \times 10^{-4} & 6.6975 \times 10^{-3} \\ 1.0289 \times 10^{-2} & 4.1659 \times 10^{-2} & 9.3634 \times 10^{-1} & 5.6272 \times 10^{-2} \\ 4.2563 \times 10^{-4} & 4.1668 \times 10^{-2} & 1.2873 \times 10^{-2} & 9.7969 \times 10^{-1} \end{bmatrix},$$

$$B = \begin{bmatrix} 4.8642 \times 10^{-6} & -1.0342 \times 10^{-3} \\ 5.5629 \times 10^{-2} & 1.5330 \times 10^{-6} \\ 1.2511 \times 10^{-2} & -3.0444 \times 10^{-2} \\ 1.2511 \times 10^{-2} & -2.0537 \times 10^{-4} \end{bmatrix}, \quad D = \begin{bmatrix} 4.268 \times 10^{-3} \\ -2.9652 \times 10^{-3} \\ 2.718 \times 10^{-3} \\ -4.9912 \times 10^{-3} \end{bmatrix},$$

$$C = \begin{bmatrix} 1 & 0 & 1 & -1 \\ 0 & 1 & 0 & 0 \end{bmatrix}, \quad E = \begin{bmatrix} 0.6 \\ 0.4 \end{bmatrix}, \quad M = \begin{bmatrix} 0.3 & -0.3 & 0.4 & -0.3 \end{bmatrix}.$$

The matrices $\mathscr{P}_1(k)$ and $\mathscr{P}_2(k)$ are taken to be

$$\mathscr{P}_1(k) = \begin{bmatrix} 0.5 & 0.5 \\ 0.3 & 0.7 \end{bmatrix}, \quad \mathscr{P}_2(k) = \begin{bmatrix} 0.6 & 0.4 \\ 0.9 & 0.1 \end{bmatrix}.$$

In this example, the $\mathscr{H}_\infty$ performance level γ, positive definite matrix W and time-horizon N are selected as 0.85, $30I$ and 100, respectively. The scalars ε_1 and ε_2 are selected as $\varepsilon_1 = 0.01$ and $\varepsilon_2 = 0.01$, respectively. The exogenous disturbance inputs are selected as

$$\nu(k) = 0.9\cos(0.7k^{1.2}), \quad \omega(k) = 0.5\cos(1.2k).$$

Based on the above parameters, the desired controller gain matrices can be calculated by solving the backward RDEs in Theorem 8.1. The simulation results are shown in Figs. 8.5, 8.6 and 8.7, where Figs. 8.5 and 8.6 plot the possible realizations

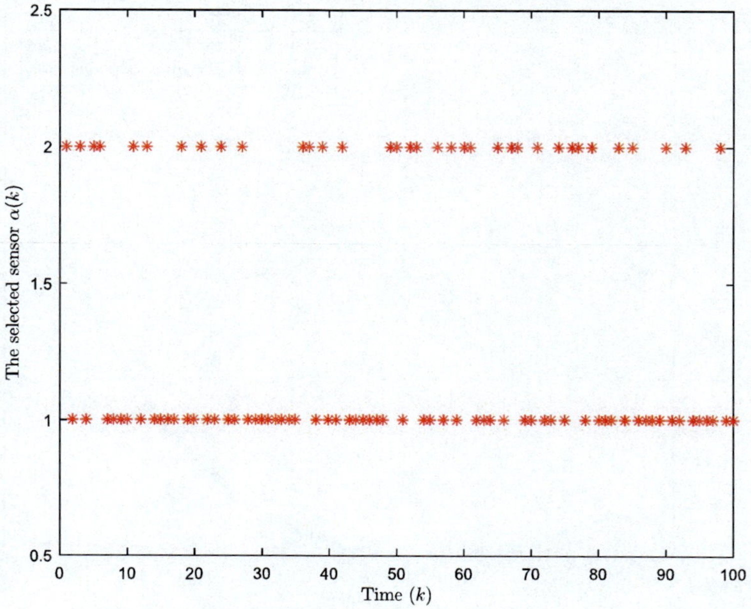

Fig. 8.5 The selected sensor obtaining access to the S/C network

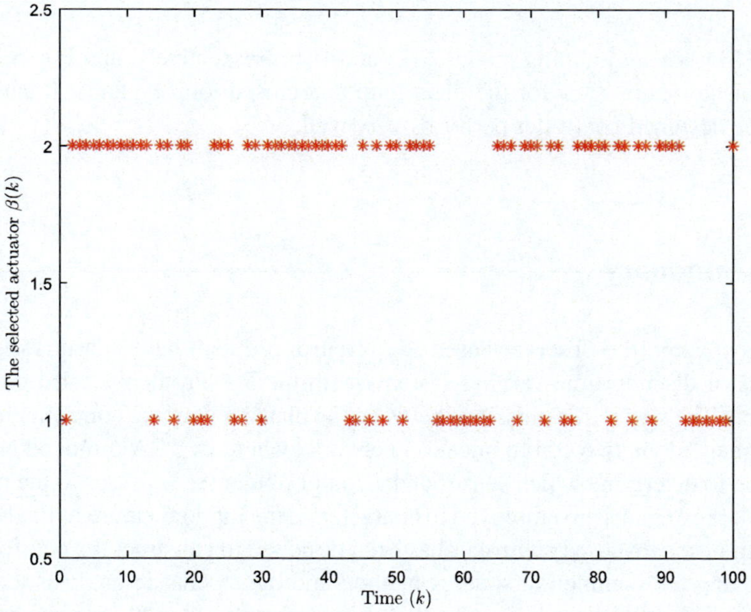

Fig. 8.6 The selected actuator obtaining access to the C/A network

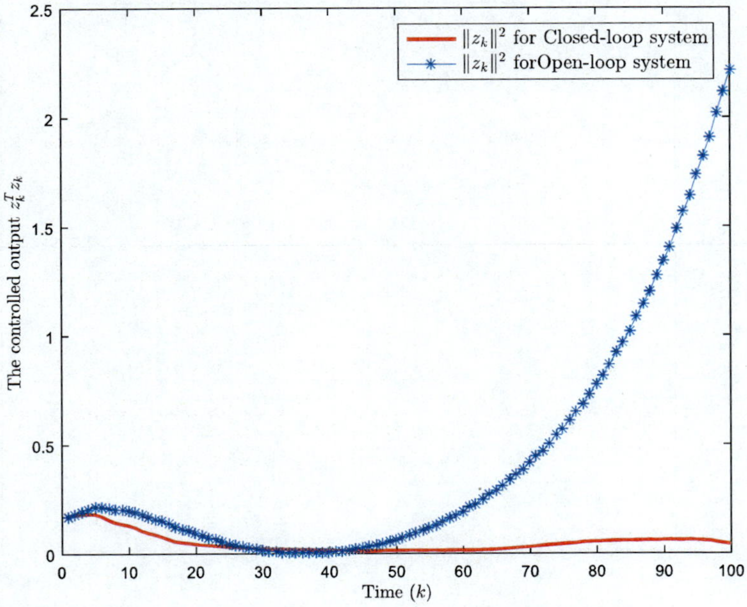

Fig. 8.7 The output trajectories of the open-loop and closed-loop system

of the Markovian jumping modes $\alpha(k)$ and $\beta(k)$, respectively, and Fig. 8.7 depicts the trajectories of $z_k^T z_k$ for the open-loop and closed-loop systems. It can be seen that the designed controller performs very well.

8.4 Summary

In this chapter, the observer-based $\mathscr{H}_\infty$ control problem has been investigated for a class of discrete time-varying systems with the RA Protocol over a given finite horizon. The signal transmissions between the plant and remote controller have been implemented via two communication networks where two RA protocols have been applied to determine which sensor and actuator obtain the accesses to the networks. The Markov chains are employed to characterize the random nature of the RA Protocol. An observer-based controller has been designed to construct the control law. By employing the completing squares method and the stochastic analysis techniques, the sufficient conditions have been derived to guarantee the $\mathscr{H}_\infty$ performance of the closed-loop system. Moreover, the desired controller parameters have been achieved by solving two coupled backward RDEs. Then, the corresponding analysis on the structure of the controller parameters has been conducted subject to the RA Protocol scheduling. Finally, two illustrative examples have been given to highlight the effectiveness of the proposed design strategy.

References

1. Xu, Y., Su, H., Pan, Y., Wu, Z., Xu, W.: Stability analysis of network control systems with round-robin scheduling and packet dropouts. J. Frankl. Inst. **350**(8), 2013–2027 (2013)
2. Penrose, R., Todd, J.A.: On best approximate solutions of linear matrix equations. Math. Proc. Camb. Philos. Soc. **52**(1), 17–19 (1956)
3. Walsh, G.C., Ye, H.: Scheduling of networked control systems. IEEE Control Syst. Mag. **21**(1), 57–65 (2001)

Chapter 9
Ultimately Bounded Control of Nonlinear Systems with Try-Once-Discard Protocol

The filtering and control problems subject to signal quantization effects have attracted considerable research interest in the past two decades. As one of the mostly studied networked-induced phenomena, signal quantization is regarded as an important source which has a significant impact on the performance of the networked system. There are generally two kinds of popular modeling methods for the quantization phenomenon that have been adopted in the literature, namely, the uniform quantization [1] and the logarithmic quantization [2], which belong to the fixed-point quantization and the floating-point quantization, respectively. From a practical point of view, because of the wide utilization of the fixed-point-number scheme in both hardware and software implementations, the uniform quantization scheme is more widely used in networked systems.

This chapter is concerned with the ultimately bounded control problem for a class of networked nonlinear systems subject to the TOD protocol scheduling and uniform quantization effects. More specifically, the objective of this chapter is to design a nonlinear-observer-based controller for the networked system subject to the uniform quantization effects and TOD scheduling such that the dynamics of the closed-loop system is ultimately bounded in mean square. The TOD protocol is utilized to regulate the signal transmission over the communication network under which the scheduling behavior is described by a special switching function. On the other hand, the uniform quantization effects of the network are characterized by a round function (i.e. the nearest integer function). Sufficient conditions are established to guarantee the ultimate boundedness of the dynamics of the closed-loop system in mean square by applying the stochastic analysis approach. Furthermore, the desired controller gains are derived by solving a convex optimization problem.

The rest of this chapter is organized as follows. In Sect. 9.1, the networked nonlinear system with the TOD scheduling and uniform quantization effects is introduced and the nonlinear-observer-based controller is proposed. In Sect. 9.2, the ultimate boundedness control issue of the closed-loop system is studied based on the stochas-

L. Zou et al., *Communication-Protocol-Based Filtering and Control of Networked Systems*, Studies in Systems, Decision and Control 430,
https://doi.org/10.1007/978-3-030-97512-8_9

tic analysis approach. Furthermore, the required controller gain matrices are obtained by solving a minimization problem subject to some matrix inequality constraints. In Sect. 9.3, the effectiveness of the proposed control scheme is demonstrated by a numerical simulation example. Finally, conclusions are drawn in Sect. 9.4.

9.1 Problem Formulation

9.1.1 The System Model

Consider an NCS shown in Fig. 9.1, where the signal transmission between the sensors and controller is implemented via a network with certain network-induced effect. In the following, we introduce the plant, communication network and the controller in the state space.

The plant shown in Fig. 9.1 is a nonlinear system of the form:

$$\begin{cases} x(k+1) = Ax(k) + f(Gx(k)) + Bu(k) + E\omega(k) \\ y(k) = Cx(k) + D\omega(k) \\ z(k) = Mx(k) \end{cases} \tag{9.1}$$

where $x(k) \in \mathbb{R}^{n_x}$, $u(k) \in \mathbb{R}^{n_u}$, $y(k) \in \mathbb{R}^{n_y}$ and $z(k) \in \mathbb{R}^{n_z}$ are, respectively, the state vector, control input, measurement output before transmitted through the communication network, and the output vector to be controlled. $\omega(k) \in \mathbb{R}^{n_\omega}$ is the disturbance noise with zero-mean and known variance $R = \bar{R}\bar{R}^T \geq 0$. $f(\cdot)$ is a nonlinear vector-valued function satisfying certain condition to be given later. A, B, C, D, E, G and M are constant matrices of appropriate dimensions.

Assumption 9.1 The nonlinear function $f(\cdot)$ satisfies $f(0) = 0$ and the following condition:

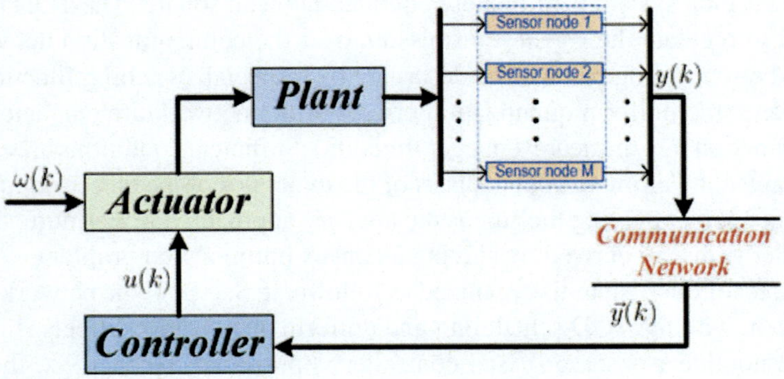

Fig. 9.1 Structure of the control system with a communication network

$$\|f(x) - f(z)\| \leq \|F(x - z)\| \tag{9.2}$$

where the matrix F is a real matrix of appropriate dimensions.

9.1.2 The Description of the Communication Network

We discuss the effects induced by the communication network. In this chapter, two network-induced constraints are considered, namely, the TOD protocol scheduling and uniform quantization effects. Without loss of generality, we first assume that the sensors of the plant are grouped into N ($N > 1$) sensor nodes according to their spatial distribution. As such, for technical analysis, the measurement output $y_N(k)$ can be rewritten as follows:

$$y(k) = \left[y_1^T(k)\ y_2^T(k)\ \cdots\ y_N^T(k)\right]^T \tag{9.3}$$

where $y_i(k)$ ($i \in \{1, 2, \ldots, N\}$) is the measurement of the ith sensor node before transmitted.

As shown in Fig. 9.2, the signal transmitted through the network is quantized first and then scheduled by the communication network subject to the communication protocol. First, we introduce the quantization effects of the network. Let the quantized measurement be $\overrightarrow{y}(k) \triangleq \left[\overrightarrow{y}_1^T(k)\ \overrightarrow{y}_2^T(k)\ \cdots\ \overrightarrow{y}_N^T(k)\right]^T$. In this chapter, the signal would be quantized by an uniform quantizer $\mathcal{Q}(\cdot)$ when it is transmitted via a communication network. Suppose the saturation level of quantization is sufficiently large. Hence, the uniform quantizer could be written as follows [3]:

$$\overrightarrow{y}(k) = \mathcal{Q}(y(k)) = \begin{bmatrix} \varepsilon\mathscr{R}(y_{(1)}(k)/\varepsilon) \\ \varepsilon\mathscr{R}(y_{(2)}(k)/\varepsilon) \\ \vdots \\ \varepsilon\mathscr{R}(y_{(n_y)}(k)/\varepsilon) \end{bmatrix} \tag{9.4}$$

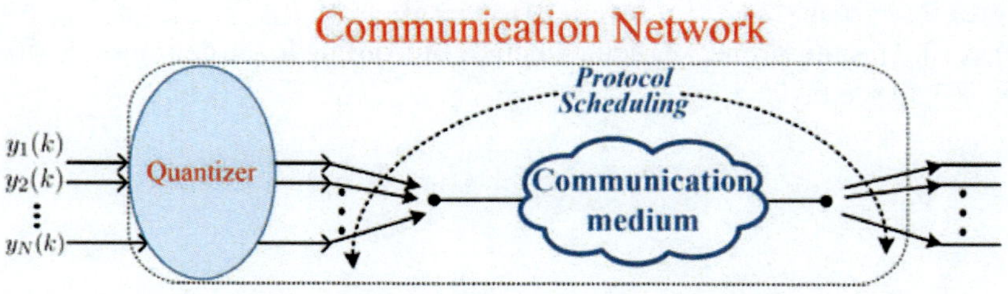

Fig. 9.2 Effects induced by the communication network

where ε is called the quantizing level. $y_{(i)}(k)$ denotes the ith element of the vector $y(k)$. The function $\mathscr{R}(\cdot)$ rounds a number to the nearest integer. Let $\Delta(k) \triangleq \mathcal{Q}(y(k)) - y(k)$ be the quantization error. Hence, it is easy to see that

$$\|\Delta(k)\|_{\infty} \leq \varepsilon/2. \tag{9.5}$$

Next, we shall introduce the effects induced by the communication protocol scheduling. In the considered networked system, the TOD protocol is utilized to schedule the signal transmission between sensors and the controller. In networked communication circumstances, communication protocols are always employed to determine which node (or nodes) obtains access to the network at each time instant. The main idea of the protocol scheduling considered in this chapter is that *only one sensor node* is permitted to send data via the communication network at each transmission instant. Let $\xi(k) \in \{1, 2, \ldots, M\}$ denote the selected sensor node obtaining access to the communication network at time instant k. Then, as shown in [4], due to the scheduling of the TOD protocol, $\xi(k)$ can be characterized by the following selection principle:

$$\xi(k) \triangleq \arg \max_{1 \leq i \leq N} \|y_i(k) - \bar{y}_i(k-1)\|^2_{Q_i} \tag{9.6}$$

where $\bar{y}_i(k-1)$ represents the previously transmitted signal before time instant k (excluding k) associating with the sensor node i, and Q_i ($i \in \{1, 2, \ldots, N\}$) is a known positive definite matrix denoting the weight matrix of the ith sensor node.

By defining

$$\begin{cases} \bar{y}(k-1) \triangleq \left[\bar{y}_1^T(k-1)\ \bar{y}_2^T(k-1) \cdots \bar{y}_N^T(k-1)\right]^T \\ \bar{Q} \triangleq \mathrm{diag}\{Q_1, Q_2, \ldots, Q_N\} \end{cases},$$

the selection principle (9.6) could be rewritten as follows:

$$\xi(k) = \arg \max_{1 \leq i \leq N} \|y(k) - \bar{y}(k-1)\|^2_{\bar{Q}\Phi_i} \tag{9.7}$$

where $\Phi_i = \mathrm{diag}\{\delta(i-1)I, \delta(i-2)I, \ldots, \delta(i-N)I\}$ ($i \in \{1, 2, \ldots, N\}$) and $\delta(\cdot) \in \{0, 1\}$ is the Kronecker delta function. According to the definition of $\bar{y}(k-1)$, it is easy to see that

$$\bar{y}(k) = \Phi_{\xi(k)} \overrightarrow{y}(k) + \big(I - \Phi_{\xi(k)}\big)\bar{y}(k-1). \tag{9.8}$$

9.1.3 The Nonlinear-Observer-Based Controller

Inspired by [5], we adopt the following nonlinear-observer-based control scheme for the networked system (9.1) subject to the TOD scheduling and uniform quantization effects:

$$\begin{cases} \hat{x}(k+1) = A\hat{x}(k) + f\big(G\hat{x}(k) + L_{2,\xi(k)}(\bar{y}(k) - C\hat{x}(k))\big) + Bu(k) \\ \qquad\qquad + L_{1,\xi(k)}(\bar{y}(k) - C\hat{x}(k)) \\ u(k) = K_{\xi(k)}\hat{x}(k) \end{cases} \tag{9.9}$$

where $\hat{x}(k) \in \mathbb{R}^{n_x}$ is the state estimate of the system (9.1), and the matrices $L_{1,\xi(k)}$, $L_{2,\xi(k)}$ and $K_{\xi(k)}$ are controller parameters to be designed.

Let the observer error be $e(k) \triangleq x(k) - \hat{x}(k)$. The error dynamics can be obtained from (9.1) and (9.9) as follows:

$$\begin{aligned} e(k+1) =& (A - L_{1,\xi(k)}C)e(k) + \tilde{f}(e(k), \omega(k), \Delta(k)) + L_{1,\xi(k)}(I - \Phi_{\xi(k)})\big(Cx(k) \\ &- \bar{y}(k-1)\big) + (E - L_{1,\xi(k)}\Phi_{\xi(k)}D)\omega(k) - L_{1,\xi(k)}\Phi_{\xi(k)}\Delta(k) \end{aligned} \tag{9.10}$$

where $\tilde{f}(e(k), \omega(k), \Delta(k)) = f(Gx(k)) - f(G\hat{x}(k) + L_{2,\xi(k)}(\bar{y}(k) - C\hat{x}(k)))$.

Denoting variable $\eta(k) = \big[x^T(k)\ \bar{y}^T(k-1)\ e^T(k)\big]^T \in \mathbb{R}^{n_\eta}$ with $n_\eta = 2n_x + n_y$, we obtain the closed-loop system as follows:

$$\begin{cases} \eta(k+1) = \vec{A}_{\xi(k)}\eta(k) + \vec{f}(\eta(k)) + \vec{E}_{\xi(k)}\omega(k) + \vec{D}_{\xi(k)}\Delta(k) \\ z(k) = \vec{M}\eta(k) \end{cases} \tag{9.11}$$

where

$$\vec{A}_{\xi(k)} = \begin{bmatrix} A + BK_{\xi(k)} & 0 & -BK_{\xi(k)} \\ \Phi_{\xi(k)}C & I - \Phi_{\xi(k)} & 0 \\ \mathcal{L}_{1,\xi(k)}C & -\mathcal{L}_{1,\xi(k)} & A - L_{1,\xi(k)}C \end{bmatrix}, \quad \vec{E}_{\xi(k)} = \begin{bmatrix} E \\ \Phi_{\xi(k)}D \\ E - L_{1,\xi(k)}\Phi_{\xi(k)}D \end{bmatrix},$$

$$\vec{D}_{\xi(k)} = \begin{bmatrix} 0 \\ \Phi_{\xi(k)} \\ -L_{1,\xi(k)}\Phi_{\xi(k)} \end{bmatrix}, \quad \vec{M} = \begin{bmatrix} M & 0 & 0 \\ 0 & 0 & 0 \\ 0 & 0 & 0 \end{bmatrix}, \quad \vec{f}(\eta(k)) = \begin{bmatrix} f(Gx(k)) \\ 0 \\ \tilde{f}(e(k), \omega(k), \Delta(k)) \end{bmatrix},$$

$$\mathcal{L}_{1,\xi(k)} = L_{1,\xi(k)}(I - \Phi_{\xi(k)}).$$

We are now in a position to state the problem addressed in this chapter as follows. The purpose of this chapter is to design a nonlinear-observer-based controller for the dynamical system (9.1) such that the following requirements are met simultaneously:

(1) the dynamics of the closed-loop system (9.11) is ultimately bounded in mean square subject to the disturbance noise $\omega(k)$ and the quantization error $\Delta(k)$;

(2) an asymptotic upper bound of the controlled output $\mathbb{E}\{\|z(k)\|^2\}$ exists and is minimized by the designed controller gain matrices $L_{1,\xi(k)}$, $L_{2,\xi(k)}$ and $K_{\xi(k)}$.

9.2 Main Results

In this section, a sufficient condition is presented to guarantee the ultimate boundedness of the closed-loop system (9.11) in mean square. Furthermore, the desired controller parameters are derived by solving a minimization problem subject to a set of matrix inequality constraints.

Theorem 9.1 *Let the controller parameters $L_{1,i}$, $L_{2,i}$ and K_i $(i = 1, 2, \ldots, N)$ be given. Assume that there exist three positive definite matrices $P \in \mathbb{R}^{(n_\eta+n_\omega)\times(n_\eta+n_\omega)}$, $\Upsilon_i \in \mathbb{R}^{n_\omega \times n_\omega}$ $(i = 1, 2)$, $N^2 + N + 1$ non-negative scalars λ_{ij} $(i, j = 1, 2, \ldots, N)$, γ_i $(i = 1, 2, \ldots, N)$, α and a positive scalar θ such that the following matrix inequalities hold for $i = 1, 2, \ldots, N$:*

$$\Xi_i = \begin{bmatrix} \Xi_{11}^i & \Xi_{12}^i & \Xi_{13}^i & \Xi_{14}^i \\ * & \Xi_{22}^i & \Xi_{23}^i & \Xi_{24}^i \\ * & * & \Xi_{33}^i & \Xi_{34}^i \\ * & * & * & \Xi_{44}^i \end{bmatrix} < 0 \tag{9.12}$$

$$\sum_{j=1}^{N} \lambda_{ij} = 1 \tag{9.13}$$

where

$$\vec{Q}_i = \bar{Q} - \bar{Q}\Phi_i,\ \vec{C} = \begin{bmatrix} C & -I & 0_{n_y \times n_x} \end{bmatrix},\ \mathcal{C} = \begin{bmatrix} \vec{C} & D \end{bmatrix},\ \bar{\Upsilon} = \mathrm{diag}\{0_{n_\eta}, \Upsilon_1\},$$

$$\mathcal{A}_i = \begin{bmatrix} \vec{A}_i & \vec{E}_i \end{bmatrix},\ \mathcal{I}_1 = \begin{bmatrix} I \\ 0_{n_\omega \times n_\eta} \end{bmatrix},\ \mathcal{D}_i = \begin{bmatrix} \vec{D}_i \\ 0_{n_\omega \times n_y} \end{bmatrix},\ \bar{\mathcal{A}}_i = \begin{bmatrix} \mathcal{A}_i^T & 0_{(n_\eta+n_\omega)\times n_\omega} \end{bmatrix}^T,$$

$$\mathcal{I}_2 = \begin{bmatrix} 0_{n_\eta \times n_\eta} \\ I \end{bmatrix},\ \mathscr{P}_i = P + \sum_{j=1}^{N} \lambda_{ij} \mathcal{C}^T \vec{Q}_j \mathcal{C},\ \mathcal{F}_i = \begin{bmatrix} \mathcal{L}_{2,i}C & -\mathcal{L}_{2,i} & G - L_{2,i}C & -L_{2,i}\Phi_i D \end{bmatrix},$$

$$\begin{aligned}
&\Xi_{11}^i = \bar{\mathcal{A}}_i^T \mathscr{P}_i \bar{\mathcal{A}}_i - (1-\theta)\big(P + \mathcal{C}^T \vec{Q}_i \mathcal{C}\big) - \bar{\Upsilon} + \gamma_i\big(\mathcal{F}_i^T F^T F \mathcal{F}_i + \mathcal{I}_3^T G^T F^T F G \mathcal{I}_3\big),\\
&\Xi_{12}^i = \bar{\mathcal{A}}_i^T \mathscr{P}_i \mathcal{I}_1,\ \Xi_{13}^i = \bar{\mathcal{A}}_i^T \mathscr{P}_i \mathcal{D}_i - \gamma_i \mathcal{F}_i^T F^T F L_{2,i}\Phi_i,\\
&\mathcal{I}_3 = \begin{bmatrix} I_{n_x} & 0_{n_x \times n_y} & 0_{n_x \times n_x} & 0_{n_x \times n_\omega} \end{bmatrix},\ \Xi_{14}^i = \bar{\mathcal{A}}_i^T \mathscr{P}_i \mathcal{I}_2,\ \Xi_{22}^i = \mathcal{I}_1^T \mathscr{P}_i \mathcal{I}_1 - \gamma_i I,\\
&\Xi_{24}^i = \mathcal{I}_1^T \mathscr{P}_i \mathcal{I}_2,\ \mathcal{L}_{2,i} = L_{2,i}(I - \Phi_i),\ \Xi_{33}^i = \mathcal{D}_i^T \mathscr{P}_i \mathcal{D}_i + \gamma_i \Phi_i^T L_{2,i}^T F^T F L_{2,i} \Phi_i - \alpha I,\\
&\Xi_{34}^i = \mathcal{D}_i^T \mathscr{P}_i \mathcal{I}_2,\ \Xi_{23}^i = \mathcal{I}_1^T \mathscr{P}_i \mathcal{D}_i,\ \Xi_{44}^i = \mathcal{I}_2^T \mathscr{P}_i \mathcal{I}_2 - \Upsilon_2.
\end{aligned}$$

Then, the dynamics of the closed-loop system (9.11) is ultimately bounded in mean square subject to the disturbance noise $\omega(k)$ and the quantization error $\Delta(k)$.

Proof To analyze the ultimate boundedness of the dynamics of (9.11), we choose the following Lyapunov-like function:

$$V(k) = \begin{bmatrix} \eta(k) \\ \omega(k) \end{bmatrix}^T (P + \mathcal{C}^T \overrightarrow{Q}_{\xi(k)} \mathcal{C}) \begin{bmatrix} \eta(k) \\ \omega(k) \end{bmatrix}. \tag{9.14}$$

According to the definition of $\overrightarrow{Q}_i$ and the selection principle (9.7), it is easy to see that the following inequality holds for all $i \in \{1, 2, \ldots, N\}$

$$\begin{bmatrix} \eta(k+1) \\ \omega(k+1) \end{bmatrix}^T \mathcal{C}^T \overrightarrow{Q}_{\xi(k+1)} \mathcal{C} \begin{bmatrix} \eta(k+1) \\ \omega(k+1) \end{bmatrix} \le \begin{bmatrix} \eta(k+1) \\ \omega(k+1) \end{bmatrix}^T \mathcal{C}^T \overrightarrow{Q}_i \mathcal{C} \begin{bmatrix} \eta(k+1) \\ \omega(k+1) \end{bmatrix} \tag{9.15}$$

for $i \in \{1, 2, \ldots, N\}$. Then, it can be concluded from (9.15) that

$$\begin{aligned} V(k+1) &= \begin{bmatrix} \eta(k+1) \\ \omega(k+1) \end{bmatrix}^T (P + \mathcal{C}^T \overrightarrow{Q}_{\xi(k+1)} \mathcal{C}) \begin{bmatrix} \eta(k+1) \\ \omega(k+1) \end{bmatrix} \\ &\le \begin{bmatrix} \eta(k+1) \\ \omega(k+1) \end{bmatrix}^T \left(P + \sum_{j=1}^{N} \lambda_{\xi(k)j} \mathcal{C}^T \overrightarrow{Q}_j \mathcal{C} \right) \begin{bmatrix} \eta(k+1) \\ \omega(k+1) \end{bmatrix}. \end{aligned} \tag{9.16}$$

Letting $i = \xi(k)$, the corresponding difference of $V(k)$ along the trajectory of system (9.11) can be calculated as follows:

$$\begin{aligned} \Delta V(k) \triangleq V(k+1) - V(k) \le & \begin{bmatrix} \eta(k+1) \\ \omega(k+1) \end{bmatrix}^T \left(P + \sum_{j=1}^{N} \lambda_{\xi(k)j} \mathcal{C}^T \overrightarrow{Q}_j \mathcal{C} \right) \begin{bmatrix} \eta(k+1) \\ \omega(k+1) \end{bmatrix} \\ & - \begin{bmatrix} \eta(k) \\ \omega(k) \end{bmatrix}^T (P + \mathcal{C}^T \overrightarrow{Q}_{\xi(k)} \mathcal{C}) \begin{bmatrix} \eta(k) \\ \omega(k) \end{bmatrix}. \end{aligned} \tag{9.17}$$

On the other hand, by denoting $\bar{\eta}(k) = \left[\eta^T(k) \ \bar{\omega}^T(k) \right]^T$, we have

$$\bar{\eta}(k+1) = \left[\bar{A}_i \ \mathcal{I}_1 \ \mathcal{D}_i \ \mathcal{I}_2 \right] \begin{bmatrix} \bar{\eta}(k) \\ \overrightarrow{f}(\eta(k)) \\ \Delta(k) \\ \omega(k+1) \end{bmatrix}. \tag{9.18}$$

Hence, it follows from (9.17) to (9.18) that

$$\Delta V(k) \le \begin{bmatrix} \bar{\eta}(k) \\ \overrightarrow{f}(\eta(k)) \\ \Delta(k) \\ \omega(k+1) \end{bmatrix}^T \begin{bmatrix} \bar{\mathcal{A}}_i^T \\ \mathfrak{I}_1^T \\ \mathfrak{D}_i^T \\ \mathfrak{I}_2^T \end{bmatrix} \Big(P + \sum_{j=1}^{N} \lambda_{ij} \mathfrak{C}^T \overrightarrow{Q}_j \mathfrak{C} \Big) \begin{bmatrix} \bar{\mathcal{A}}_i^T \\ \mathfrak{I}_1^T \\ \mathfrak{D}_i^T \\ \mathfrak{I}_2^T \end{bmatrix}^T \begin{bmatrix} \bar{\eta}(k) \\ \overrightarrow{f}(\eta(k)) \\ \Delta(k) \\ \omega(k+1) \end{bmatrix}$$
$$- \bar{\eta}^T(k)\big(P + \mathfrak{C}^T \overrightarrow{Q}_i \mathfrak{C}\big)\bar{\eta}(k)$$
$$= \xi^T(k) \begin{bmatrix} \hbar_i & \bar{\mathcal{A}}_i^T \mathscr{P}_i \mathfrak{I}_1 & \bar{\mathcal{A}}_i^T \mathscr{P}_i \mathfrak{D}_i & \bar{\mathcal{A}}_i^T \mathscr{P}_i \mathfrak{I}_2 \\ * & \mathfrak{I}_1^T \mathscr{P}_i \mathfrak{I}_1 & \mathfrak{I}_1^T \mathscr{P}_i \mathfrak{D}_i & \mathfrak{I}_1^T \mathscr{P}_i \mathfrak{I}_2 \\ * & * & \mathfrak{D}_i^T \mathscr{P}_i \mathfrak{D}_i & \mathfrak{D}_i^T \mathscr{P}_i \mathfrak{I}_2 \\ * & * & * & \mathfrak{I}_2^T \mathscr{P}_i \mathfrak{I}_2 \end{bmatrix} \xi(k) \tag{9.19}$$

where $\xi(k) = \Big[\bar{\eta}^T(k)\ \overrightarrow{f}^T(\eta(k))\ \Delta^T(k)\ \omega^T(k+1)\Big]^T$ and $\hbar_i = \bar{\mathcal{A}}_i^T \mathscr{P}_i \bar{\mathcal{A}}_i - \big(P + \mathfrak{C}^T \overrightarrow{Q}_i \mathfrak{C}\big)$.

Furthermore, one can infer from Assumption 9.1 that

$$\|\tilde{f}(e(k), \omega(k), \Delta(k))\|^2$$
$$\le \big(G(x(k) - \hat{x}(k)) - L_{2,i}(\bar{y}(k) - C\hat{x}(k))\big)^T F^T F \big(G(x(k) - \hat{x}(k)) L_{2,i}(\bar{y}(k)$$
$$- C\hat{x}(k))\big)$$
$$= \begin{bmatrix} \bar{\eta}(k) \\ \Delta(k) \end{bmatrix}^T \begin{bmatrix} \mathfrak{F}_i^T \\ -\Phi_i^T L_{2,i}^T \end{bmatrix} F^T F \begin{bmatrix} \mathfrak{F}_i & -L_{2,i}\Phi_i \end{bmatrix} \begin{bmatrix} \bar{\eta}(k) \\ \Delta(k) \end{bmatrix}. \tag{9.20}$$

Hence, we have

$$\|\overrightarrow{f}(\eta(k))\|^2 = \|f(Gx(k))\|^2 + \|\tilde{f}(e(k), \omega(k), \Delta(k))\|^2$$
$$\le x^T G^T F^T F G x(k) + \begin{bmatrix} \bar{\eta}(k) \\ \Delta(k) \end{bmatrix}^T \begin{bmatrix} \mathfrak{F}_i^T \\ -\Phi_i^T L_{2,i}^T \end{bmatrix} F^T F \begin{bmatrix} \mathfrak{F}_i & -L_{2,i}\Phi_i \end{bmatrix} \begin{bmatrix} \bar{\eta}(k) \\ \Delta(k) \end{bmatrix}$$
$$= \begin{bmatrix} \bar{\eta}(k) \\ \Delta(k) \end{bmatrix}^T \begin{bmatrix} \bar{\hbar}_i & -\mathfrak{F}_i^T F^T F L_{2,i}\Phi_i \\ * & \Phi_i^T L_{2,i}^T F^T F L_{2,i}\Phi_i \end{bmatrix} \begin{bmatrix} \bar{\eta}(k) \\ \Delta(k) \end{bmatrix} \tag{9.21}$$

where $\bar{\hbar}_i = \mathfrak{F}_i^T F^T F \mathfrak{F}_i + \mathfrak{I}_3^T G^T F^T F G \mathfrak{I}_3$. Therefore

$$\Delta V(k) \le \xi^T(k)\Xi_i\xi(k) - \theta V(k) + \alpha\|\Delta(k)\|^2 + \omega^T(k)\Upsilon_1\omega(k)$$
$$+ \omega^T(k+1)\Upsilon_2\omega(k+1)$$
$$\le -\theta V(k) + \alpha\|\Delta(k)\|^2 + \omega^T(k)\Upsilon_1\omega(k) + \omega^T(k+1)\Upsilon_2\omega(k+1) \tag{9.22}$$

which implies that

$$\mathbb{E}\{\Delta V(k)\} \leq -\theta \mathbb{E}\{V(k)\} + \rho \tag{9.23}$$

where $\rho = \alpha n_y \frac{\varepsilon^2}{4} + \text{tr}\{\bar{R}^T(\Upsilon_1 + \Upsilon_2)\bar{R}\}$. To this end, it can be derived that, for any scalar $\sigma > 0$,

$$\begin{aligned} &\mathbb{E}\{\sigma^{k+1}V(k+1)\} - \mathbb{E}\{\sigma^k V(k)\} \\ =&\sigma^{k+1}\big(\mathbb{E}\{V(k+1)\} - \mathbb{E}\{V(k)\}\big) + \sigma^k(\sigma - 1)\mathbb{E}\{V(k)\} \\ \leq&\sigma^k\big(\sigma - 1 - \sigma\theta\big)\mathbb{E}\{V(k)\} + \sigma^{k+1}\rho. \end{aligned} \tag{9.24}$$

Letting $\sigma = \sigma_0 = \frac{1}{1-\theta}$ and summing up both sides of (9.24) from 0 to $t-1$ with respect to k, we obtain

$$\mathbb{E}\{\sigma^t V(t)\} - \mathbb{E}\{V(0)\} \leq \frac{\sigma_0(1-\sigma_0^t)}{1-\sigma_0}\rho \tag{9.25}$$

which implies that

$$\mathbb{E}\{V(k)\} \leq (1-\theta)^k\big(\mathbb{E}\{V(0)\} - \frac{\rho}{\theta}\big) + \frac{\rho}{\theta}. \tag{9.26}$$

Then, it is easy to see that

$$\mathbb{E}\{\|\eta(k)\|^2\} \leq \frac{1}{\lambda_{\min}\{P\}}\mathbb{E}\{\bar{\eta}^T(k)P\bar{\eta}(k)\} \leq \frac{(1-\theta)^k}{\lambda_{\min}\{P\}}\big(\mathbb{E}\{V(0)\} - \frac{\rho}{\theta}\big) + \frac{\rho}{\theta\lambda_{\min}\{P\}}. \tag{9.27}$$

Hence, it can be concluded that the dynamics of the closed-loop system (9.11) is ultimately bounded in mean square subject to the disturbance noise $\omega(k)$ and the quantization error $\Delta(k)$. The proof is complete.

Remark 9.1 Theorem 9.1 provides a sufficient condition under which the dynamics of the closed-loop system (9.11) is ultimately bounded in mean square subject to the disturbance noise $\omega(k)$ and the quantization error. It can be observed from Theorem 9.1 that the ultimate boundedness of (9.11) is dependent only on the system parameters of the plant (9.1) and the weight matrix of the TOD protocol. The information of the disturbance noise $\omega(k)$ and the quantization error have been reflected in the asymptotic upper bound of $\mathbb{E}\{\|\eta(k)\|^2\}$.

Note that the decay rate of the closed-loop system (9.11) is an important performance index for the control performance. Next, we focus our attention on the design of the controller gain matrices $L_{1,i}$, $L_{2,i}$ and K_i $(i = 1, 2, \ldots, N)$ based on a given decay rate. In what follows, we aim to provide a design scheme of the controller parameters L_1, L_2 and K. For this purpose, by defining

$$\mathscr{A}_i = \begin{bmatrix} A & 0 & 0 \\ \Phi_i C & I - \Phi_i & 0 \\ 0 & 0 & A \end{bmatrix}, \ \mathscr{B} = \begin{bmatrix} B \\ 0 \\ 0 \end{bmatrix}, \ \mathscr{E}_i = \begin{bmatrix} E \\ \Phi_i D \\ E \end{bmatrix}, \ \bar{\mathscr{A}_i} = \begin{bmatrix} \mathscr{A}_i & \mathcal{E}_i \\ 0 & 0 \end{bmatrix}, \ \bar{\mathscr{B}} = \begin{bmatrix} \mathscr{B} \\ 0 \end{bmatrix},$$

$$\bar{\mathscr{I}_2} = \begin{bmatrix} \mathscr{I}_2 \\ 0 \end{bmatrix}, \ \mathscr{I}_1 = \begin{bmatrix} I \ 0 \ -I \end{bmatrix}, \ \bar{\mathscr{I}_1} = \begin{bmatrix} \mathscr{I}_1 \ 0 \end{bmatrix}, \ \mathscr{C}_i = \begin{bmatrix} (I - \Phi_i)C \ -(I - \Phi_i) \ -C \end{bmatrix},$$

$$\bar{\mathscr{C}_i} = \begin{bmatrix} \mathscr{C}_i \ -\Phi_i D \end{bmatrix}, \ \check{\Phi}_i = \begin{bmatrix} 0_{n_y \times n_x} \ \Phi_i^T \ 0_{n_y \times n_x} \ 0_{n_y \times n_\omega} \end{bmatrix}^T, \ \mathscr{I}_2 = \begin{bmatrix} 0 \ 0 \ I \end{bmatrix}^T,$$

we obtain

$$\bar{\mathcal{A}}_i = \bar{\mathscr{A}_i} + \bar{\mathscr{B}} K_i \bar{\mathscr{I}_1} + \bar{\mathscr{I}_2} L_{1,i} \bar{\mathscr{C}_i}, \quad \mathcal{D}_i = \check{\Phi}_i - \bar{\mathscr{I}_2} L_{1,i} \Phi_i. \tag{9.28}$$

Then, we have the following corollary.

Corollary 9.1 *For the closed-loop system (9.11), under Assumption 9.1, let a scalar $0 < \theta < 1$ be given. Suppose that there exist three positive definite matrices $P \in \mathbb{R}^{(n_\eta + n_\omega) \times (n_\eta + n_\omega)}$, $\Upsilon_j \in \mathbb{R}^{n_\omega \times n_\omega}$ $(j = 1, 2)$, $N^2 + 1$ non-negative scalars λ_{ij} $(i = 1, 2, \ldots, N;\ j = 1, 2, \ldots, N - 1)$, γ_i $(i = 1, 2, \ldots, N)$ and α, $3N$ matrices $L_{1,i}$, $\bar{L}_{2,i}$ and K_i $(i = 1, 2, \ldots, N)$ satisfying the following matrix inequalities for $i = 1, 2, \ldots, N$:*

$$\vec{\Xi}_i = \begin{bmatrix} \bar{\Xi}_i & \Pi_1 & \Pi_2 \\ * & -\Omega_i & 0 \\ * & * & -\gamma_i I \end{bmatrix} < 0 \tag{9.29}$$

$$P + \bar{\lambda}_i \mathcal{C}^T \vec{Q}_N \mathcal{C} + \sum_{j=1}^{N-1} \lambda_{ij} \mathcal{C}^T \vec{Q}_j \mathcal{C} - 2I \leq -\Omega_i \tag{9.30}$$

$$\mathscr{M}^T \mathscr{M} \leq P \tag{9.31}$$

$$\bar{\lambda}_i \geq 0 \tag{9.32}$$

where

$$\bar{\Xi}_i = \text{diag}\{\vec{\hbar}_i, -\gamma_i I, -\alpha I, -\Upsilon_2\}, \ \Pi_1 = \left[(\Pi_1^1)^T \ (\Pi_1^2)^T \ (\Pi_1^3)^T \ (\Pi_1^4)^T \right]^T,$$

$$\vec{\hbar}_i = -(1 - \theta)(P + \mathcal{C}^T \vec{Q}_i \mathcal{C}) - \bar{\Upsilon} + \gamma_i \mathfrak{I}_3^T G^T F^T F G \mathfrak{I}_3,$$

$$\vec{Q}_i = \bar{Q} - \bar{Q}\Phi_i, \ \Pi_1^1 = (\bar{\mathscr{A}_i} + \bar{\mathscr{B}} K_i \bar{\mathscr{I}_1} + \bar{\mathscr{I}_2} L_{1,i} \bar{\mathscr{C}_i})^T, \ \Pi_1^2 = \mathfrak{I}_1^T,$$

$$\Pi_1^3 = \check{\Phi}_i^T - \Phi_i L_{1,i}^T \bar{\mathscr{I}_2}^T, \ \Pi_1^4 = \mathfrak{I}_2^T, \ \Pi_2^2 = -\Phi_i \bar{L}_{2,i}^T F^T, \ \Pi_2^1 = \bar{\mathfrak{F}}_i^T F^T,$$

$$\bar{\mathfrak{F}}_i = \left[\bar{\mathscr{L}}_{2,i} C \ -\bar{\mathscr{L}}_{2,i} \ \gamma_i G - \bar{L}_{2,i} C \ -\bar{L}_{2,i} \Phi_i D \right],$$

$$\bar{\mathscr{L}}_{2,i} = \bar{L}_{2,i}(I - \Phi_i), \ \bar{\lambda}_i = 1 - \sum_{j=1}^{N-1} \lambda_{ij}, \ \Pi_2 = \left[(\Pi_2^1)^T \ 0 \ (\Pi_2^2)^T \ 0 \right]^T,$$

$$\vec{C} = \left[C \ -I \ 0_{n_y \times n_x} \right], \ \mathcal{C} = \left[\vec{C} \ D \right], \ \bar{\Upsilon} = \text{diag}\{0_{n_\eta}, \Upsilon_1\}, \ \mathscr{M} = \text{diag}\{\vec{M}, 0_{n_\omega \times n_\omega}\},$$

$$\mathcal{J}_1 = \begin{bmatrix} I \\ 0_{n_\omega \times n_\eta} \end{bmatrix}, \ \mathcal{J}_2 = \begin{bmatrix} 0_{n_\eta \times n_\eta} \\ I \end{bmatrix}, \ \mathcal{J}_3 = \begin{bmatrix} I_{n_x} & 0_{n_x \times n_y} & 0_{n_x \times n_x} & 0_{n_x \times n_\omega} \end{bmatrix}.$$

Then, the dynamics of the closed-loop system (9.11) is ultimately bounded in mean square subject to the disturbance noise $\omega(k)$ and the quantization error $\Delta(k)$. Furthermore, the decay rate of the output vector $z(k)$ is less than $1-\theta$, and the minimum of the asymptotic upper bound of $\mathbb{E}\{\|z(k)\|^2\}$ can be derived by solving the following minimization problem

$$\begin{cases} \min \left\{ \operatorname{tr}\{\bar{R}^T(\Upsilon_1 + \Upsilon_2)\bar{R}\} + \alpha n_y \frac{\varepsilon^2}{4} \right\} \\ \textit{subject to the constraints (9.29)–(9.32).} \end{cases} \tag{9.33}$$

The desired controller gain matrices are given by K_i, $L_{1,i}$ and $L_{2,i} = \gamma_i^{-1}\bar{L}_{2,i}$.

Proof By denoting $\bar{\lambda}_i = 1 - \sum_{j=1}^{N-1} \lambda_{ij}$, $\mathscr{P}_i$ defined in Theorem 9.1 is replaced by $\mathscr{P}_i = P + \bar{\lambda}_i \mathcal{C}^T \vec{Q}_N \mathcal{C} + \sum_{j=1}^{N-1} \lambda_{ij} \mathcal{C}^T \vec{Q}_j \mathcal{C}$. Then, Ξ_i could be rewritten as

$$\Xi_i = \bar{\Xi}_i + \begin{bmatrix} \bar{\mathcal{A}}_i^T \\ \mathcal{J}_1^T \\ \mathcal{D}_i^T \\ \mathcal{J}_2^T \end{bmatrix} \mathscr{P}_i \begin{bmatrix} \bar{\mathcal{A}}_i & \mathcal{J}_1 & \mathcal{D}_i & \mathcal{J}_2 \end{bmatrix} + \gamma_i \begin{bmatrix} \mathcal{F}_i^T F^T \\ 0 \\ -\Phi_i L_{2,i}^T F^T \\ 0 \end{bmatrix} \begin{bmatrix} F\mathcal{F}_i & 0 & -FL_{2,i}\Phi_i & 0 \end{bmatrix}. \tag{9.34}$$

Noting $\bar{L}_{2,i} = \gamma_i L_{2,i}$, (9.28) and (9.30), it follows from (9.34) that

$$\Xi_i \le \bar{\Xi}_i + \Pi_1 \Omega_i^{-1} \Pi_1^T + \gamma_i^{-1} \Pi_2 \Pi_2^T. \tag{9.35}$$

Then, by using Schur Complement lemma, it can be seen from (9.35) to (9.29) that $\Xi < 0$. Hence, it follows from Theorem 9.1 that the dynamics of the closed-loop system (9.11) is ultimately bounded in mean square.

Next, it can be derived from (9.31) to (9.26) that

$$\begin{aligned} \mathbb{E}\{\|z(k)\|^2\} &\le \mathbb{E}\left\{ \begin{bmatrix} \eta(k) \\ \omega(k) \end{bmatrix}^T P \begin{bmatrix} \eta(k) \\ \omega(k) \end{bmatrix} \right\} \le \mathbb{E}\{V(k)\} \\ &\le (1-\theta)^k \left(\mathbb{E}\{V(0)\} - \frac{\rho}{\theta} \right) + \frac{\rho}{\theta} \end{aligned} \tag{9.36}$$

where $\rho = \operatorname{tr}\{\bar{R}^T(\Upsilon_1 + \Upsilon_2)\bar{R}\} + \alpha n_y \frac{\varepsilon^2}{4}$. Therefore, it can be concluded that the asymptotic bound of $\mathbb{E}\{\|z(k)\|^2\}$ is $\frac{\rho}{\theta}$ and the minimum of this asymptotic bound can be derived by minimizing ρ, which is equivalent to the optimization problem (9.33). The proof is complete.

Remark 9.2 Compared with the traditional control scheme (i.e. the Luenberger-observer-based control scheme, where the observer is constructed by the following dynamical system: $\hat{x}(k+1) = A\hat{x}(k) + f(G\hat{x}(k)) + Bu(k) + L_{1,\xi(k)}(\bar{y}(k) - C\hat{x}(k))$), our developed controller (nonlinear-observer-based controller) is more flexible due to the extra controller parameters $L_{2,\xi(k)}$. It can be observed that the controller design scheme established in this chapter could be specialized to the design of Luenberger-observer-based controller by setting $L_{2,i} = 0\ (1 \le i \le N)$. In such a case, the minimum value of ρ would be larger than that obtained in Corollary 9.1. In other words, the nonlinear-observer-based control scheme could provide a better performance with a smaller asymptotic upper bound of $\mathbb{E}\{\|z(k)\|^2\}$ as compared with the Luenberger-observer-based control.

9.3 Illustrative Examples

In this section, we present two simulation examples to illustrate the effectiveness of the proposed controller design scheme for networked systems with TOD protocols.

Example 9.1 Consider the system (9.1) with the following parameters:

$$A = \begin{bmatrix} 0.93 & 0.11 \\ 0.07 & 0.92 \end{bmatrix},\ B = \begin{bmatrix} 1.1 & 0.2 \\ -0.3 & 0.9 \end{bmatrix},\ E = \begin{bmatrix} 0.2 \\ -0.1 \end{bmatrix},\ C = \begin{bmatrix} 0.6 & 0.4 \\ 0.3 & 0.7 \end{bmatrix},\ D = \begin{bmatrix} -0.1 \\ 0.1 \end{bmatrix}$$

and $M = 0.8I$.

The nonlinear function is taken to be

$$f(x(k)) = 0.1 \begin{bmatrix} \sin(1.2x_1(k)) \\ \sin(2.2x_2(k)) \end{bmatrix}$$

with which the matrices F and G could be selected as $F = 0.1I$ and $G = \mathrm{diag}\{2, 2.2\}$, respectively. The variance of the disturbance noise is $R = 0.81I$. Set the decay rate be 0.99 (i.e. $\theta = 0.01$). Then, according to Corollary 9.1, the minimization problem (9.33) can be solved via the MATLAB software (YALMIP 3.0). The controller gain matrices are derived as follows:

$$K_1 = \begin{bmatrix} -0.2685 & -0.1381 \\ -0.4190 & -0.5074 \end{bmatrix},\ K_2 = \begin{bmatrix} -0.1253 & -0.0628 \\ -0.2707 & -0.6112 \end{bmatrix},\ L_{1,1} = \begin{bmatrix} 0.7936 & 0.3049 \\ 0.6393 & 0.3683 \end{bmatrix},$$

$$L_{1,2} = \begin{bmatrix} -0.0155 & 0.2307 \\ 0.0241 & 0.2720 \end{bmatrix},\ L_{2,1} = \begin{bmatrix} 0.0213 & 0.0809 \\ 0.0183 & 0.2953 \end{bmatrix},\ L_{2,2} = \begin{bmatrix} 0.0129 & 0.1762 \\ 0.0006 & 0.3381 \end{bmatrix}.$$

Set the simulation run length be 160, and let the initial state be $x(0) = \begin{bmatrix} 20 & 12 \end{bmatrix}^T$. Based on the derived controller parameters, numerical simulation results are given in Figs. 9.3 and 9.4. Figure 9.3 plots the state trajectories of the open-loop and closed-loop system. Figure 9.4 shows the selected sensor node obtaining access to the com-

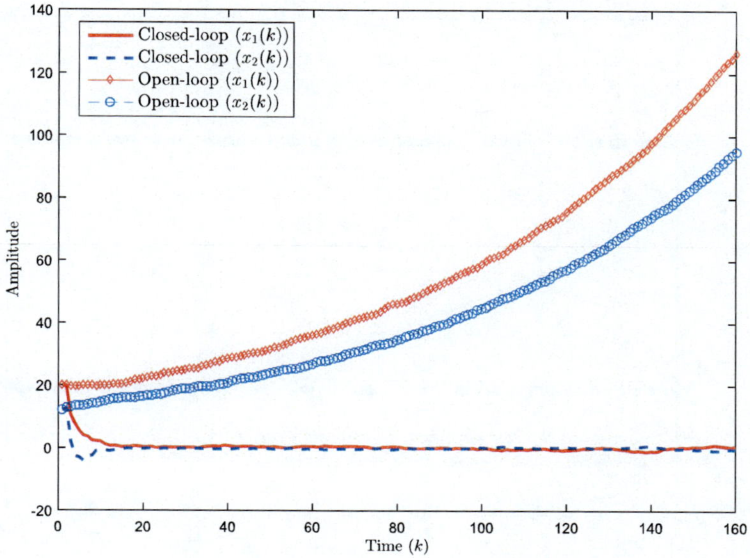

Fig. 9.3 The state trajectories of the open-loop and closed-loop system

munication network under the TOD scheduling. It can be observed from the state trajectories of the open-loop system that the plant under consideration is an unstable system. By employing our designed controller with the above-derived parameters, the state trajectories would eventually enter an ultimate boundedness region. The simulation results have confirmed that the designed ultimate boundedness controller performs very well.

For the purpose of examining the advantages of our proposed control scheme compared with the Luenberger-observer-based control scheme, we shall provide a simulation comparison between these two control approaches subject to different variance of the noise. Simulation results are shown in Table 9.1, where $\tilde{\rho}_1$ and $\tilde{\rho}_2$ represent the minimum of ρ under the nonlinear-observer-based control and the Luenberger-observer-based control, respectively. It can be seen that our developed control method performs better than the Luenberger-observer-based control method.

Table 9.1 Comparison to the Luenberger-observer-based control scheme

R	$0.25I$	$0.36I$	$0.49I$	$0.64I$	$0.81I$
$\tilde{\rho}_1$	1.3566	1.5499	1.8100	2.1559	2.6073
$\tilde{\rho}_2$	1.3663	1.5605	1.8225	2.1701	2.6244

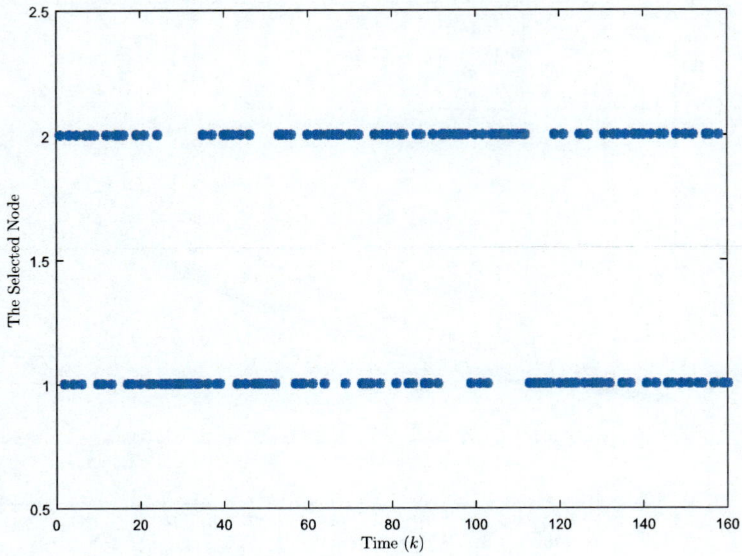

Fig. 9.4 The selected sensor node obtaining access to the network

Example 9.2 Consider the nonlinear system (9.1) with the following parameters:

$$A = \begin{bmatrix} 0.53 & 0.12 & 0.33 \\ 0.17 & 0.32 & 0.52 \\ 0.22 & 0.37 & 0.46 \end{bmatrix}, \ B = \begin{bmatrix} 1 & 0.2 \\ -0.3 & 1 \\ 0.8 & 0.3 \end{bmatrix}, \ E = \begin{bmatrix} 0.1 \\ -0.1 \\ 0.2 \end{bmatrix}, \ C = \begin{bmatrix} 0.6 & 0.4 & 0.2 \\ 0.3 & 0.7 & -0.2 \end{bmatrix},$$
$$D = \begin{bmatrix} -0.1 \\ 0.1 \end{bmatrix}, \ M = \begin{bmatrix} 0.4 & 0.3 & 0.3 \end{bmatrix}.$$

The nonlinear function is taken to be

$$f(x(k)) = 0.1 \begin{bmatrix} \sin(0.8x_1(k)) \\ \sin(1.1x_2(k)) \\ \sin(0.9x_2(k)) \end{bmatrix}$$

with which the matrices G and F could be selected as $G = \text{diag}\{0.8, 1.1, 0.9\}$ and $F = 0.1I$, respectively. The variance of the disturbance noise is $R = 0.36I$. Set the decay rate be 0.99 (i.e. $\theta = 0.01$). Then, according to Corollary 9.1, the minimization problem (9.33) can be solved via the MATLAB software (YALMIP 3.0). The controller gain matrices are derived as follows:

$$K_1 = \begin{bmatrix} -0.0003 & -0.2324 & 0.0102 \\ -0.3709 & -0.1548 & -0.3408 \end{bmatrix}, \ K_2 = \begin{bmatrix} -0.0817 & -0.3196 & 0.1267 \\ -0.2768 & -0.0939 & -0.2113 \end{bmatrix},$$
$$L_{1,1} = \begin{bmatrix} 0.5926 & 0.0512 \\ 0.6589 & 0.0375 \\ 0.3126 & -0.0263 \end{bmatrix}, \ L_{1,2} = \begin{bmatrix} 0.4587 & 0.3324 \\ 0.3822 & 0.3473 \\ -0.0381 & 0.1889 \end{bmatrix},$$

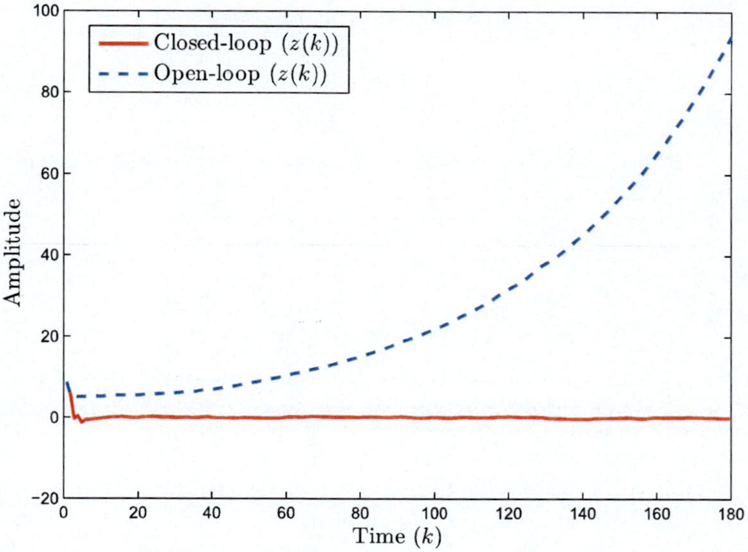

Fig. 9.5 The trajectories of $z(k)$ for the open-loop and closed-loop system

$$L_{2,1} = \begin{bmatrix} 0.0245 & -0.0031 \\ 0.0298 & 0.0002 \\ 0.0327 & 0.0019 \end{bmatrix}, \quad L_{2,2} = \begin{bmatrix} 0.0225 & 0.0107 \\ 0.0377 & 0.0169 \\ 0.0178 & 0.0170 \end{bmatrix}.$$

Set the simulation run length to be 180 and let the initial state be $x(0) = [20\ 12\ -10]^T$. Based on the derived controller parameters, numerical simulation results are given in Figs. 9.5 and 9.6. Figure 9.5 plots the trajectories of $z(k)$ for the open-loop and closed-loop system. Figure 9.6 shows the selected sensor node obtaining access to the communication network under the TOD scheduling. Obviously, it is easy to see from the trajectory of the open-loop system that the plant under consideration is unstable. Our developed control scheme is capable of guaranteeing the ultimate boundedness of the controlled output. All the simulation results have confirmed that the designed ultimate boundedness controller performs very well.

9.4 Summary

In this chapter, the ultimately bounded control problem has been addressed for a class of networked nonlinear systems subject to the so-called TOD protocol scheduling and uniform quantization effects. The communication between sensor nodes and the controller is implemented via a constraint network under which only one node is allowed to send data at each time instant. The TOD protocol has been employed to determine which node obtains access to the network. A novel nonlinear-observer-

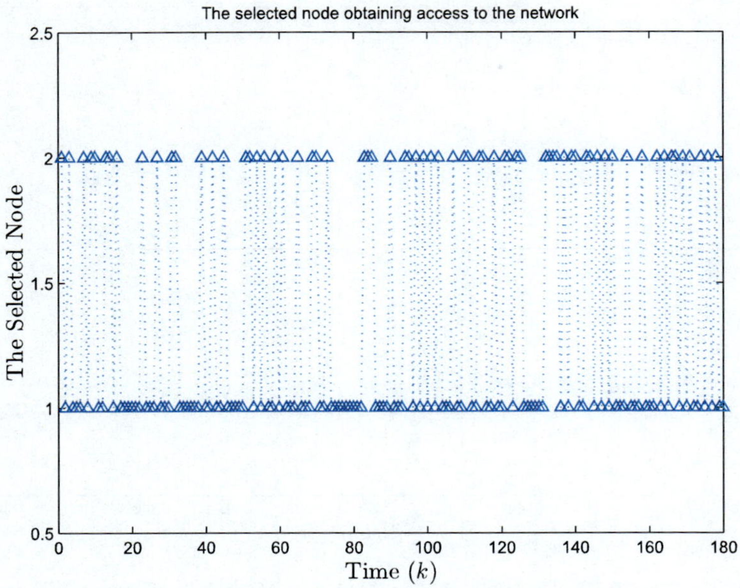

Fig. 9.6 The selected sensor node obtaining access to the network

based controller has been presented to guarantee that the dynamics of the closed-loop system is ultimately bounded in mean square subject to the disturbance noise and the quantization error. The simulation examples have been presented to demonstrate the effectiveness of the proposed controller design scheme.

References

1. Mohamed EI Mongi, B.G., Çela, A.: Trading quantization precision for update rates for systems with limited communication in the uplink channel. Automatica **46**(7), 1210–1214 (2010)
2. Liu, M., Ho, D.W.C., Niu, Y.: Robust filtering design for stochastic system with mode-dependent output quantization. IEEE Trans. Signal Process. **58**(12), 6410–6416 (2010)
3. Yun, S.W., Choi, Y.J., Park, P.: Dynamic output-feedback guaranteed cost control for linear systems with uniform input quantization. Nonlinear Dyn. **62**(1), 95–104 (2010)
4. Donkers, M.C.F., Heemels, W.P.M.H., van de Wouw, N., Hetel, L.: Stability analysis of networked control systems using a switched linear systems approach. IEEE Trans. Autom. Control **56**(9), 2101–2115 (2011)
5. Arcak, M., Kokotović, P.: Nonlinear observers: a circle criterion design and robustness analysis. Automatica **37**(12), 1923–1930 (2001)

Chapter 10
Finite-Horizon Consensus Control of Multi-agent Systems with Random Access Protocol

A multi-agent system is a special complex system composed of multiple interacting "agents". The past decades have witnessed a surge of an increasing research interest on the multi-agent systems due primarily to their extensive applications in various fields such as the multi-vehicle systems, satellite clusters, mobile ad hoc networks and coupled oscillators. Because of the clear physical and engineering insights of multi-agent systems, the cooperative control problems for various kinds of multi-agent systems have gained considerable attention. In general, the approaches to study multi-agent systems can be classified as centralized and distributed ones. Compared with the centralized approach, the distributed approach has proven to be an effective method for the cooperative control due to its advantages including low operational costs, less system requirements, high robustness, strong adaptivity and flexible scalability. Accordingly, considerable research attention has been devoted to distributed cooperative control problems for multi-agent systems.

The consensus problem is one of the critical problems for cooperative control issues of multi-agent systems. The objective of the consensus control is to design a controller for each agent by using its neighboring information such that all the agents asymptotically reach an agreement on a consistent state. Since the pioneering works proposed in [1, 2], there has been an ongoing research interest in the consensus problem of multi-agent systems, see, e.g. [3–11]. Most available results are concerned with *time-invariant* models. Unfortunately, in real-world applications, almost all practical systems are subject to certain type of time-varying parameter variations including temperature fluctuations, operating point shifting and graduate aging of system components, which motivates researchers to study the consensus control problems of time-varying multi-agent systems.

In this chapter, we shall consider the $\mathscr{H}_\infty$ consensus problem for a class of time-varying multi-agent systems subject to the RA protocol. A directed graph is used to characterize the communication topology of the multi-agent network. The data transmission between each agent and the neighboring ones is implemented via a

L. Zou et al., *Communication-Protocol-Based Filtering and Control of Networked Systems*, Studies in Systems, Decision and Control 430,
https://doi.org/10.1007/978-3-030-97512-8_10

constrained communication channel where only one neighboring agent is allowed to transmit data at each time instant. The RA protocol is applied to schedule the signal transmission of the multi-agent system. A sequence of random variables is utilized to capture the scheduling behavior of the RA protocol. The purpose of the addressed problem is to design a cooperative controller for each agent such that, for all probabilistic scheduling behaviors, the $\mathscr{H}_\infty$ consensus performance is achieved over a given finite horizon for the closed-loop multi-agent system.

The rest of this chapter is organized as follows. In Sect. 10.1, the time-varying multi-agent system with the RA protocol scheduling is introduced and some preliminaries are briefly outlined. In Sect. 10.2, the design problem of the cooperative controllers is solved in terms of the solution of two coupled backward RDEs. Furthermore, a numerical simulation example is given in Sect. 10.3 to illustrate the effectiveness of the proposed controller design scheme. Finally, we conclude the chapter in Sect. 10.4.

10.1 Problem Formulation and Preliminaries

10.1.1 Graph Topology

In this chapter, the communication topology of the multi-agent system is described by a directed graph $\mathscr{G} = (\mathscr{V}, \mathscr{E}, \mathscr{H})$ of order N with the set of agents $\mathscr{V} = \{1, 2, \ldots, N\}$, the set of edges $\mathscr{E} = \mathscr{V} \times \mathscr{V}$, and the weighted adjacency matrix $\mathscr{H} = [h_{ij}]$ with non-negative adjacency element h_{ij}. An directed edge of $\mathscr{G}$ denoted by (i, j) means that the agent i can receive information from the agent j. The elements of the weighted adjacency matrix $\mathscr{H}$ are defined such that $h_{ij} > 0 \Longleftrightarrow (i, j) \in \mathscr{E}$, while $h_{ij} = 0 \Longleftrightarrow (i, j) \notin \mathscr{E}$. Throughout this chapter, self-edges (i, i) are not allowed, i.e. $(i, i) \notin \mathscr{E}$ and $h_{ii} = 0$ for any $i \in \mathscr{V}$. The Laplacian of the directed graph $\mathscr{G}$ is defined as $\bar{\mathscr{H}} = \mathscr{H} - \mathscr{D}$ where $\mathscr{D} = \mathrm{diag}\{d_1, d_2, \ldots, d_N\}$ is the degree matrix with $d_i = \sum_{j\in\mathscr{V}_i} h_{ij}$. For each agent $i \in \mathscr{V}$, the neighboring agents of the agent i are denoted by the set $\mathscr{V}_i \triangleq \{j \in \mathscr{V} : (i, j) \in \mathscr{E}\}$.

10.1.2 Random Access Protocol of Multi-agent Systems

Consider a multi-agent system with N identical agents labeled by $\mathscr{V} = \{1, 2, \ldots, N\}$. For an agent $l \in \mathscr{V}$, only one neighboring agent $i \in \mathscr{V}_l$ is allowed to transmit data to agent l at each transmission instant for the purpose of preventing the data from collisions. Let $\xi_l(k) \in \mathscr{V}_l$ be the selected neighboring agent of agent l which obtains the opportunity to transmit data at time instant k. The occurrence probability of $\xi_l(k) = i$ is given by

$$\text{Prob}\{\xi_l(k) = i\} = p_l^i(k), \quad \forall l \in \{1, 2, \ldots, N\} \tag{10.1}$$

where $p_l^i(k) \geq 0$ $(i \in \mathscr{V}_l)$ is the occurrence probability that agent i is selected to transmit data to agent l at time instant k and $\sum_{i \in \mathscr{V}_l} p_l^i(k) = 1$ $(i \in \mathscr{V}_l)$. Without loss of generality, in this chapter, we assume that $p_l^i(k) = 0$ for all the $i \notin \mathscr{V}_l$.

10.1.3 Problem Formulation

Consider a multi-agent system with N identical agents labeled by $\mathscr{V} = \{1, 2, \ldots, N\}$ in which the dynamics of agent $l \in \mathscr{V}$ is described by the following discrete time-varying system defined on the finite-horizon ($k \in [0, \mathscr{T} - 1]$):

$$\begin{cases} x_l(k+1) = A(k)x_l(k) + B(k)u_l(k) + E(k)\omega_l(k) \\ \quad y_l(k) = C(k)x_l(k) + D(k)\nu_l(k) \\ \quad z_l(k) = M(k)x_l(k) \end{cases} \tag{10.2}$$

where $x_l(k) \in \mathbb{R}^{n_x}$, $y_l(k) \in \mathbb{R}^{n_y}$, $u_l(k) \in \mathbb{R}^{n_u}$ and $z_l(k) \in \mathbb{R}^{n_z}$ denote the system state, measurement output, control input and the controlled output of agent i, respectively. $\omega_i(k) \in l_2([0, \mathscr{T} - 1]; \mathbb{R}^{n_\omega})$ and $\nu_i(k) \in l_2([0, \mathscr{T} - 1]; \mathbb{R}^{n_\nu})$ are the external disturbances. $A(k)$, $B(k)$, $C(k)$, $D(k)$, $E(k)$ and $M(k)$ are known time-varying matrices with compatible dimensions.

In this chapter, the finite-horizon $\mathscr{H}_\infty$ consensus problem is taken into account to alleviate the interferences from the time-varying parameters and the external disturbances on the desired agreement among the agents. Now, let us introduce the following definition about the finite-horizon $\mathscr{H}_\infty$ consensus performance.

Definition 10.1 Let a disturbance attenuation level $\gamma > 0$ and a positive definite matrix $W = W^T > 0$ be given. The multi-agent system (10.2) with a prefixed connected topology is said to satisfy the $\mathscr{H}_\infty$ consensus performance constraint over the finite-horizon $[0, \mathscr{T} - 1]$ if the following inequality

$$\mathbb{E}\left\{\sum_{l=1}^{N}\sum_{k=0}^{\mathscr{T}-1}\left(\|\bar{z}_l(k)\|^2 - \gamma^2\left(\|\omega_l(k)\|^2 + \|\nu_l(k)\|^2\right)\right) - \gamma^2\sum_{l=1}^{N}\bar{x}_l^T(0)W\bar{x}_l(0)\right\} < 0 \tag{10.3}$$

holds, where $\bar{z}_l(k) = z_l(k) - \frac{1}{N}\sum_{l=1}^{N} z_l(k)$ and $\bar{x}_l(0) = x_l(0) - \frac{1}{N}\sum_{i=1}^{N} x_i(0)$. Obviously, the consensus performance of the multi-agent system (10.2) is affected by the RA protocol scheduling behavior, which is modeled by a sequence of random variables with the occurrence probabilities given by (10.1). Hence, the consensus definition (10.3) is defined in a probabilistic sense for the purpose of capturing the consensus performance subject to the random nature of the RA protocol scheduling.

10.1.4 Cooperative Controllers Design

We now consider the information exchange among the agents. In this chapter, the RA protocol is utilized to schedule the data transmissions between each agent and its neighboring agents. For the agent $l \in \mathscr{V}$, let $\xi_l(k) \in \mathscr{V}_l$ denote the selected neighboring agent transmitting data to agent l at the time instant k. Then, under the RA protocol scheduling, $\xi_l(k)$ can be governed by a random sequence with the known probability distribution given by (10.1).

According to the RA protocol scheduling, the cooperative controller of the agent l is designed as

$$\begin{aligned} u_l(k) &= K(k)h_{l\xi_l(k)}\big(y_{\xi_l(k)} - y_l(k)\big) \\ &= K(k)\sum_{i\in\mathscr{V}_l} h_{li}\lambda_i^l(k)\big(y_i(k) - y_l(k)\big) \end{aligned} \tag{10.4}$$

where $K(k) \in \mathbb{R}^{n_x\times n_y}$ $(0 \le k \le \mathscr{T}-1)$ are feedback gain matrices to be determined, $\lambda_i^l(k) = \delta(i - \xi_l(k))$ $(i \in \mathscr{V}_l)$ and $\delta(\cdot)$ is the Kronecker delta function.

Defining

$$u(k) \triangleq \left[u_1^T(k)\ u_2^T(k)\ \cdots\ u_N^T(k)\right]^T,\ y(k) \triangleq \left[y_1^T(k)\ y_2^T(k)\ \cdots\ y_N^T(k)\right]^T,$$

it can be concluded from (10.4) that

$$u(k) = \mathscr{K}(k)\big(\overrightarrow{\mathscr{H}}(\bar{\xi}(k)) \otimes I_{n_y}\big)y(k) \tag{10.5}$$

where $\bar{\xi}(k) = \left[\xi_1(k)\ \xi_2(k)\ \cdots\ \xi_N(k)\right]$ and

$$\begin{aligned} &\overrightarrow{\mathscr{H}}(\bar{\xi}(k)) = \mathscr{H} \circ \overrightarrow{\Lambda}(k) - \overrightarrow{\mathscr{D}}(k), \quad \overrightarrow{\Lambda}(k) = \left[\lambda_j^i(k)\right]_{N\times N}, \quad \mathscr{K}(k) = I_N \otimes K(k), \\ &\overrightarrow{d}_i(k) = \sum_{j\in\mathscr{V}_i} h_{ij}\lambda_j^i(k), \quad \overrightarrow{\mathscr{D}}(k) = \operatorname{diag}\{\overrightarrow{d}_1(k), \overrightarrow{d}_2(k), \ldots, \overrightarrow{d}_N(k)\}. \end{aligned}$$

Obviously, the value of the matrix $\overrightarrow{\mathscr{H}}(\bar{\xi}(k))$ depends on the sequences $\xi_i(k)$ $(i \in \{1, 2, \ldots, N\},\ k = 0, 1, \ldots, \mathscr{T}-1)$, which renders $\overrightarrow{\mathscr{H}}(\bar{\xi}(k))$ a *stochastic parameter matrix*. For analysis convenience, we now reformulate the controllers (10.5) by mapping the sequences of random variables $\xi_i(k)$ $(i \in \{1, 2, \ldots, N\}$, $k = 0, 1, \ldots, \mathscr{T}-1)$ to one random sequence.

Proposition 10.1 *The RA protocol scheduling behaviors governed by $\xi_i(k)$ $(i = 1, 2, \ldots, N)$ of the controllers (10.5) can be mapped to the sequence $r(k) \in \mathscr{R} \triangleq \left\{1, 2, \ldots, N^N\right\}$ by the following mapping $\mathcal{R}(\cdot)$:*

$$r(k) = \mathcal{R}\big(\bar{\xi}(k)\big) \triangleq \sum_{l=1}^{N} N^{l-1}\big(\xi_l(k) - 1\big) + 1 \tag{10.6}$$

in which $\prod_{s=1}^{0} n_s$ *is set to be* $\prod_{s=1}^{0} n_s = 1$*. Moreover, if* $r(k)$ *is given, the values of* $\xi_l(k)$ *can be derived by* $\phi_l(r(k))$ $(l = 1, 2, \ldots, N)$*:*

$$\xi_l(k) = \phi_l(r(k)) \triangleq \text{mod}\left(\left\lfloor \frac{r(k) - 1}{N^{l-1}} \right\rfloor, N\right) + 1. \tag{10.7}$$

Proof First, it is easy to see that the value of $r(k)$ obtained by (10.6) satisfies $r(k) \in \mathscr{R}$. Next, we shall prove that $\xi_l(k)$ can be derived by (10.7). For any given $r(k) \in \mathscr{R}$, we have

$$\begin{aligned}
\phi_l(r(k)) &= \text{mod}\left(\left\lfloor \frac{r(k) - 1}{N^{l-1}} \right\rfloor, N\right) + 1 \\
&= \text{mod}\left(\sum_{l=1}^{N} N^{l-1}\big(\xi_l(k) - 1\big), N\right) + 1 \\
&= \xi_l(k) - 1 + 1 = \xi_l(k).
\end{aligned}$$

The proof is complete.

It can be seen from Proposition 10.1 that there is a *one-to-one correspondence* between the variable $r(k)$ and the vector $\bar{\xi}(k)$. Hence, the characteristics of the new random sequence $r(k)$ $(k = 0, 1, \ldots, \mathscr{T} - 1)$ is determined by the random sequences $\xi_l(k)$ $(l \in \{1, 2, \ldots, N\}, k = 0, 1, \ldots, \mathscr{T} - 1)$. The probability distribution of the random sequence $r(k)$ is proposed in the following proposition.

Proposition 10.2 *For any* $k \in [0, \mathscr{T} - 1]$*, the random sequence* $r(k)$ *takes values in* $\mathscr{R}$ *and the occurrence probability of* $r(k) = i \in \mathscr{R}$ *is calculated as follows:*

$$\bar{p}_i(k) \triangleq \text{Prob}\{r(k) = i\} = \Pi_{l=1}^{N}\text{Prob}\,(\xi_l(k) = \phi_l(i)) = \Pi_{l=1}^{N} p_l^{\phi_l(i)}(k). \tag{10.8}$$

Proof The proof is quite straightforward and is therefore omitted for conciseness.

10.1.5 The Closed-Loop System

The follow lemmas will be used in obtaining our main results.

Lemma 10.1 *[12] For arbitrary matrices A, B, C and D with appropriate dimensions, the Kronecker product* $\otimes$ *satisfies*

(i) $(A+B)\otimes C = A\otimes C + B\otimes C,$
(ii) $(A\otimes B)(C\otimes D) = (AC)\otimes(BD),$
(iii) $(A\otimes B)^T = A^T\otimes B^T.$

Lemma 10.2 *[13] For arbitrary matrices $A\in\mathbb{R}^{n\times n}$ and $B\in\mathbb{R}^{n\times n}$, the row-sums of $A\circ B$ are the diagonal elements of AB^T:*

$$\sum_{j=1}^{N}\left(A\circ B\right)_{i,j} = \left(AB^T\right)_{i,i}$$

where the notation $\left(A\right)_{i,j}$ denotes the entry in the ith row and jth column of the matrix A.

For the purpose of simplicity, we introduce the following notations:

$$x(k) = \left[x_1^T(k)\; x_2^T(k)\;\cdots x_N^T(k)\right]^T,\; z(k) = \left[z_1^T(k)\; z_2^T(k)\;\cdots z_N^T(k)\right]^T,$$
$$\omega(k) = \left[\omega_1^T(k)\; \omega_2^T(k)\;\cdots \omega_N^T(k)\right]^T,\; \nu(k) = \left[\nu_1^T(k)\; \nu_2^T(k)\;\cdots \nu_N^T(k)\right]^T.$$

Then, the closed-loop system resulting from (10.2) to (10.5) can be written as

$$\begin{cases} x(k+1) = \big(I_N\otimes A(k)\big)x(k) + \big(I_N\otimes B(k)\big)u(k) + \big(I_N\otimes E(k)\big)\omega(k) \\ \quad = \Big(\big(I_N\otimes A(k)\big) + \big(I_N\otimes B(k)\big)\mathscr{K}(k)\big(\check{\Lambda}(r(k))\otimes I_{n_y}\big)\big(I_N\otimes C(k)\big)\Big)x(k) \\ \qquad + \big(I_N\otimes E(k)\big)\omega(k) + \big(I_N\otimes B(k)\big)\mathscr{K}(k)\big(\check{\Lambda}(r(k))\otimes I_{n_y}\big)\big(I_N\otimes D(k)\big)\nu(k) \\ \qquad + \Big(\big(\check{\Lambda}(r(k))\otimes B(k)K(k)C(k)\big) + \big(I_N\otimes A(k)\big)\Big)x(k) + \big(I_N\otimes E(k)\big)\omega(k) \\ \qquad + \big(\check{\Lambda}(r(k))\otimes B(k)K(k)D(k)\big)\nu(k) \\ z(k) = \big(I_N\otimes M(k)\big)x(k) \end{cases} \tag{10.9}$$

where $\check{\Lambda}(r(k)) = \overrightarrow{\mathscr{H}}(\xi(k))$.

Let

$$\bar{x}(k) = \left[\bar{x}_1^T(k)\; \bar{x}_2^T(k)\;\cdots\; \bar{x}_N^T(k)\right]^T,\; \bar{z}(k) = \left[\bar{z}_1^T(k)\; \bar{z}_2^T(k)\;\cdots\; \bar{z}_N^T(k)\right]^T$$

with $\bar{x}_l(k) = x_l(k) - \frac{1}{N}\sum_{i=0}^{N}x_i(k)$. Noticing that $\bar{x}(k) = \big(\mathfrak{I}\otimes I_{n_x}\big)x(k)$ and $\bar{z}(k) = \big(\mathfrak{I}\otimes I_{n_z}\big)z(k)$ where $\mathfrak{I} = I_N - \frac{1}{N}\mathbf{1}_N\mathbf{1}_N^T$, it can be derived that

$$\begin{cases} \bar{x}(k+1) = \Big(\big(\mathfrak{I}\check{\Lambda}(r(k))\otimes B(k)K(k)C(k)\big) + \big(\mathfrak{I}\otimes A(k)\big)\Big)x(k) \\ \qquad\qquad + \big(\mathfrak{I}\otimes E(k)\big)\omega(k) + \big(\mathfrak{I}\check{\Lambda}(r(k))\otimes B(k)K(k)D(k)\big)\nu(k) \\ \bar{z}(k) = \big(\mathfrak{I}\otimes M(k)\big)x(k). \end{cases} \tag{10.10}$$

On the other hand, it can be concluded from Lemma 10.2 that

$$\sum_{j=1}^{N}\big(\check{\Lambda}(r(k))\big)_{i,j}=\sum_{j=1}^{N}\big(\mathscr{H}\circ\overrightarrow{\Lambda}(k)\big)_{i,j}-\overrightarrow{d}_i(k)$$
$$=\big(\mathscr{H}\overrightarrow{\Lambda}^T(k)\big)_{i,i}-\sum_{j\in\mathscr{V}_i}h_{ij}\lambda_j^i(k)=0. \tag{10.11}$$

Then, it can be seen from (10.11) and the property of the matrix $\mathfrak{I}$ that $\check{\Lambda}(r(k))\mathfrak{I}=\check{\Lambda}(r(k))$. Apparently, (10.10) can be converted into the form

$$\begin{cases}\bar{x}(k+1)=\big(\mathscr{A}(k)+\mathscr{B}_{r(k)}(k)\big)\bar{x}(k)+\mathscr{D}_{r(k)}(k)\bar{\omega}(k)\\ \bar{z}(k)=\mathscr{M}(k)\bar{x}(k)\end{cases} \tag{10.12}$$

where

$$\mathscr{A}(k)=I_N\otimes A(k),\ \mathscr{M}(k)=I_N\otimes M(k),\ \mathscr{B}_{r(k)}(k)=\mathfrak{I}\check{\Lambda}(r(k))\otimes B(k)K(k)C(k),$$
$$\mathscr{D}_{r(k)}(k)=\big[\mathfrak{I}\otimes E(k)\ \mathfrak{I}\check{\Lambda}(r(k))\otimes B(k)K(k)D(k)\big],\quad \bar{\omega}(k)=\big[\omega^T(k)\ \nu^T(k)\big]^T.$$

Then, according to the probability distribution of $r(k)$ given by Proposition 10.2, we have

$$\bar{\Lambda}(k)\triangleq\mathbb{E}\{\check{\Lambda}(r(k))\}=\sum_{i\in\mathscr{R}}\bar{p}_i(k)\check{\Lambda}(i). \tag{10.13}$$

Defining $\tilde{\Lambda}(r(k))\triangleq\check{\Lambda}(r(k))-\bar{\Lambda}(k)$, the time-varying system (10.12) can be rewritten as follows:

$$\begin{cases}\bar{x}(k+1)=\big(\mathscr{A}(k)+\bar{\mathscr{B}}(k)+\tilde{\mathscr{B}}_{r(k)}(k)\big)\bar{x}(k)+\big(\bar{\mathscr{D}}(k)+\tilde{\mathscr{D}}_{r(k)}(k)\big)\bar{\omega}(k)\\ \bar{z}(k)=\mathscr{M}(k)\bar{x}(k)\end{cases} \tag{10.14}$$

where

$$\bar{\mathscr{B}}(k)=\mathfrak{I}\bar{\Lambda}(k)\otimes B(k)K(k)C(k),\ \tilde{\mathscr{D}}_{r(k)}(k)=\big[0\ \mathfrak{I}\tilde{\Lambda}(r(k))\otimes B(k)K(k)D(k)\big],$$
$$\bar{\mathscr{D}}(k)=\big[\mathfrak{I}\otimes E(k)\ \mathfrak{I}\bar{\Lambda}(k)\otimes B(k)K(k)D(k)\big],\ \tilde{\mathscr{B}}_{r(k)}(k)=\mathfrak{I}\tilde{\Lambda}(r(k))\otimes B(k)K(k)C(k).$$

Remark 10.1 Due to the RA protocol scheduling behavior (10.1) and the cooperative controllers (10.4), the closed-loop system (10.14) is reformulated to be a time-varying system with a *stochastic parameter matrix* $\tilde{\Lambda}(r(k))$ whose statistical properties are given by (10.13). The main difficulty encountered in this chapter would be the design of the controller parameters $K(k)$ ($0\le k\le\mathscr{T}-1$) for the time-varying system (10.14) with such a stochastic parameter matrix $\tilde{\Lambda}(r(k))$. Note that the state estimation problem with stochastic parameter matrices has been investigated in [14, 15] for networked systems. This chapter represents one of the first attempts to study the consensus control problem for multi-agent systems with stochastic parameter matrices and we endeavor to derive the necessary and sufficient condition under which the desired controller parameters exist.

We are now in the position to state the problem addressed in this chapter. We aim to design the controller parameters $K(k)$ $(0 \leq k \leq \mathscr{T} - 1)$ for the time-varying system (10.12) such that, for all possible realizations of the random sequence $r(k)$, the following $\mathscr{H}_\infty$ consensus performance is achieved over the given finite-horizon $[0, \mathscr{T} - 1]$:

$$J_1 = \mathbb{E}\left\{ \sum_{k=0}^{\mathscr{T}-1} \left(\|\bar{z}(k)\|^2 - \gamma^2 \|\bar{\omega}(k)\|^2 \right) - \gamma^2 \bar{x}^T(0)\left(I_N \otimes W\right)\bar{x}(0) \right\} < 0, \quad \forall(\bar{\omega}(k), \bar{x}(0)) \neq 0 \tag{10.15}$$

where W is a given positive definite matrix.

10.2 Main Results

Lemma 10.3 *[16] Let $\mathscr{U}$, $\mathscr{V}$ and $\mathscr{W}$ be known nonzero matrices with appropriate dimensions. The solution X to $\min_X \|\mathscr{U} X \mathscr{W} - \mathscr{V}\|_F$ is $\mathscr{U}^\dagger \mathscr{V} \mathscr{W}^\dagger$.*

Lemma 10.4 *Consider the time-varying system (10.14) with the stochastic parameter matrix $\tilde{\Lambda}(r(k))$ whose statistical properties is given in (10.13). Let the cooperative controller parameters $\{K(k)\}_{0 \leq k \leq \mathscr{T}-1}$, the disturbance attenuation level $\gamma > 0$ and the positive definite matrix W be given. For any disturbance sequence $\{\bar{\omega}(k)\}_{0 \leq k \leq \mathscr{T}-1}$, the time-varying system (10.14) satisfies the $\mathscr{H}_\infty$ performance requirement if and only if there exist a family of non-negative definite matrices $\{P(k)\}_{0 \leq k \leq \mathscr{T}-1}$ (with final condition $P(\mathscr{T}) = 0$) satisfying the following backward RDE:*

$$\begin{aligned} P(k) = &(\mathscr{A}(k) + \bar{\mathscr{B}}(k))^T P(k+1)(\mathscr{A}(k) + \bar{\mathscr{B}}(k)) + \Omega_1(k+1) + \mathscr{M}^T(k)\mathscr{M}(k) \\ &+ \left(\Omega_2(k+1) + (\mathscr{A}(k) + \bar{\mathscr{B}}(k))^T P(k+1)\bar{\mathscr{D}}(k)\right)\Delta^{-1}(k)\left(\Omega_2(k+1)\right. \\ &\left.+ (\mathscr{A}(k) + \bar{\mathscr{B}}(k))^T P(k+1)\bar{\mathscr{D}}(k)\right)^T \end{aligned} \tag{10.16}$$

subject to

$$\begin{cases} \Delta(k) \triangleq \gamma^2 I - \Omega_3(k+1) - \bar{\mathscr{D}}^T(k)P(k+1)\bar{\mathscr{D}}(k) > 0 \\ P(0) \leq \gamma^2(I_N \otimes W) \end{cases} \tag{10.17}$$

where

$$\Omega_1(k+1) = \sum_{i\in\mathscr{R}} \bar{p}_i(k)\tilde{\mathscr{B}}_i^T(k)P(k+1)\tilde{\mathscr{B}}_i(k),\ \mathfrak{C} = B(k)K(k)C(k),$$

$$\Omega_2(k+1) = \sum_{i\in\mathscr{R}} \bar{p}_i(k)\tilde{\mathscr{B}}_i^T(k)P(k+1)\tilde{\mathscr{D}}_i(k),\ \mathfrak{D} = B(k)K(k)D(k),$$

$$\Omega_3(k+1) = \sum_{i\in\mathscr{R}} \bar{p}_i(k)\tilde{\mathscr{D}}_i^T(k)P(k+1)\tilde{\mathscr{D}}_i(k)$$

and the nonempty finite set $\mathscr{R}$ is defined in Proposition 10.1.

Proof *Sufficiency:* Define $V(k) = \bar{x}^T(k)P(k)\bar{x}(k)$. Then, calculate $V(k+1) - V(k)$ along the system trajectory (10.14) and take expectation as follows:

$$\begin{aligned}
&\mathscr{J}_1(k) \triangleq \mathbb{E}\{V(k+1) - V(k)\}\\
=&\mathbb{E}\{\bar{x}^T(k+1)P(k+1)\bar{x}(k+1) - \bar{x}^T(k)P(k)\bar{x}(k)\}\\
=&\mathbb{E}\Big\{\big((\mathscr{A}(k) + \bar{\mathscr{B}}(k) + \tilde{\mathscr{B}}_{r(k)}(k))\bar{x}(k) + (\bar{\mathscr{D}}(k) + \tilde{\mathscr{D}}_{r(k)}(k))\bar{\omega}(k)\big)^T P(k+1)\big((\mathscr{A}(k)\\
&+ \bar{\mathscr{B}}(k) + \tilde{\mathscr{B}}_{r(k)}(k))\bar{x}(k) + (\bar{\mathscr{D}}(k) + \tilde{\mathscr{D}}_{r(k)}(k))\bar{\omega}(k)\big) - \bar{x}^T(k)P(k)\bar{x}(k)\Big\}\\
=&\mathbb{E}\Big\{\big((\mathscr{A}(k) + \bar{\mathscr{B}}(k))\bar{x}(k) + \bar{\mathscr{D}}(k)\bar{\omega}(k)\big)^T P(k+1)\big((\mathscr{A}(k) + \bar{\mathscr{B}}(k))\bar{x}(k)\\
&+ \bar{\mathscr{D}}(k)\bar{\omega}(k)\big) + \bar{x}^T(k)\tilde{\mathscr{B}}_{r(k)}^T(k)P(k+1)\tilde{\mathscr{B}}_{r(k)}(k)\bar{x}(k) - \bar{x}^T(k)P(k)\bar{x}(k)\\
&+ 2\bar{x}^T(k)\tilde{\mathscr{B}}_{r(k)}^T(k)P(k+1)\tilde{\mathscr{D}}_{r(k)}(k)\bar{\omega}(k) + \bar{\omega}^T(k)\tilde{\mathscr{D}}_{r(k)}^T P(k+1)\tilde{\mathscr{D}}_{r(k)}\bar{\omega}(k)\Big\}.
\end{aligned} \tag{10.18}$$

In view of the statistical properties of $\tilde{\Lambda}(r(k))$, it is easy to see that

$$\begin{cases}
\mathbb{E}\{\tilde{\mathscr{B}}_{r(k)}^T(k)P(k+1)\tilde{\mathscr{B}}_{r(k)}(k)\} = \Omega_1(k+1)\\
\mathbb{E}\{\tilde{\mathscr{B}}_{r(k)}^T(k)P(k+1)\tilde{\mathscr{D}}_{r(k)}(k)\} = \Omega_2(k+1)\\
\mathbb{E}\{\tilde{\mathscr{D}}_{r(k)}^T(k)P(k+1)\tilde{\mathscr{D}}_{r(k)}(k)\} = \Omega_3(k+1).
\end{cases} \tag{10.19}$$

Therefore, it can be concluded from (10.18) to (10.19) that

$$\begin{aligned}
\mathscr{J}_1(k) =& \mathbb{E}\Big\{\bar{x}^T(k)\big((\mathscr{A}(k) + \bar{\mathscr{B}}(k))^T P(k+1)(\mathscr{A}(k) + \bar{\mathscr{B}}(k)) + \Omega_1(k+1)\\
&- P(k)\big)\bar{x}(k) + 2\bar{x}^T(k)\big(\Omega_2(k+1) + (\mathscr{A}(k) + \bar{\mathscr{B}}(k))^T P(k+1)\bar{\mathscr{D}}(k)\big)\bar{\omega}(k)\\
&+ \bar{\omega}^T(k)\big(\Omega_3(k+1) + \bar{\mathscr{D}}^T(k)P(k+1)\bar{\mathscr{D}}(k)\big)\bar{\omega}(k)\Big\}.
\end{aligned} \tag{10.20}$$

Adding the following zero term

$$\|\bar{z}(k)\|^2 - \gamma^2\|\bar{\omega}(k)\|^2 - (\|\bar{z}(k)\|^2 - \gamma^2\|\bar{\omega}(k)\|^2) \tag{10.21}$$

to both sides of (10.20), we have

$$\begin{aligned}\mathscr{J}_1(k) =&\mathbb{E}\Big\{\gamma^2\|\bar{\omega}(k)\|^2 + \bar{x}^T(k)\big(\mathscr{M}^T(k)\mathscr{M}(k) + (\mathscr{A}(k) + \bar{\mathscr{B}}(k))^T P(k+1)(\mathscr{A}(k)\\&+ \bar{\mathscr{B}}(k)) + \Omega_1(k+1) - P(k)\big)\bar{x}(k) - \|\bar{z}(k)\|^2 + 2\bar{x}^T(k)\big(\Omega_2(k+1)\\&+ (\mathscr{A}(k) + \bar{\mathscr{B}}(k))^T P(k+1)\bar{\mathscr{D}}(k)\big)\bar{\omega}(k) - \bar{\omega}^T(k)\big(\gamma^2 I - \Omega_3(k+1)\\&- \bar{\mathscr{D}}^T(k)P(k+1)\bar{\mathscr{D}}(k)\big)\bar{\omega}(k)\Big\}.\end{aligned} \tag{10.22}$$

Applying the completing squares method, we have

$$\begin{aligned}\mathscr{J}_1(k) =&\mathbb{E}\Big\{\bar{x}^T(k)\big(\mathscr{M}^T(k)\mathscr{M}(k) + (\mathscr{A}(k) + \bar{\mathscr{B}}(k))^T P(k+1)(\mathscr{A}(k) + \bar{\mathscr{B}}(k))\\&+ \Omega_1(k+1) - P(k)\big)\bar{x}(k) + (\bar{\omega}^*(k))^T \Delta(k)(\bar{\omega}^*(k)) - \big(\bar{\omega}(k) - \bar{\omega}^*(k)\big)^T\\&\times \Delta(k)\big(\bar{\omega}(k) - \bar{\omega}^*(k)\big)\Big\} - \mathbb{E}\big\{\|\bar{z}(k)\|^2 - \gamma^2\|\bar{\omega}(k)\|^2\big\}\end{aligned} \tag{10.23}$$

where $\bar{\omega}^*(k) = \Delta^{-1}(k)\big(\bar{\mathscr{D}}^T(k)P(k+1)(\mathscr{A}(k) + \bar{\mathscr{B}}(k)) + \Omega_2^T(k+1)\big)\bar{x}(k)$.

Taking the sum on both sides of (10.23) from 0 to $\mathscr{T} - 1$, we obtain

$$\begin{aligned}&\mathbb{E}\{\bar{x}^T(\mathscr{T})P(\mathscr{T})\bar{x}(\mathscr{T}) - \bar{x}^T(0)P(0)\bar{x}(0)\} = \sum_{k=0}^{\mathscr{T}-1} \mathscr{J}_1(k)\\&=\mathbb{E}\left\{-\sum_{k=0}^{\mathscr{T}-1} \big(\bar{\omega}(k) - \bar{\omega}^*(k)\big)^T \Delta(k)\big(\bar{\omega}(k) - \bar{\omega}^*(k)\big)\right\}\\&\quad - \mathbb{E}\left\{\sum_{k=0}^{\mathscr{T}-1} \big(\|\bar{z}(k)\|^2 - \gamma^2\|\bar{\omega}(k)\|^2\big)\right\}.\end{aligned} \tag{10.24}$$

Since $\Delta(k) > 0$, $P(0) \le \gamma^2(I_N \otimes W)$ and $P(\mathscr{T}) = 0$, we have

$$\begin{aligned}J_1 =&\mathbb{E}\Big\{-\sum_{k=0}^{\mathscr{T}-1} \big(\bar{\omega}(k) - \bar{\omega}^*(k)\big)^T \Delta(k)\big(\bar{\omega}(k) - \bar{\omega}^*(k)\big)\\&+ \bar{x}^T(0)\big(P(0) - \gamma^2(I_N \otimes W)\big)\bar{x}(0)\Big\}\end{aligned} \tag{10.25}$$

which means the pre-specified $\mathscr{H}_\infty$ performance (10.15) is satisfied. The proof of sufficiency is complete.

Necessity: The proof follows directly from that of Lemma 3 in [15] and is therefore omitted.

So far, we have analyzed the $\mathscr{H}_\infty$ performance in terms of the solvability of a backward Riccati equation in Lemma 10.4. In the next stage, we like to develop an approach for dealing with the design issue of the cooperative controllers (10.5) under the worst situation, i.e. $\bar{\omega}(k) = \bar{\omega}^*(k) = \Delta^{-1}(k)\big(\bar{\mathscr{D}}^T(k)P(k+1)(\mathscr{A}(k)+\bar{\mathscr{B}}(k)) + \Omega_2^T(k+1)\big)\bar{x}(k)$. In this case, the closed-loop system (10.14) can be rewritten as follows:

$$\begin{aligned}\bar{x}(k+1) = &\Big(\mathscr{A}(k) + \bar{\mathscr{D}}(k)\mathscr{W}(k)\Big)\bar{x}(k) + \mathfrak{B}(k)\bar{u}(k) + \big(\tilde{\mathscr{B}}_{r(k)}(k) \\ &+ \tilde{\mathscr{D}}_{r(k)}(k)\mathscr{W}(k)\big)\bar{x}(k)\end{aligned} \tag{10.26}$$

where

$$\begin{aligned}&\mathscr{W}(k) = \Delta^{-1}(k)\big(\bar{\mathscr{D}}^T(k)P(k+1)(\mathscr{A}(k)+\bar{\mathscr{B}}(k)) + \Omega_2^T(k+1)\big),\\ &\mathfrak{B}(k) = I_N \otimes B(k), \quad \bar{u}(k) = (I_N \otimes K(k))(\mathfrak{I}\bar{\Lambda}(k)\otimes C(k))\bar{x}(k).\end{aligned}$$

Then, we define the following cost functional:

$$J_2 \triangleq \mathbb{E}\left\{\sum_{k=0}^{\mathscr{T}-1}\big(\|\bar{z}(k)\|^2 + \varepsilon\|\bar{u}(k)\|^2\big)\right\} \tag{10.27}$$

where ε is a known constant introduced for more flexibility in the controller parameter design. Obviously, such a cost functional is in the form of the Linear-Quadratic functional, where $\sum_{k=0}^{\mathscr{T}-1}\|\bar{z}(k)\|^2$ represents the control performance, $\sum_{k=0}^{\mathscr{T}-1}\|\bar{u}(k)\|^2$ stands for the control effort. The values of the coefficient ε could be selected according to the special requirement on the cost functional.

Theorem 10.1 *Consider the multi-agent system (10.2) with the RA protocol scheduling governed by (10.1). Let the disturbance attenuation level $\gamma > 0$, the positive scalar $\varepsilon > 0$ and the positive definite matrix W be given. There exist controller parameters $\{K(k)\}_{0\le k\le \mathscr{T}-1}$ for the cooperative controllers (10.4) such that the closed-loop system (10.14) satisfies the $\mathscr{H}_\infty$ performance requirement (10.15) if there exist solutions $\{P(k), Q(k), K(k)\}_{0\le k\le\mathscr{T}-1}$ satisfying the backward RDE (10.16) as well as the following backward RDE:*

$$\begin{cases} Q(k) = \big(\mathscr{A}(k) + \bar{\mathscr{D}}(k)\mathscr{W}(k)\big)^T Q(k+1)\big(\mathscr{A}(k)+\bar{\mathscr{D}}(k)\mathscr{W}(k)\big) + \mathscr{M}^T(k)\mathscr{M}(k) \\ \qquad + \Upsilon_1(k+1) + \Upsilon_2(k+1)\mathscr{W}(k) + \mathscr{W}^T(k)\Upsilon_2^T(k+1) + \mathscr{W}^T(k)\Upsilon_3(k+1) \\ \qquad \times \mathscr{W}(k) + \mathscr{W}^T(k)\bar{\mathscr{D}}^T(k)Q(k+1)\bar{\mathscr{B}}(k) + \bar{\mathscr{B}}^T(k)Q(k+1) \\ \qquad \times \bar{\mathscr{D}}(k)\mathscr{W}(k) - \mathscr{A}^T(k)Q(k+1)\mathfrak{B}(k)\Theta^{-1}(k)\mathfrak{B}^T(k)Q(k+1)\mathscr{A}(k) \\ Q(N) = 0 \end{cases} \tag{10.28}$$

subject to

$$\begin{cases} P(0) \leq \gamma^2(I_N \otimes W), \quad P(\mathscr{T}) = Q(\mathscr{T}) = 0 \\ \Delta(k) = \gamma^2 I - \Omega_3(k+1) - \bar{\mathscr{D}}^T(k)P(k+1)\bar{\mathscr{D}}(k) > 0 \\ \Theta(k) = \varepsilon I + \mathfrak{B}^T(k)Q(k+1)\mathfrak{B}(k) > 0 \end{cases} \tag{10.29}$$

with the controller parameters given as follows:

$$K(k) = \arg\min_{K(k)} \left\| (I_N \otimes K(k))(\mathfrak{I}\bar{\Lambda}(k) \otimes C(k)) + \Theta^{-1}(k)\mathfrak{B}^T(k)Q(k+1)\mathscr{A}(k) \right\|_F \tag{10.30}$$

where

$$\Upsilon_1(k+1) = \sum_{i\in\mathscr{R}} \bar{p}_i(k)\tilde{\mathscr{B}}_i^T(k)Q(k+1)\tilde{\mathscr{B}}_i(k),$$
$$\Upsilon_2(k+1) = \sum_{i\in\mathscr{R}} \bar{p}_i(k)\tilde{\mathscr{B}}_i^T(k)Q(k+1)\tilde{\mathscr{D}}_i(k),$$
$$\Upsilon_3(k+1) = \sum_{i\in\mathscr{R}} \bar{p}_i(k)\tilde{\mathscr{D}}_i^T(k)Q(k+1)\tilde{\mathscr{D}}_i(k)$$

and the other corresponding matrix parameters are defined as in Lemma 10.4.

Proof First, it follows from Lemma 10.4 that, if there exist solutions $P(k)$ to (10.16) such that $\Delta(k) > 0$ and $P(0) < \gamma^2(I_N \otimes W)$, then system (10.14) achieves the pre-specified $\mathscr{H}_\infty$ performance. In this case, the worst-case disturbance can be expressed as $\bar{\omega}^*(k) = \mathscr{W}(k)\bar{x}(k)$. In what follows, by employing the worst-case disturbance, we aim to provide a design scheme of the controller parameters $K(k)$. For this purpose, it follows from (10.26) that

$$\begin{aligned}
&\mathbb{E}\{\bar{x}^T(k+1)Q(k+1)\bar{x}(k+1) - \bar{x}^T(k)Q(k)\bar{x}(k)\} \\
=&\mathbb{E}\Big\{\Big(\big(\mathscr{A}(k) + \tilde{\mathscr{B}}_{r(k)}(k) + \big(\bar{\mathscr{D}}(k) + \tilde{\mathscr{D}}_{r(k)}(k)\big)\mathscr{W}(k)\big)\bar{x}(k) + \mathfrak{B}(k)\bar{u}(k)\Big)^T \\
&\times Q(k+1)\Big(\big(\mathscr{A}(k) + \tilde{\mathscr{B}}_{r(k)}(k) + \big(\tilde{\mathscr{D}}_{r(k)}(k) + \bar{\mathscr{D}}(k)\big)\mathscr{W}(k)\big)\bar{x}(k) + \mathfrak{B}(k)\bar{u}(k)\Big) \\
&- \bar{x}^T(k)Q(k)\bar{x}(k)\Big\} \\
=&\mathbb{E}\Big\{\bar{x}^T(k)\Big(\Big(\big(\mathscr{A}(k) + \bar{\mathscr{D}}(k)\mathscr{W}(k)\big) + \big(\tilde{\mathscr{D}}_{r(k)}(k)\mathscr{W}(k) + \tilde{\mathscr{B}}_{r(k)}(k)\big)\Big)^T Q(k+1) \\
&\times \Big(\big(\mathscr{A}(k) + \bar{\mathscr{D}}(k)\mathscr{W}(k)\big) + \big(\tilde{\mathscr{B}}_{r(k)}(k) + \tilde{\mathscr{D}}_{r(k)}(k)\mathscr{W}(k)\big)\Big) - Q(k)\Big)\bar{x}(k) \\
&+ 2\bar{x}^T(k)\big(\mathscr{A}(k) + \bar{\mathscr{D}}(k)\mathscr{W}(k)\big)^T Q(k+1)\mathfrak{B}(k)\bar{u}(k) + \bar{u}^T(k)\mathfrak{B}^T(k)Q(k+1) \\
&\times \mathfrak{B}(k)\bar{u}(k)\Big\}
\end{aligned}$$

$$
\begin{aligned}
=&\mathbb{E}\Big\{\bar{x}^T(k)\Big(\big(\mathscr{A}(k)+\bar{\mathscr{D}}(k)\mathscr{W}(k)\big)^T Q(k+1)\big(\mathscr{A}(k)+\bar{\mathscr{D}}(k)\mathscr{W}(k)\big)\\
&+2\tilde{\mathscr{B}}_{r(k)}^T(k)Q(k+1)\tilde{\mathscr{D}}_{r(k)}(k)\mathscr{W}(k)+\mathscr{W}^T(k)\tilde{\mathscr{D}}_{r(k)}^T(k)Q(k+1)\tilde{\mathscr{D}}_{r(k)}(k)\mathscr{W}(k)\\
&-Q(k)\Big)\bar{x}(k)+2\bar{x}^T(k)\big(\mathscr{A}(k)+\bar{\mathscr{D}}(k)\mathscr{W}(k)\big)^T Q(k+1)\mathfrak{B}(k)\bar{u}(k)\\
&+\bar{u}^T(k)\mathfrak{B}^T(k)Q(k+1)\mathfrak{B}(k)\bar{u}(k)+\tilde{\mathscr{B}}_{r(k)}^T(k)Q(k+1)\tilde{\mathscr{B}}_{r(k)}(k)\Big\}.
\end{aligned}
\tag{10.31}
$$

Similar to the proof of Lemma 10.4, one has

$$
\begin{cases}
\mathbb{E}\{\tilde{\mathscr{B}}_{r(k)}^T(k)Q(k+1)\tilde{\mathscr{B}}_{r(k)}(k)\}=\Upsilon_1(k+1)\\
\mathbb{E}\{\tilde{\mathscr{B}}_{r(k)}^T(k)Q(k+1)\tilde{\mathscr{D}}_{r(k)}(k)\}=\Upsilon_2(k+1)\\
\mathbb{E}\{\tilde{\mathscr{D}}_{r(k)}^T(k)Q(k+1)\tilde{\mathscr{D}}_{r(k)}(k)\}=\Upsilon_3(k+1).
\end{cases}
\tag{10.32}
$$

By substituting (10.32) into (10.31), we have

$$
\begin{aligned}
&\mathbb{E}\{\bar{x}^T(k+1)Q(k+1)\bar{x}(k+1)-\bar{x}^T(k)Q(k)\bar{x}(k)\}\\
=&\mathbb{E}\Big\{\bar{x}^T(k)\Big(\big(\mathscr{A}(k)+\bar{\mathscr{D}}(k)\mathscr{W}(k)\big)^T Q(k+1)\big(\mathscr{A}(k)+\bar{\mathscr{D}}(k)\mathscr{W}(k)\big)+\Upsilon_1(k+1)\\
&+2\Upsilon_2(k+1)\mathscr{W}(k)+\mathscr{W}^T(k)\Upsilon_3(k+1)\mathscr{W}(k)-Q(k)\Big)\bar{x}(k)+2\bar{x}^T(k)\big(\mathscr{A}(k)\\
&+\bar{\mathscr{D}}(k)\mathscr{W}(k)\big)^T Q(k+1)\mathfrak{B}(k)\bar{u}(k)+\bar{u}^T(k)\mathfrak{B}^T(k)Q(k+1)\mathfrak{B}(k)\bar{u}(k)\Big\}\\
=&\mathbb{E}\Big\{\bar{x}^T(k)\Big(\big(\mathscr{A}(k)+\bar{\mathscr{D}}(k)\mathscr{W}(k)\big)^T Q(k+1)\big(\mathscr{A}(k)+\bar{\mathscr{D}}(k)\mathscr{W}(k)\big)+\Upsilon_1(k+1)\\
&+2\Upsilon_2(k+1)\mathscr{W}(k)+2\mathscr{W}^T(k)\bar{\mathscr{D}}^T(k)Q(k+1)\bar{\mathscr{B}}(k)+\mathscr{W}^T(k)\Upsilon_3(k+1)\mathscr{W}(k)\\
&-Q(k)\Big)\bar{x}(k)+2\bar{x}^T(k)\mathscr{A}^T(k)Q(k+1)\mathfrak{B}(k)\bar{u}(k)+\bar{u}^T(k)\mathfrak{B}^T(k)Q(k+1)\\
&\mathfrak{B}(k)\bar{u}(k)\Big\}.
\end{aligned}
\tag{10.33}
$$

Then, it follows from (10.28) and the cost functional (10.27) that

$$
\begin{aligned}
J_2=&\mathbb{E}\Big\{\|\bar{z}(k)\|^2+\sum_{k=0}^{\mathscr{T}-1}\Big(\bar{x}^T(k+1)Q(k+1)\bar{x}(k+1)-\bar{x}^T(k)Q(k)\bar{x}(k)\Big)\\
&+\bar{x}^T(0)Q(0)\bar{x}(0)+\varepsilon\|\bar{u}(k)\|^2\Big\}\\
=&\sum_{k=0}^{\mathscr{T}-1}\mathbb{E}\Big\{\big(\bar{u}(k)+\bar{u}^*(k)\big)^T\Theta(k)\big(\bar{u}(k)+\bar{u}^*(k)\big)\Big\}+\mathbb{E}\{\bar{x}^T(0)Q(0)\bar{x}(0)\}\\
\leq&\mathbb{E}\{\bar{x}^T(0)Q(0)\bar{x}(0)\}+\sum_{k=0}^{\mathscr{T}-1}\mathbb{E}\Big\{\big\|\mathscr{K}(k)(\mathfrak{I}\bar{\Lambda}(k)\otimes C(k))\\
&+\Theta^{-1}(k)\mathfrak{B}^T(k)Q(k+1)\mathscr{A}(k)\big\|_F^2\|\bar{x}(k)\|^2\Big\}
\end{aligned}
\tag{10.34}
$$

where $\bar{u}^*(k) = \Theta^{-1}(k)\mathfrak{B}^T(k)Q(k+1)\mathscr{A}(k)\bar{x}(k)$.

For the purpose of suppressing the cost functional (10.27), the best choice of the controller parameters is the $K(k)$ given by (10.30), and this ends the proof.

In general, it is difficult to solve the optimization problem (10.30) due to the special structure of $\mathscr{K}(k)$. For the convenience in application, the expression of the parameter $K(k)$ can be acquired by using the Moore–Penrose pseudo inverse combined with the partitioning matrix method in the following theorem.

Theorem 10.2 *Consider the multi-agent system (10.2) with the RA protocol scheduling governed by (10.1). Let the disturbance attenuation level $\gamma > 0$, the positive scalar $\varepsilon > 0$ and the positive definite matrix W be given. There exist controller parameters $\{K(k)\}_{0\le k\le M}$ for the cooperative controllers (10.4) such that the closed-loop system (10.14) satisfies the $\mathscr{H}_\infty$ performance requirement (10.15) if there exist solutions $\{P(k), Q(k), K(k)\}_{0\le k\le \mathscr{T}-1}$ satisfying the coupled backward RDEs (10.16) and (10.28) subject to (10.29) with the controller parameters given by:*

$$K(k) = -\left[\vartheta_k^{(1)}\ \vartheta_k^{(2)}\ \cdots\ \vartheta_k^{(N)}\right]\left[\psi_k^{(1)}\ \psi_k^{(2)}\ \cdots\ \varphi_k^{(N)}\right]^{\dagger} \tag{10.35}$$

where

$$\mathfrak{I}\bar{A}(k)\otimes C(k) \triangleq \left[\left(\psi_k^{(1)}\right)^T\ \left(\psi_k^{(2)}\right)^T\ \cdots\ \left(\psi_k^{(N)}\right)^T\right]^T,$$
$$\Theta^{-1}(k)\mathfrak{B}^T(k)Q(k+1)\mathscr{A}(k) \triangleq \left[\left(\vartheta_k^{(1)}\right)^T\ \left(\vartheta_k^{(2)}\right)^T\ \cdots\ \left(\vartheta_k^{(N)}\right)^T\right]^T.$$

Proof In Theorem 10.1, the controller parameters $K(k)$ are determined by (10.30), which can be rearranged by

$$\begin{aligned} K(k) &= \arg\min_{K(k)} \left\|(I_N\otimes K(k))(\mathfrak{I}\bar{A}(k)\otimes C(k)) + \Theta^{-1}(k)\mathfrak{B}^T(k)Q(k+1)\mathscr{A}(k)\right\|_F \\ &= \arg\min_{K(k)} \left\|K(k)\left[\psi_k^{(1)}\ \psi_k^{(2)}\ \cdots\ \psi_k^{(N)}\right] + \left[\vartheta_k^{(1)}\ \vartheta_k^{(2)}\ \cdots\ \vartheta_k^{(N)}\right]\right\|_F. \end{aligned} \tag{10.36}$$

Therefore, it follows from Lemma 10.3 that (10.35) is the solution of the optimization problem (10.30). With the designed $K(k)$, it follows from Theorem 10.1 that the multi-agent system achieves the pre-specified $\mathscr{H}_\infty$ performance constraint (10.15) over the finite-horizon $[0, \mathscr{T}-1]$. The proof is now complete.

By means of Theorem 10.2, we can summarize the Finite-Horizon Consensus Control Design (FHCCD) algorithm as follows:

Algorithm FHCCD:

Step 1. Let scalars $\gamma > 0$, $\varepsilon > 0$ and the positive definite matrix $W > 0$ be given. Set $k = \mathscr{T} - 1$ and $P(\mathscr{T}) = Q(\mathscr{T}) = 0$.

Step 2. Calculate the matrix $\Theta(k)$ by (10.29). If $\Theta(k)$ is positive definite, then the controller parameter $K(k)$ can be solved by (10.35), and go to the next step, else jump to *Step 6*.

Step 3. Compute the matrix $\Delta(k)$ by (10.17). If $\Delta(k)$ is positive definite, then step to the next procedure, else jump to *Step 6*.

Step 4. Solve the backward RDEs of (10.16) and (10.28) to get $P(k)$ and $Q(k)$, respectively. If $k \neq 0$, set $k = k - 1$ and go back to *Step 2*, else turn to the next step.

Step 5. If the condition $\{\Theta(k) > 0,\ \Delta(k) > 0,\ P(0) \leq \gamma^2(I_N \otimes W)\}$ is not satisfied, this algorithm is infeasible. Stop.

Remark 10.2 In this chapter, we examine how the RA protocol scheduling behavior influences the performance of the consensus control for multi-agent systems with a fixed topology. It can be observed from *Algorithm FHCCD* that, in the controller design procedure, all the important factors contributing to the system complexity have been reflected, which include: (1) the time-varying system parameters; (2) the occurrence probabilities about the node transmission opportunities under the RA protocol scheduling; and (3) the prescribed $\mathscr{H}_\infty$ consensus performance index. Compared with the existing literature, the results obtained in this chapter have the following three distinguishing features: (1) this chapter represents one of the first attempts to consider the consensus control problem for a time-varying multi-agent system with a communication protocol scheduling; (2) the stochastic parameter matrix is employed to characterize the RA protocol scheduling of agents; and (3) a *necessary and sufficient condition* is obtained for handling the $\mathscr{H}_\infty$ consensus control of a time-varying multi-agent system with a stochastic parameter matrix induced by the RA protocol scheduling.

10.3 An Illustrative Example

In order to illustrate the effectiveness of the proposed $\mathscr{H}_\infty$ cooperative controllers design scheme, we present a numerical example to illustrate the effectiveness of the obtained theoretical results in this section.

Consider a network of five agents whose topology is represented by a directed communication graph $\mathscr{G}$ (shown in Fig. 10.1) with the set of nodes $\mathscr{V} = \{1, 2, 3, 4, 5\}$ and the associated adjacency matrix $\mathscr{H}$ given as follows:

$$\mathscr{H} = \begin{bmatrix} 0\,1\,1\,0\,0 \\ 1\,0\,1\,0\,0 \\ 0\,1\,0\,1\,1 \\ 1\,0\,1\,0\,1 \\ 1\,1\,0\,1\,0 \end{bmatrix}.$$

Fig. 10.1 The directed communication graph $\mathscr{G}$

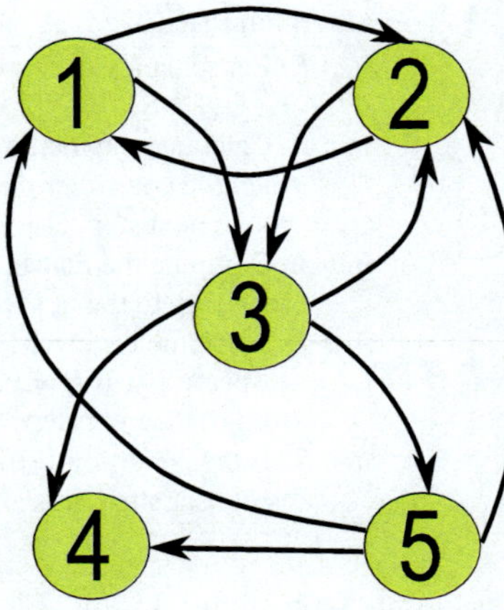

The individual dynamics of agents are given by (10.2) with the following parameters:

$$A(k) = \begin{bmatrix} 0.99 + \frac{\cos(0.4k)}{20} & -0.45 \\ -0.10 & -0.73 - \frac{\cos(0.5k)}{10} \end{bmatrix}, \quad B(k) = \begin{bmatrix} 0.2 - \frac{\sin(0.3k)}{20} \\ 0.12 \end{bmatrix},$$
$$C(k) = \begin{bmatrix} 1.05 \ 0.82 \end{bmatrix}, \quad D(k) = -0.3, \quad E(k) = \begin{bmatrix} 0.06 \ 0.24 \end{bmatrix}^T, \quad M(k) = 0.1I.$$

In this example, the $\mathscr{H}_\infty$ consensus performance level γ, the positive definite matrix W and time-horizon $\mathscr{T}$ are taken as 0.8, 9.875I and 101, respectively. The occurrence probabilities $p_i^j(k)$ is given by

$p_i^j(k)$	$j=1$	$j=2$	$j=3$	$j=4$	$j=5$
$i=1$	0	0.4	0.6	0	0
$i=2$	0.5	0	0.5	0	0
$i=3$	0	0.4	0	0.3	0.3
$i=4$	0.4	0	0.3	0	0.3
$i=5$	0.2	0.3	0	0.5	0

By applying Algorithm FHCCD, the coupled backward RDEs (10.16) and (10.28) can be solved recursively, and the controller parameters $\{K(k)\}_{0\le k\le \mathscr{T}-1}$ are listed in Table 10.1

Table 10.1 The controller parameters $\{K(k)\}_{0\le k\le \mathscr{T}-1}$

k	0	1	2	3	4	…
$K(k)$	0.4439	0.4344	0.4026	0.3594	0.3169	…

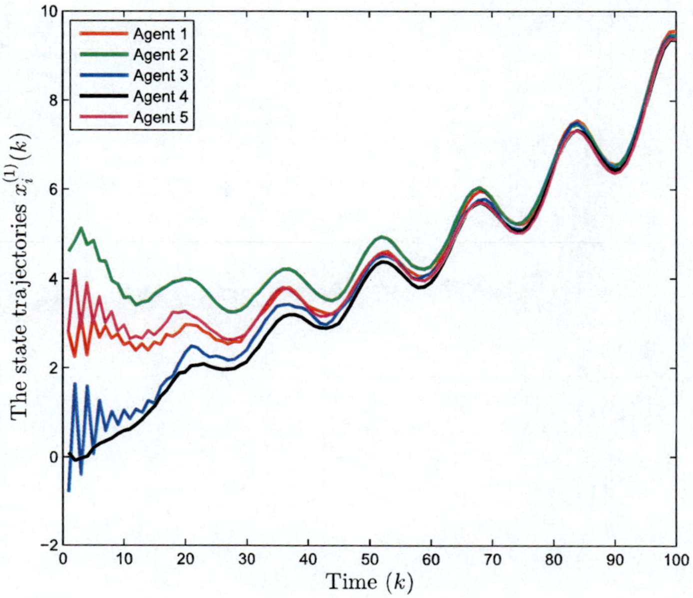

Fig. 10.2 The state trajectories $x_i^{(1)}(k)$, $i = 1, 2, 3, 4, 5$

In the simulation, the disturbance inputs $\omega_i(k)$ and $\nu_i(k)$ are selected as follows:

$$\begin{aligned}
&\omega_1(k) = 0.8\cos(1.2k),\ \ \omega_2(k) = 0.8\sin(0.5k),\ \ \omega_3(k) = 0.8\sin(1.3k),\\
&\omega_4(k) = 0.8\cos(0.5k),\ \ \omega_5(k) = 0.4\sin(0.4k) + 0.4\cos(0.7k),\\
&\nu_1(k) = 0.8\sin(1.2k),\ \ \nu_2(k) = 0.8\sin(0.7k),\ \ \nu_3(k) = 0.8\cos(k),\\
&\nu_4(k) = 0.8\cos(0.5k),\ \ \nu_5(k) = 0.4\sin(0.8k) + 0.4\cos(0.6k).
\end{aligned}$$

The initial values $x_i(0)$ $(i = 1, 2, \ldots, 5)$ are selected as $x_1(0) = \begin{bmatrix}2.8 & 1.6\end{bmatrix}^T$, $x_2(0) = \begin{bmatrix}4.6 & -0.2\end{bmatrix}^T$, $x_3(0) = \begin{bmatrix}-0.8 & -3.8\end{bmatrix}^T$, $x_4(0) = \begin{bmatrix}0.1 & -0.2\end{bmatrix}^T$, $x_5(0) = \begin{bmatrix}2.8 & -2.9\end{bmatrix}^T$. Simulation results are presented in Figs. 10.2, 10.3, 10.4 and 10.5, where the state trajectories of $x_i(k)$ $(i = 1, 2, \ldots, 5)$ are depicted by Figs. 10.2 and 10.3. Clearly, each individual agent has unstable dynamics. The output consensus errors $\bar{z}_i(k)$ $(i = 1, 2, \ldots, 5)$ are shown in Figs. 10.4 and 10.5. The simulation results have confirmed that the designed $\mathscr{H}_\infty$ cooperative controllers perform very well.

10.4 Conclusion

In this chapter, the $\mathscr{H}_\infty$ consensus control problem has been addressed for a time-varying multi-agent system subject to the scheduling of the so-called RA protocol. The scheduling behavior of the RA protocol has been modeled by a sequence of

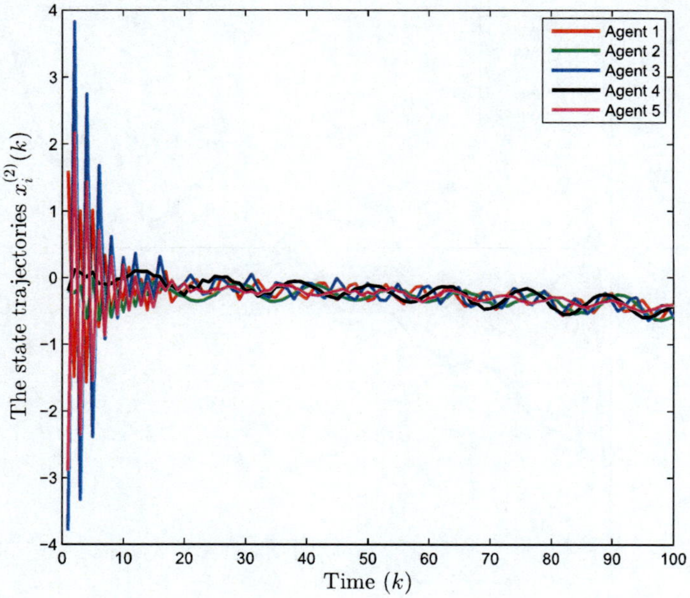

Fig. 10.3 The state trajectories $x_i^{(2)}(k)$, $i = 1, 2, 3, 4, 5$

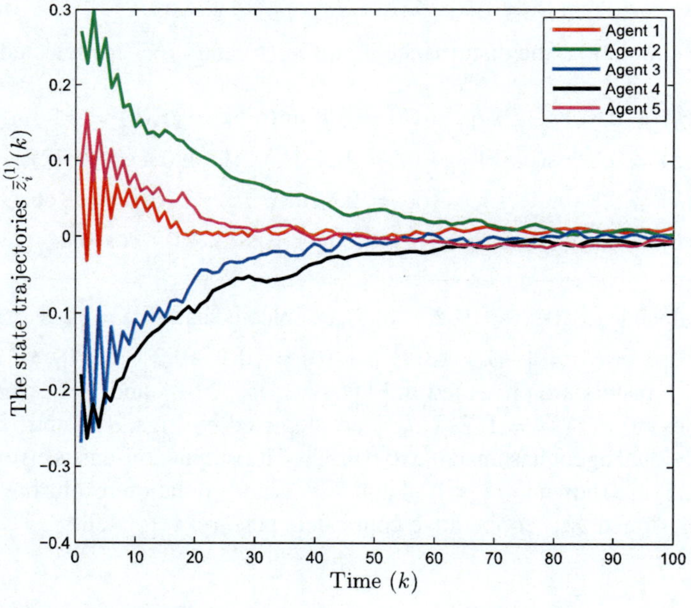

Fig. 10.4 The output consensus error $\bar{z}_i^{(1)}(k)$, $i = 1, 2, 3, 4, 5$

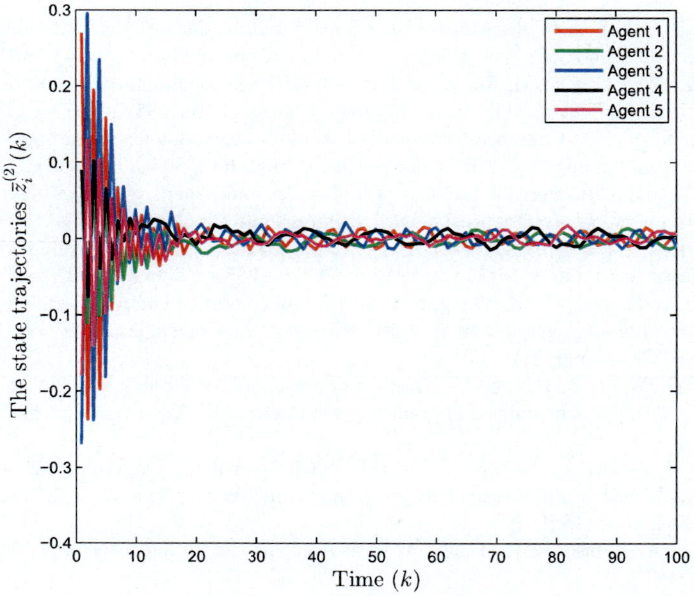

Fig. 10.5 The output consensus error $\bar{z}_i^{(2)}(k)$, $i = 1, 2, 3, 4, 5$

random variables with certain known occurrence probabilities. The cooperative controllers have been designed to generate the control signals according to the RA protocol scheduling as well as the topology of the multi-agent systems. By using the *mapping technology* and the *Hadamard product*, the closed-loop system has been described by a time-varying system with a *stochastic parameter matrix*. By employing the completing squares method and the stochastic analysis technique, a necessary and sufficient condition has been established to guarantee the prescribed $\mathscr{H}_\infty$ consensus performance constraint. Furthermore, the controller gains have been explicitly characterized by means of the solutions of two coupled backward RDEs. Finally, an illustrative example has been provided to highlight the effectiveness of the developed consensus control approach. It is worth mentioning that the proposed consensus control scheme in this chapter can be applied to some practical applications, such as the vehicle formations [17].

References

1. Jadbabaie, A., Lin, J., Morse, A.: Coordination of groups of mobile autonomous agents using nearest neighbor rules. IEEE Trans. Autom. Control **48**(6), 988–1001 (2003)
2. Olfati-Saber, R., Murray, R.: Consensus problems in networks of agents with switching topology and time-delays. IEEE Trans. Autom. Control **49**(9), 1520–1533 (2004)
3. Liu, C., Liu, F.: Stationary consensus of heterogeneous multi-agent systems with bounded communication delays. Automatica **47**(9), 2130–2133 (2011)

4. Zhang, Y., Tian, Y.: Consensus of data-sampled multi-agent systems with random communication delay and packet loss. IEEE Trans. Autom. Control **55**(4), 939–943 (2010)
5. Yu, W., Chen, G., Cao, M.: Some necessary and sufficient conditions for second-order consensus in multi-agent dynamical systems. Automatica **46**(6), 1089–1095 (2010)
6. Wang, S., Xie, D.: Consensus of second-order multi-agent systems via sampled control: undirected fixed topology case. IET Control Theory Appl. **6**(7), 893–899 (2012)
7. Liu, Y., Jia, Y.: Consensus problem of high-order multi-agent systems with external disturbances: an $\mathscr{H}_\infty$ analysis approach. Int. J. Robust Nonlinear Control **20**(14), 1579–1593 (2010)
8. Patterson, S., Bamieh, B., Abbadi, A.: Convergence rates of distributed average consensus with stochastic link failures. IEEE Trans. Autom. Control **55**(4), 880–892 (2010)
9. Gao, C., Wang, Z., He, X., Dong, H: Fault-tolerant consensus control for multi-agent systems: an encryption-decryption scheme. IEEE Trans. Autom. Control https://doi.org/10.1109/TAC.2021.3079407 (in press)
10. Ding, D., Wang, Z., Han, Q.-L.: Neural-network-based consensus control for multi-agent systems with input constraints: the event-triggered case. IEEE Trans. Cybern. **50**(8), 3719–3730 (2020)
11. Wang, L., Wang, Z., Wei, G., Alsaadi, F.E.: Observer-based consensus control for discrete-time multiagent systems with coding-decoding communication protocol. IEEE Trans. Cybern. **49**(12), 4335–4345 (2019)
12. Horn, R.A., Johnson, C.R.: Topics in Matrix Analysis. Cambridge University Press, Cambridge (1994)
13. Styan, G.P.H.: Hadamard products and multivariate statistical analysis. Linear Algebr. Appl. **6**, 217–240 (1973)
14. Caballero-Águila, R., Hermoso-Carazo, A., Linares-Pérez, J.: Optimal state estimation for networked systems with random parameter matrices, correlated noises and delayed measurements. Int. J. Gen. Syst. **44**(2), 142–154 (2015)
15. Ding, D., Wang, Z., Dong, H., Shu, H.: Distributed $\mathscr{H}_\infty$ state estimation with stochastic parameters and nonlinearities through sensor networks: the finite-horizon case. Automatica **48**(8), 1575–1585 (2012)
16. Penrose, R., Todd, J.A.: On best approximate solutions of linear matrix equations. Math. Proc. Camb. Philos. Soc. **52**(1), 17–19 (1956)
17. Alexander Fax, J., Murray, R.M.: Information flow and cooperative control of vehicle formations. IEEE Trans. Autom. Control **49**(9), 1465–1476 (2004)

Chapter 11
Conclusions and Future Work

This chapter draws conclusions on this book and points out some possible research directions related to the results done in this book.

11.1 Conclusions

The focus of the book has been placed on the filtering and control issues for networked systems under the effects of communication protocols. First, the concepts of networked systems and communication protocols have been introduced and the up-to-date research results on analysis and synthesis of networked systems subject to communication protocols have been reviewed. Then, the following communication-protocol-based filtering and control problems have been discussed in detail.

- In Chap. 2, we have discussed the ultimately bounded filtering problem for a class of time-delay complex networks subject to the RR protocol. First, we have established the transmission model for the complex network under the RR protocol effects. Then, the filtering error dynamics has been derived based on the introduced transmission model. By using the switch-based framework, the mean-square ultimate boundedness of the filtering error has been analyzed by constructing a novel Lyapunov-like functional and employing the stochastic analysis approach. The desired filter gains have been subsequently calculated by solving a convex optimization problem. Finally, we have proposed an illustrative example to demonstrate the effectiveness of the developed filtering scheme.
- In Chap. 3, we have considered the finite-horizon $\mathscr{H}_\infty$ filtering problem for a class of nonlinear time-varying systems with high-rate communication network and the RA protocol scheduling effects. A Markov chain has been adopted to model the scheduling behaviors of the RA protocol. The signal transmissions between the

L. Zou et al., *Communication-Protocol-Based Filtering and Control of Networked Systems*, Studies in Systems, Decision and Control 430,
https://doi.org/10.1007/978-3-030-97512-8_11

plant and filter have been implemented over a so-called high-rate communication network, which would give rise to the oversampling on transmission signals (i.e. *multiple transmissions* will occur between each two adjacent sampling instants of the sensors). A mapping technique has been employed to describe the transmission behaviors under the effects of RA protocol and high-rate communication network. The desired filter parameters have then been obtained by solving a set of recursive matrix inequalities to ensure the required $\mathscr{H}_\infty$ performance of the resultant filtering error dynamics. Finally, two numerical examples have been presented to illustrate the effectiveness of the proposed finite-horizon $\mathscr{H}_\infty$ filtering scheme.

- In Chap. 4, we have extended the results in Chap. 3 to the finite-horizon $\mathscr{H}_\infty$ fault estimation of time-varying systems subject to the RA protocol. Different from the scheduling model employed in Chap. 3, the signal transmission behaviors are described by using a sequence of i.i.d. random variables. The fault signals studied in this chapter have been assumed to be either incipient faults or abrupt faults, which are the most common faults in industrial processes. By using the similar recursive-matrix-inequality-based approach, the desired time-varying fault estimator gain has been calculated recursively. The illustrative examples have finally shown the effectiveness of the developed fault estimation method.
- In Chap. 5, we have studied the communication-protocol-based set-membership filtering technique, which is another well-known filtering algorithm dealing with time-varying systems. This chapter has concerned with the set-membership filtering problem for time-varying systems with mixed-time-delays under two deterministic protocols (i.e. the RR protocol and TOD protocol). By resorting to the recursive-matrix-inequality-based approach, sufficient conditions have been obtained for the time-varying filter to guarantee that the derived state estimate of the given system can be confined to certain ellipsoidal region at each time instant. Within the established theoretical framework, we have provided the design methods for the time-varying filter gain matrix for different underlying communication protocols in terms of two online optimization problems.
- In Chap. 6, we have dealt with the communication-protocol-based recursive filtering problem for linear time-varying systems. A two-step MMSE filter has been conducted to generate the desired state estimates through minimizing the trace of the filtering error covariance. Under the scheduling behavior of RA protocol, the resultant filtering error has been characterized by a time-varying system with a stochastic parameter matrix. Sufficient conditions have been proposed to guarantee the uniform boundedness of the filtering error covariance. Finally, we have addressed two numerical examples to verify the effectiveness and correctness of the developed communication-protocol-based recursive filtering algorithm.
- In Chap. 7, we have considered the practical application of the studied filtering techniques subject to communication protocols. This chapter has been concerned with the filter design issue for the so-called communication-based train control systems subject to the p-persistent CSMA protocol. The train under consideration has been modeled by n cars linked by couplers, whose dynamic behavior is characterized by a sequence of Newton's motion equations. The transmission behavior under the p-persistent CSMA protocol has been described by a Bernoulli dis-

tributed variable. The desired filter gains have been calculated based on different requirements through solving two optimization problems. The illustrative example has shown the effectiveness of our proposed filtering approach.

- In Chap. 8, we have extended the results about the communication-protocol-based finite-horizon $\mathscr{H}_\infty$ filtering to the communication-protocol-based finite-horizon $\mathscr{H}_\infty$ control problem, where the signal transmissions over the S/C channel and C/A channel have been scheduled by two RA protocols. An observer-based controller has been developed to achieve the desired control performance. By using the "completing the square" technique, the desired controller parameters have been derived via the solution of two coupled backward RDEs. Furthermore, the impact of the RA protocol on the characteristic of the controller parameters has also been analyzed. Finally, two illustrative examples have been addressed to demonstrate the effectiveness of the designed controller.
- Chap. 9 has considered the ultimately bounded control problem for nonlinear systems subject to the TOD protocol scheduling and uniform quantization effects. As a typical dynamical protocol, the scheduling behavior of the TOD protocol is governed by a special selection principle according to a quadratic function. A novel nonlinear-observer-based controller has been constructed to ensure that the dynamics of the closed-loop system is ultimately bounded in mean square subject to the norm-bounded disturbance and quantization error. By using the similar techniques proposed in Chap. 2, we have derived the values of the desired controller gains by resorting to the linear-matrix-inequality-based approach.
- Chapter 10 has discussed the finite-horizon $\mathscr{H}_\infty$ consensus control for time-varying multi-agent systems subject to the RA protocol scheduling. The results in this chapter can be regarded as extension of Chap. 8. First, we have established the scheduling model for multi-agent systems by extending the original RA protocol scheduling model for centralized networked systems. Then, the dynamics of the closed-loop system has been proposed by using the mapping technology and the Hadamard product. According to the RDE-based method utilized in Chap. 8, an improved controller design method has been developed by using the "completing the square" method and graph-theory-based analysis approach. Finally, the numerical example has shown the effectiveness of the developed finite-horizon $\mathscr{H}_\infty$ consensus controller.

11.2 Future Work

Generally speaking, this book has established a unified theoretical framework for the filtering and control of networked systems subject to communication protocols. Some of the related research future research topics are listed below

- In real-world applications, the signal transmissions of a networked system might be implemented via multiple communication networks with different protocols. The corresponding filtering and control processes of such a networked system would be affected by the combined effects of multiple communication protocols. In this sense, such filtering and control problems are referred to as the filtering and control problems subject to hybrid communication protocols. The dynamical behaviors subject to the hybrid communication protocol would be quite different from those subject to the single communication protocol. Hence, it would be a promising research topic to develop the suitable approaches to handle the corresponding filtering and control problems.

- The studies on the adaptive filtering and control problems subject to communication protocols are very few. Adaptive strategies are capable of handling the systems with unknown parameters. In practical applications, sometimes the parameters of the underlying communication protocols are fully accessible for a variety of reasons. Hence, the adaptive filtering and control techniques might fit well with such a case. Obviously, it leads to a particularly attractive area for investigating the adaptive filtering and control problems subject to communication protocols.
- It should be noted that the extensively studied communication protocols (the RR, TOD, RA protocols) are in fact some widely utilized time division multiple access (TDMA) protocols in modern industry. Code division multiple access (CDMA) protocol is another well-known communication protocol that is widely adopted in cellular telephone systems. The transmission behaviors under the CDMA protocol are quite distinguished from those under the TDMA protocol. Accordingly, the existing research frameworks are incapable of dealing with the analysis and synthesis issues under the CDMA protocol. As such, a trend for future research is to develop certain new research frameworks to deal with the analysis and synthesis of networked systems with CDMA protocols.

Index

L. Zou et al., *Communication-Protocol-Based Filtering and Control of Networked Systems*, Studies in Systems, Decision and Control 430,
https://doi.org/10.1007/978-3-030-97512-8